Constructal Theory of Social Dynamics

Constructal Theory of Social Dynamics

Edited by

Adrian Bejan

Department of Mechanical Engineering and Materials Science
Duke University
Durham, NC, USA

and

Gilbert W. Merkx

Department of Sociology
Duke University
Durham, NC, USA

Springer

Adrian Bejan
Department of Mechanical Engineering and Materials Science
Duke University
Durham, NC, USA

Gilbert W. Merkx
Department of Sociology
Duke University
Durham, NC, USA

Constructal Theory of Social Dynamics

Library of Congress Control Number: 2006939575

ISBN-13: 978-0-387-47680-3 e-ISBN-13: 978-0-387-47681-0

Printed on acid-free paper.

9 8 7 6 5 4 3 2 1

springer.com

Contents

Preface

Society is a "live" flow system, perhaps the most complex and puzzling we know. It is a jungle of flow systems—a vast multiscale system of systems—with organization, pattern, hierarchies, and usefulness (design). It is the most difficult to comprehend because we, the individuals who try to make sense of it, are inside the flow system. Difficult, because each of us is like an alveolus in the lung, an eddy in a turbulent river, or a leaf on a tree branch. From such a position of singularity, which is *identical* in rank to the positions of enormous numbers of individuals, it is a formidable task to see and describe the big picture—the lung, the river basin, and the forest.

Man's great fortune has been the fact that Nature has shape, structure, configuration, pattern, rhythm, and similarity. From this stroke of luck, science was born and developed to the present day, where it is responsible for our physical and intellectual well-being. The puzzling architecture and history of society has many things in common with the architecture and evolution of other complex (but simpler) flow systems: blood vascularization, river basins and deltas, animal movement, turbulence, respiration, dendritic solidification, etc. Coincidences that occur in the billions are loud hints that a universal phenomenon is in play. Is there a single physics principle from which the phenomenon of configuration and rhythm can be *deduced* without recourse to empiricism?

In this book we show that there is such a principle, and it is based on the common observation that if a flow system (e.g., river basin, vascularized tissue, city traffic) is endowed with sufficient freedom to change its configuration, the system exhibits configurations that provide progressively better access routes for the currents that flow. Observations of this kind come in the billions, and they mean one thing: a time arrow is associated with the sequence of flow configurations that constitutes the existence—the survival—of the system. Existing drawings are replaced by easier-flowing drawings. This physics principle is the *constructal law* of the generation of configuration in Nature: "For a finite size flow system to persist in time (to survive) its configuration must evolve in such a way that it provides easier and easier access to the currents that flow through it."

At Duke, where constructal theory began by accident in 1996 as a thermodynamics principle that unites physics with biology and engineering, we have stumbled upon another accident: scientists and sociologists view the generation of design in societies based on the same principle. Duke is a wonderful place not because of beautiful gardens and basketball, but because of *freedom*. Freedom is good for all design, from the better-flowing river basins to the faster, cheaper, and safer flowing rivers of people and goods (our society with all its live tree flows), all the way to the design called "better science."

Freedom brought the two of us together, a sociologist and an engineering scientist, and we were soon joined in this fertile discussion by our prominent colleagues Ed Tiryakian and Ken Land. Together we decided that the élan that constructal theory had generated in science is so contagious, and the theory itself so commonsense, concise, and useful, that it deserves to be discussed more broadly with colleagues throughout social sciences. We proposed this vision to the Human and Social Dynamics program of the National Science Foundation, which gave us an exploratory grant to "develop a community of scholars around the constructal theory of social dynamics."

This book is the first of its kind in this new field. It is the first account of the ideas, results, and future plans that came out of putting scientists, sociologists, and engineers together. The chapters of this book are based on the contributions made by prominent invited speakers at the First International Workshop on the Constructal Theory of Social Dynamics, which was held on 4–5 April 2006 at Duke University. We wish to thank the authors for their contributions to the workshop and to this book:

Prof. Sylvie Lorente
Prof. Heitor Reis
Prof. Antonio Miguel
Mr. Stephen Périn
Prof. Edward Tiryakian
Prof. John Staddon
Prof. Anthony Oberschall
Prof. Kenneth Land
Prof. Carter Butts
Ms. Miruna Petrescu-Prahova
Ms. Lorien Jasny
Dr. Franca Morroni
Dr. John Angle
Mr. Cyrus Amoozegar
Mr. Jean-Christophe Danaës

The constructal theory of social dynamics developed in this book surprises even us with the breadth and freshness of the territory that it covers. Major threads of this emerging theory of social organization are as follows:

- The organized multiscale distribution of living settlements. The idea is to place the community–community access in geometric terms, and to optimize it everywhere, subject to space constraints. Allocation of territory to movement (people, goods, information) is the fundamental idea.
- The occurrence of multiscale structure inside a settlement. In a city, for example, we see a compounding of scales, and each flowing thing has its

own hierarchy of scales. One example is how small streets coexist with larger (fewer) streets, and how the latter sustain a single artery. There are macroscopic features that appear in the largest cities (finger-shaped growth, beltways) that may be attributed to the same global principle of maximization of access.

- Development, and the connection between "flowing" societies, advancement, and prosperity. There is an opportunity to exploit the constructal idea of the need to be free to change the flow configuration, and connect it with the Darwinian view that the living constructs that prosper are those that possess the greatest ability to change.
- Migration patterns on the globe, in space and in time. Where and when people settle may be random individually, but the society appears to be the result of global optimization.
- Globalization, and the problematic aspects of overcoming obstacles to efficient flows, e.g., investment funds from the public and private sectors.

In sum, this book is about the tearing down of fences that are presumed to exist between the most central fields of human thought. To tear down fences means the opposite of "to destroy." It means *to construct* a far bigger tent that covers the designs (the bodies of knowledge) of historically separate fields.

Science is our knowledge of how nature works. Nature is *everything*, including engineering and society. Our knowledge is condensed in simple statements (thoughts, connections), which evolve in time by being replaced by simpler statements. We "know more" because of this evolution in time, not because brains become bigger and neurons smaller and more numerous. Our finite-size brains keep up with the steady inflow of new information through a process of simplification by replacement: in time, and stepwise, bulky catalogs of empirical information (e.g., measurements, observations, data, complex empirical models) are replaced by much simpler summarizing statements (e.g., concepts, formulas, constitutive relations, laws). A hierarchy of statements emerges along the way: it emerges naturally, because it is better than what existed before.

The simplest and most universal are the laws. The bulky and the laborious are being replaced by the compact and the fast. In time, science optimizes and organizes itself in the same way that a river basin evolves: toward configurations (links, connections, design) that provide faster access, or easier flowing.

The hierarchy that science exhibited at every stage in the history of its development is an expression of its never-ending struggle to optimize and redesign itself. Hierarchy means that measurements, ad hoc assumptions, and empirical models come in huge number, a "continuum" above which the compact statements (the laws) rise as needle-shaped peaks. Both are needed, the numerous and the singular. One class of flows (information links) sustains the other.

Civilization with all its constructs (science, religion, language, writing, etc.) is this never-ending physics of generation of new configurations, from the flow of mass, energy, and knowledge to the world migration of the special persons to whom ideas occur (the creative). Good ideas travel and persist. Better-flowing configurations replace existing configurations. Empirical facts are extremely numerous, like the hill slopes of a river basin. The laws are the extremely few big rivers, the Seine and the Danube. This book is about the big river of all "live" flow systems, including social dynamics: the constructal law.

Adrian Bejan
Gilbert W. Merkx

Duke University
December 2006

List of Contributors

Adrian Bejan
Department of Mechanical Engineering and Materials Science
Duke University
Durham, NC 27708-0300
abejan@duke.edu

Gilbert W. Merkx
Department of Sociology
Duke University
Durham, NC 27708-0006
gilbert.merkx@duke.edu

Sylvie Lorente
National Institute of Applied Sciences
Department of Civil Engineering
Laboratory of Materials and Durability
of Constructions, 31077 Toulouse, France
Lorente@insa-toulouse.fr

A. Heitor Reis
Departamento de Fisica
Universidade de Évora
Colegio Luis Verney
Rua Romao Ramalho, 59
7000-67 1 Évora Portugal
ahr@uevora.pt

Antonio F. Miguel
Departamento de Fisica
Universidade de Évora
Colegio Luis Verney
Rua Romao Ramalho, 59
7000-67 1 Évora Portugal
afm@uevora.pt

Stephen Périn
NeoSYS
7 rue du Théâtre
91300 Massy, France
stephen.perin@neosys.fr

Edward A. Tiryakian
Professor Emeritus
Department of Sociology
Duke University
Durham, NC 27708-0088
durkhm@soc.duke.edu

John E. R. Staddon
James B. Duke Professor
Duke University
Durham, NC 27708-1050
Staddon@psych.duke.edu

Anthony Oberschall
Emeritus professor
University of North Carolina at Chapel Hill
Chapel Hill, NC 27514
tonob@email.unc.edu

Kenneth G. Manton, **Kenneth C. Land**, and Eric Stallard
Department of Sociology and the Center for Demographic Studies
Duke University
Durham, NC 27708-0088
kland@soc.duke.edu

Carter T. Butts
Department of Sociology
University of California, Irvine
2145 Social Science Plaza A
Irvine, CA 92697-5100
buttsc@uci.edu

Miruna Petrescu-Prahova
Department of Sociology
University of California, Irvine
2145 Social Science Plaza A
Irvine, CA 92697-5100
mpetresc@uci.edu

Lorien Jasny
Department of Sociology
University of California, Irvine
2145 Social Science Plaza A
Irvine, CA 92697-5100
ljasny@uci.edu

Franca Morroni
DEXIA Asset Management
Head of SRT Analysis
Brussels, Belgium
fmorroni@yahoo.fr

John Angle
Economic Research Service
US Department of Agriculture
1800 M Street NW, Room N4097
Washington, DC, 20036-5831
jackangle@earthlink.com

Cyrus Amoozegar
Department of Mechanical Engineering and Materials Science
Duke University
Durham, NC 27708-0300
cyrus.amoozegar@duke.edu

Jean-Christophe Danaës
Cognitive Computer Science Department
University of Quebec at Montreal
Montreal, Quebec, Canada
denaes.jean-christophe@uqam.ca

Chapter 1
The Constructal Law in Nature and Society

Adrian Bejan

1.1. The Constructal Law

Society with all its layers and features of organization is a flow system. It is a "live" system, perhaps the most complex and puzzling we know. It is the most difficult to comprehend because we, the individuals who try to make sense of it, are inside the flow system. Each of us is like an alveolus in the lung, an eddy in a turbulent river, or a leaf on a tree branch. From such a position* of singularity, which is identical in rank to the positions of enormous numbers of individuals, it is a formidable task to see and describe the big picture—the lung, the river basin, and the forest.

Nature impresses us with shape, structure, configuration, pattern, rhythm, and similarity. This was our stroke of luck. From it, science was born and developed to the present day, where it is responsible for our physical and intellectual well-being. The puzzling architecture and history of society has many things in common with the architecture and evolution of other complex (but simpler) flow systems: blood vascularization, river basins and deltas, animal movement, respiration, dendritic solidification, etc. Coincidences that occur in the billions are loud hints that a universal phenomenon is in play. Is there a single physics principle from which the phenomenon of configuration and rhythm can be *deduced* without recourse to empiricism?

There is such a principle, and it is based on the common (universal) observation that if a flow system (e.g., river basin, blood vessel) is endowed with sufficient freedom to change its configuration, the system exhibits configurations that provide progressively better access routes for the currents that flow. Observations of this kind come in billions, and they mean one thing: a time arrow is associated with the sequence of flow configurations that constitutes the existence of the system. Existing drawings are replaced by easier-flowing drawings.

---

* Here, the meaning of position is geometric. The individual is a *particular* point of view in space. That point is occupied by *this* individual (his or her view of the world) and not by anybody else.

I formulated this principle in 1996 as the *constructal law* of the generation of flow configuration (Bejan 1996, 1997a–c):

> For a finite size flow system to persist in time (to survive) its configuration must evolve in such a way that it provides easier and easier access to the currents that flow through it.

This law is the basis for the *constructal theory* of organization in nature, which was first summarized in book form in Bejan (1997c). Today this body of work represents a new extension of physics: the thermodynamics of flow systems with configuration (Bejan and Lorente 2004, 2005).

To see why the constructal law is a law of physics, ask why the constructal law is different than (i.e., distinct from, or complementary to) the other laws of thermodynamics. Think of an isolated thermodynamic system that is initially in a state of internal nonuniformity (e.g., regions of higher and lower pressures or temperature, separated by internal partitions that suddenly break). The first and second laws account for billions of observations that describe a tendency in time, a time arrow: if enough time passes, the isolated system settles into a state of equilibrium (no internal flows, maximum entropy at constant energy, etc.). The first and second laws speak of a black box. They say nothing about the configurations (the drawings) of the things that flow. Classical thermodynamics was not concerned with the configurations of its nonequilibrium (flow) systems.

This tendency, this time sequence of drawings that the flow system exhibits as it evolves, is the phenomenon covered by the constructal law: not the drawings per se, but the time direction in which they morph if given freedom. No drawing in nature is "predetermined" or "destined" to be or to become a particular image. The actual evolution or lack of evolution (rigidity) of the drawing depends on many factors, which are mostly random. One cannot count on having the freedom to morph in peace (undisturbed).

Once again, the juxtaposition of the constructal law with the laws of classical thermodynamics can be useful. No isolated system in nature is predetermined or destined to end up in a state of mathematically uniform intensive properties so that all future flows are ruled out. One cannot count on the removal of all the internal constraints. One can count even less on anything being left in peace, in isolation.

As a thought, the second law proclaims the existence of a "final" state: the concept of equilibrium in an isolated system, at sufficiently long times. Similarly, the constructal law proclaims the existence of a concept: the equilibrium flow architecture, when all possibilities of increasing morphing freedom have been exhausted.

Constructal theory is now a fast-growing field with contributions from many sources, which have been reviewed on several occasions (Poirier 2003; Lewins 2003; Rosa et al. 2004; Torre 2004; Upham and Wolo 2004; Bejan and Lorente 2006; Reis 2006). The basic idea, however, is that constructal theory is the 1996 law cited at the start of this section.

The constructal law statement is general: it does not use words such as *tree, complex* versus *simple*, and *natural* versus *engineered*. How to deduce a class of flow configurations by invoking the constructal law is an entirely different (separate, subsequent) thought, which should not be confused with the constructal law. There are several (not many) classes of flow configurations, and each class can be derived from the constructal law in several ways, analytically (pencil and paper) or numerically, approximately or more accurately, blindly (via random search) or using intelligence (strategy, short cuts), etc. Classes that we have treated in detail, and by several methods, are the cross-sectional shapes of ducts, the cross-sectional shapes of rivers, internal spacings, and tree-shaped architectures (Bejan 1997c, 2000, 2006; Bejan and Lorente 2005).

Regarding trees, our group treated them not as models* (many have published and continue to publish models), but as fundamental access-maximization problems: volume to point, area to point, line to point, and the respective reverse flow directions. Important is the geometric notion that the "volume," the "area," and the "line" represent *infinities* of points. Our theoretical discovery of trees stems from the decision to connect one point (source or sink) with an *infinity* of points (volume, area, line). It is the reality of the continuum that is routinely discarded by modelers who approximate the space as a finite number of discrete points, and then cover the space with "sticks" drawings, which (of course) cover the space incompletely (and, from this, fractal geometry). Recognition of the continuum requires a study of the interstitial spaces between the tree links. The interstices can only be bathed by high-resistivity diffusion (an invisible, disorganized flow), while the tree links serve as conduits for low-resistivity organized flow (visible streams, ducts).

The two modes of flowing with thermodynamic imperfection (i.e., with resistances), the interstices and the links, must be *balanced* so that together they contribute minimum imperfection to the global flow architecture. Choke points must be balanced and distributed. The flow architecture is the graphical expression of the balance between links and their interstices. The deduced architecture (tree, duct shape, spacing, etc.) is the *optimal distribution of imperfection*. Those who model natural trees and then draw them as black lines on white paper (while not optimizing the layout of every black line on its optimally sized and allocated white patch) miss half of the drawing. The white is as important as the black.

Our discovery of tree-shaped flow architectures was based on three approaches. In Bejan (1996), the start was an analytical short cut based on several simplifying assumptions: 90° between stem and tributaries, a construction sequence in which smaller optimized constructs are retained, constant-thickness branches, etc. (e.g., Section 1.2). Months later, we published the same problem (Ledezma et al. 1997) numerically, by abandoning most of the simplifying assumptions (e.g., the construction sequence) used in the first papers. We also

---

* The great conceptual difference between modeling and theory is spelled out in *Physics Today*, July 2005, p. 20.

did this work in an area-point flow domain with random low-resistivity blocks embedded in a high-resistivity background (Errera and Bejan 1998), by using the language of Darcy flow (permeability, instead of conductivity and resistivity). Along the way, we found better performance and "more natural looking" trees as we progressed in time; i.e., as we endowed the flow structure with more freedom to morph.

And so I end this section with the "click" that I felt as I ended the second paper on constructal trees (for the full version, see p. 813–815 in Bejan 1997a):

The commonality of these phenomena is much too obvious to be overlooked. It was noted in the past and most recently (empirically) in fractal geometry, where it was simulated based on repeated fracturing that had to be assumed and truncated. The origin of such algorithms was left to the explanation that the broken pieces (or building blocks, from the point of view of this paper) are the fruits of a process of self-optimization and self-organization. The present paper places a purely deterministic approach behind the word "self": the search for the easiest path (least resistance) when global constraints (current, flow rate, size) are imposed.

If we limit the discussion to examples of living flow systems (lungs, circulatory systems, nervous systems, trees, roots, leaves), it is quite acceptable to end with the conclusion that such phenomena are common because they are the end result of a long running process of "natural selection". A lot has been written about natural selection and the impact that efficiency has on survival. In fact, to refer to living systems as complex power plants has become routine. The tendency of living systems to become optimized in every building block and to develop optimal associations of such building blocks has not been explained: it has been abandoned to the notion that it is imprinted in the genetic code of the organism.

If this is so, then what genetic code might be responsible for the development of equivalent structures in inanimate systems such as rivers and lightning? What genetic code is responsible for man-made networks (such as the trees in this paper)? Certainly not mine, because although highly educated, neither of my parents knew heat transfer (by the way, classical thermodynamics was not needed in this paper). Indeed, whose genetic code is responsible for the societal trees that connect us, for all the electronic circuits, telephone lines, air lines, assembly lines, alleys, streets highways and elevator shafts in multistory buildings?

There is no difference between the animate and the inanimate when it comes to the opportunity to find a more direct route subject to global constraints, for example, the opportunity of getting from here to there in an easier manner. If living systems can be viewed as engines in competition for better thermodynamic performance, then inanimate systems too can be viewed as living entities (animals!) in competition for survival.

This analogy is purely empirical: we have a very large body of case-by-case observations indicating that flow configurations (animate and inanimate) evolve and persist in time, while others do not. Now we know the particular feature (maximum flow access) that sets each surviving design apart, but we have no theoretical basis on which to expect that the design that persists in time is the one that has this particular feature. This body of empirical evidence forms the basis for a new *law of nature* that can be summarized as ... [the constructal law, at the start of this section]. This new law brings life and time explicitly into thermodynamics and creates a bridge between physics and biology.

1.2. The Urge to Organize Is an Expression of Selfish Behavior

Why are streets usually arranged in clusters (patterns, grids) that look almost similar from block to block and from city to city? Why are streets and street patterns a mark of civilization? Indeed, *why do streets exist?* Constructal theory provided answers to these questions by addressing the following area-point access maximization problem.

Consider a finite-size geographical area A and a point M situated inside A or on its boundary (Fig. 1.1). Each member of the population living in A must travel between his or her point of residence P(x, y) and point M. The latter serves as common destination for all the people who live in A. The density of this traveling population—i.e., the rate at which people must travel to M—is fixed and described by $\dot{n}''$ (people/m^2 s). This also means that the rate at which people are streaming into M is constrained, $\dot{n} = \dot{n}''A$. Determine the optimal bouquet of paths that link the points P of area A with the common destination M such that the time of travel required by the entire population is the shortest.

The problem is how to connect a finite area (A) to a single point (M). Area A contains an infinite number of points, and every one of these points must be taken into account when optimizing the access from A to M and back. Time has shown that this problem was a lot tougher than the empirical game of connecting "many points": i.e., a finite number of points distributed over an area. The many-points problem can be solved on the computer using brute-force methods (random walk or Monte Carlo—more points on better computers), which are not theory.

The area A could be a flat piece of farmland populated uniformly, with M as its central market or harbor. The oldest solution to this problem was to unite with a straight line each point P and the common destination M. The straight-line solution was the preferred pattern as long as humans had only one mode

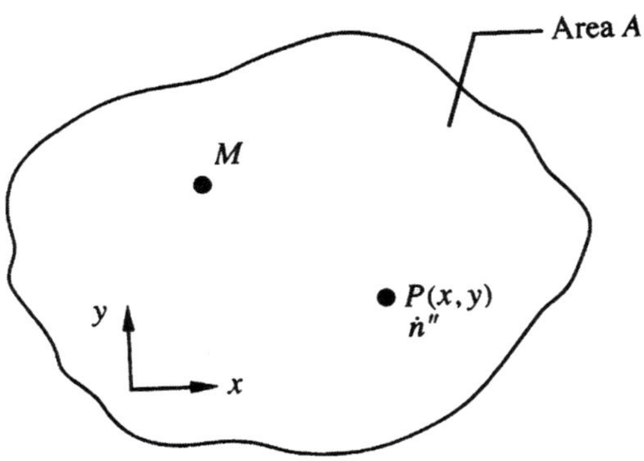

FIGURE 1.1. Finite-size area (A) covered by a uniformly distributed population ($\dot{n}''$) traveling to a common destination (M) (Bejan 1996)

of locomotion: walking, with the average speed V_0. The farmer and the hunter would walk straight to the point (farm, village, river) where the market was located.

The radial pattern disappeared naturally in areas where settlements were becoming too dense to permit straight-line access to everyone. Why the radial pattern disappeared "naturally" is the area-point access problem. Another important development was the horse-driven carriage: with it, people had two modes of locomotion, walking (V_0) and riding in a carriage with an average velocity V_1 that was significantly greater than V_0. It is as if the area A became a composite material with two conductivities, V_0 and V_1. Clearly, it would be faster for every inhabitant (P, in Fig. 1.1) to travel in straight lines to M with the speed V_1. This would be impossible, because the area A would end up being covered by beaten tracks, leaving no space for the inhabitants and their land properties.

The modern problem, then, is one of bringing the street near a small but finite-size group of inhabitants; this group would first have to walk to reach the street. The problem is one of allocating a finite length of street to each finite patch of area A_1, where $A_1 << A$. The problem is also one of connecting these street lengths in an optimal way such that the time of travel of the population is minimum.

The first analytical approach to this problem was "atomistic," from the smaller subsystem (detail) of area A to the larger subsystem, and ultimately to area A itself. The area subsystem to which a street length may be allocated cannot be smaller than the size fixed by the living conditions (e.g., property) of the people who will be using the street. This smallest area scale is labeled A_1 in Fig. 1.2. For simplicity we assume that the A_1 element is rectangular. Although A_1 is fixed, its shape or aspect ratio H_1/L_1 is not. Indeed, the first objective is to anticipate optimal form: the area shape that maximizes the access of the A_1 population to the street segment allocated to A_1.

Symmetry suggests that the best position for the street segment is along the longer of the axes of symmetry of A_1. This choice has been made in Fig. 1.2, where $L_1 > H_1$ and the street has the length L_1 and width D_1. The traveling population density $\dot{n}''$ is distributed uniformly on A_1. To get out of A_1, each person must travel from a point of residence P(x, y) to the (0,0) end of the street. The person can travel at two speeds: (1) a low speed V_0 when off the street and (2) a higher speed A_1 when on the street.

We assume that the rectangle $H_1 \times L_1$ is sufficiently slender ($L_1 > H_1$) so that the V_0 travel is approximated well by a trajectory aligned with the y axis. The time of travel between P(x, y) and (0,0) is $(x/V_1) + (y/V_0)$. The average travel time of the A_1 population is given by

$$\bar{t}_1 = \frac{1}{H_1 L_1} \int_{-H_1/2}^{H_1/2} \int_0^{L_1} \left(\frac{x}{V_1} + \frac{y}{V_0} \right) dx\, dy \qquad (1.1)$$

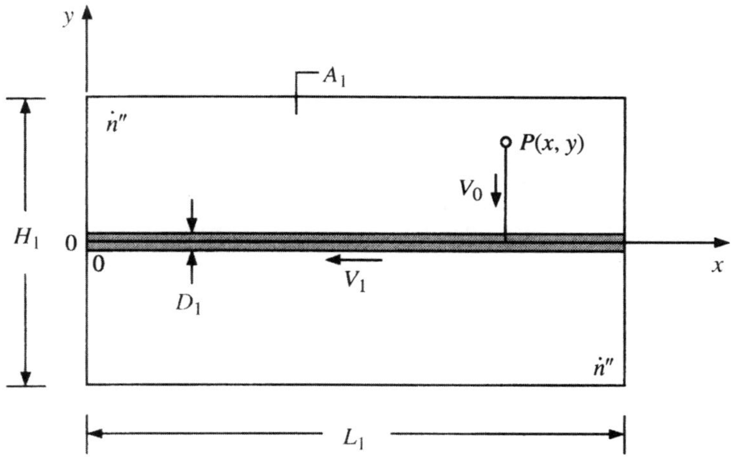

FIGURE 1.2. Smallest (innermost) elemental area, A_1, and the street segment allocated to it (Bejan 1996)

which yields

$$\bar{t}_1 = \frac{L_1}{2V_1} + \frac{H_1}{4V_0} \tag{1.2}$$

The elemental area is fixed ($A_1 = H_1 L_1$, constant); therefore, $\bar{t}_1$ can be expressed as a function of H_1, which represents the shape of A_1:

$$\bar{t}_1 = \frac{A_1}{2V_1H_1} + \frac{H_1}{4V_0} \tag{1.3}$$

The average travel time has a sharp minimum with respect to H_1. Solving $\partial\bar{t}_1/\partial H_1 = 0$, we obtain

$$H_{1,\,opt} = \left(2\frac{V_0}{V_1}A_1\right)^{1/2} \tag{1.4}$$

and subsequently,

$$L_{1,\,opt} = \left(\frac{V_1A_1}{2V_0}\right)^{1/2} \tag{1.5}$$

$$\left(\frac{H_1}{L_1}\right)_{opt} = \frac{2V_0}{V_1} < 1 \tag{1.6}$$

Equation (1.6) shows the optimal slenderness of the smallest area element A_1. This result validates the initial assumption that $H_1/L_1 < 1$; indeed, the optimal smallest rectangular area should be slender when the street velocity is sensibly greater than the lowest (walking) velocity. The rectangular area A_1 must become more slender as V_1 increases relative to V_0—i.e., as time passes and technology advances. This trend is confirmed by a comparison between the streets built in antiquity and those that are being built today. In antiquity the first streets were short, typically with two or three houses on one side. In the housing developments that are being built today, the first streets are sensibly longer, with 10 or more houses on one side. This contrast is illustrated by modern Rome (Fig. 1.3). In the center, which is the ancient city, the streets are considerably shorter than in the more recently built, peripheral areas (e.g., the upper corners in Fig. 1.3).

Important is the observation that exactly the same optimum [Eqs. (1.4–1.6)] is found by minimizing the longest travel time (t_1) instead of minimizing the area-averaged time of Eq. (1.1). The longest time is required by those who travel from one of the distant corners ($x = L_1$, $y = \pm H_1/2$) to the origin (0,0) and is given by

$$t_1 = \frac{L_1}{V_1} + \frac{H_1}{2V_0} \tag{1.7}$$

Equations (1.7) and (1.2) show that the geometric minimizations of t_1 and $\bar{t}_1$ are equivalent. It is both interesting and important that the optimization of the shape of the A_1 element is of interest to every inhabitant: *What is good for the*

FIGURE 1.3. Plan of modern Rome, showing that in the ancient city (the center) the street length scales are considerably shorter than in the newer outskirts (Bejan 1997c)

most disadvantaged person is good for every member of the community. This conclusion has profound implications in the spatial organization of all living groups, from bacterial colonies all the way to our own societies. *The urge to organize is an expression of selfish behavior.*

The time obtained by minimizing t_1 or by substituting Eqs. (1.4) and (1.5) into Eq. (1.7) is

$$t_{1,\min} = \left(\frac{2A_1}{V_0 V_1}\right)^{1/2} \tag{1.8}$$

At this minimum, the two terms that make up t_1 in Eq. (1.7) are equal. This equipartition of time principle means that the total travel time is minimum when it is divided equally between traveling along the street and traveling perpendicularly to the street. We return to this feature in Section 1.4.

In Fig. 1.2 we see the smallest loop of the traffic network that will eventually cover the given area A. The next question is how to connect the D_1 streets such that each innermost loop has access to the common destination M. One answer—the simplest, albeit approximate—is obtained by *repeating* the preceding geometric optimization several times, each time for a larger area element, until the largest scale (A) is reached. This construction is detailed in Bejan (1996, 1997c), and is explained by considering the rectangular area $A_1 = H_2L_2$ shown in Fig. 1.4. This area consists of a certain number of the smallest patches A_1. The purpose of this assembly of A_1 elements is to connect the D_1 streets so that the traveling population ($\dot{n}'' A_2$) can leave A_1 in the quickest manner. We invoke symmetry as the reason for placing the new (second) street along the long axis of the A_2 rectangle. In Fig. 1.4, the stream of travelers ($\dot{n}'' A_2$) leaves A_2 through the left end of the D_2 street. There exists an optimal shape H_2L_2, and a minimal global travel time.

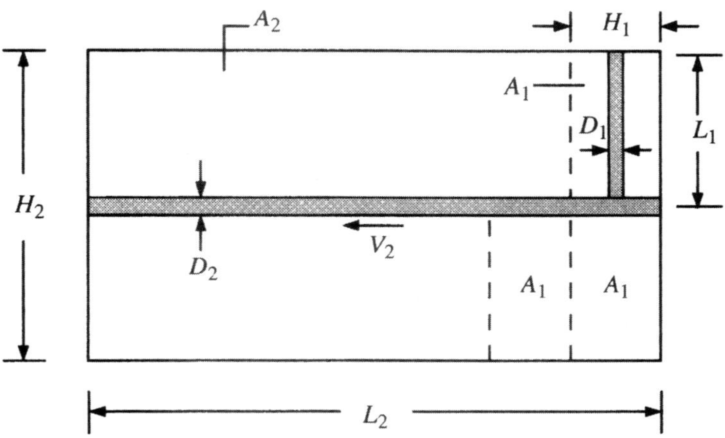

FIGURE 1.4. Area construct A_2 as an assembly of connected innermost elements A_1 (Bejan 1996)

The atomistic construction started in Figs. 1.2 and 1.4 can be continued toward larger assemblies of areas. This is not the best way to allocate streets to areas mathematically, but it is the most transparent. Its value is that it shows the emergence of a tree network (the streets) from principle (the maximization of access), not by copying from nature. In constructal theory, flow architectures such as trees are *discovered*. They are now known, observed, modeled, or copied from nature.

This sequence is shown in Fig. 1.5 (top) only for illustration, because it is unlikely to be repeated beyond the third-generation street. The reason is that as the community and the area inhabited by it grow, other common destinations (e.g., church, hospital, bank, school, train station) emerge on A in addition to the original M point (Fig. 1.1). Some of the streets that were meant to provide access to only one end of the area element must be extended all the way across the area to provide access to both ends of the street. As the destinations multiply and shift around the city, the dead ends of the streets of the first few generations disappear, and what replaces the growth pattern is a *grid* with access to both ends of each street. The multiple scales of this grid, and the self-similar structure of certain areas (neighborhoods) of the grid, however, are the fingerprints of the deterministic organization principle (the constructal law).

The area-point access problem formulated in this section was stated in two dimensions (Fig. 1.1). The corresponding problem in three dimensions is this: minimize the time of travel from all the points P of a volume V to one common destination point M, subject to the constraint that the traveling population rate is fixed. One application is the sizing and shaping of the floor plan in a multistory building, along with the selection and placement of the optimal number of elevator shafts and staircases.

The same organization theory can be extended generally to areas that are populated unevenly, or specifically to highways, railroads, telecommunications, and air routes (e.g., the organization of such connections into hubs, or centrals). A clear application of these concepts is in operations research and manufacturing, where the invention of the first auto assembly line is analogous to the appearance of the first street (Carone 2003; Carone et al. 2003; Hernandez 2001; Hernandez et al. 2003).

The atomistic construction sequence presented until now is just an approximate and simple way to illustrate how a tree of organized (channeled) flow emerges on a background covered by individual (disorganized) movement. The "exact" way to generate the tree architecture from the *same* principle is to endow the flow architecture with maximum freedom to morph (Bejan and Lorente 2004, 2005) and to use numerical simulations to morph the flow structure through all its eligible configurations. This more exact work is illustrated by relaxing the assumption (made in Figs. 1.2 and 1.4) that the paths intersect at right angles.

Assume that in Fig. 1.2 the angle between the V_0 and V_1 paths may vary. This general situation is shown in Fig. 1.6, which is set for calculating the maximum

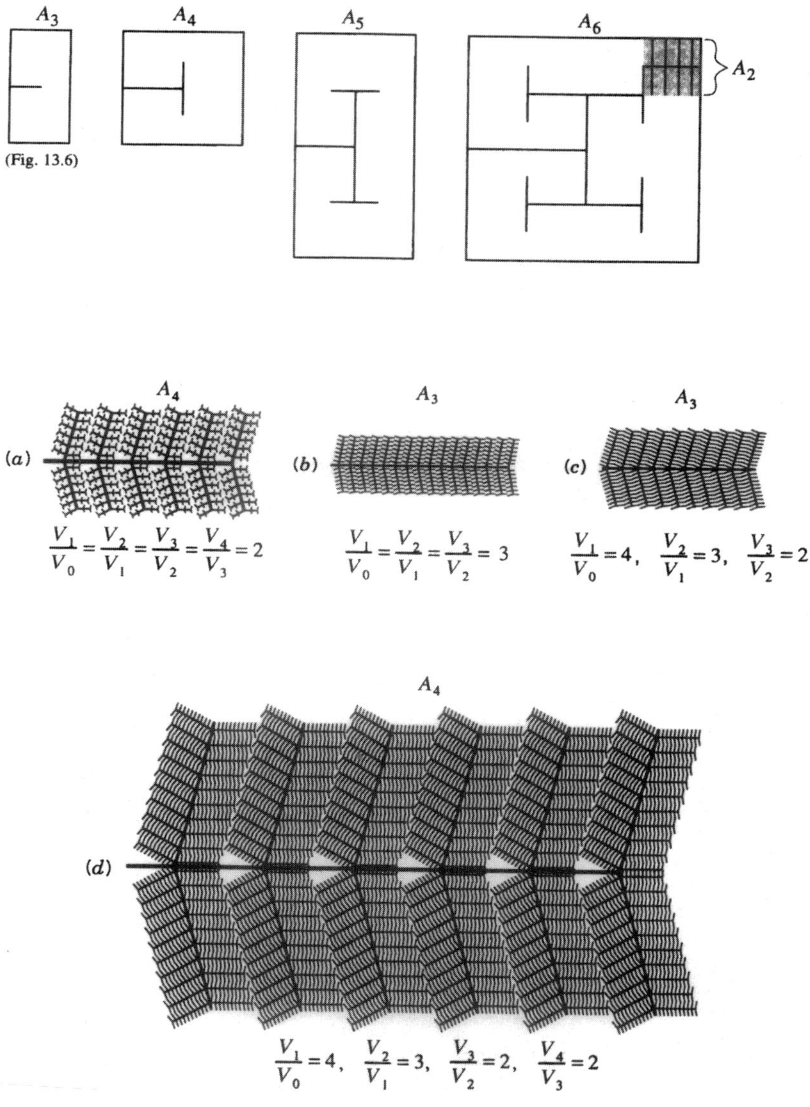

FIGURE 1.5. Top: higher-order constructs in the sequence of Figs. 1.2 and 1.4 (Bejan 1996). Bottom: urban growth patterns in which each construct was optimized for overall shape and angle of street confluence (Ledezma and Bejan 1998)

travel time between the distant corner (P) and the common destination (M). In place of Eq. (1.7), we obtain

$$t_1 = \frac{L_1}{V_1} + \frac{H_1}{2V_0}\left(\frac{1}{\cos\beta} - \frac{V_0}{V_1}\frac{\sin\beta}{\cos\beta}\right) \tag{1.9}$$

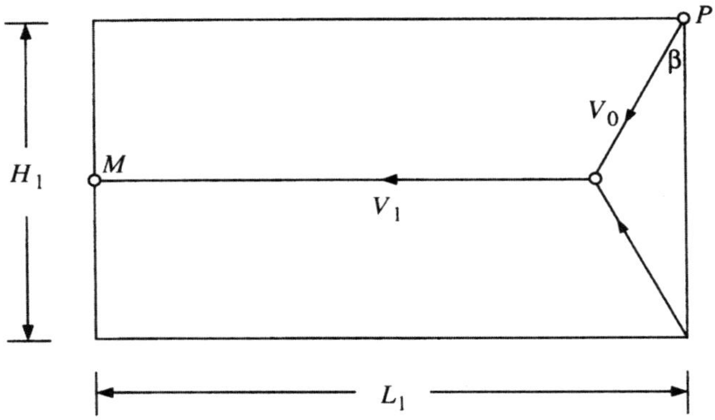

FIGURE 1.6. Smallest area (A_1) and the variable angle between the V_0 and V_1 paths

where β is the angle between the V_0 path and the H_1 side. Now the minimization of t_1 has two degrees of freedom: the geometric aspect ratio H_1/L_1 and the angle β. The optimal angle for minimum t_1

$$\beta_{opt} = \sin^{-1} \frac{V_0}{V_1} \tag{1.10}$$

confirms the statement made above Eq. (1.1) that V_0 should be perpendicular to V_1 (i.e., that $\beta = 0$) when $V_0 << V_1$. The minimization with respect to H_1/L_1 subject to $A_1 = H_1/L_1$ is the same as earlier in this section. The twice-minimized travel time is

$$t_{1,\,min} = \left(\frac{2A_1}{V_0 V_1} \cos \beta_{opt} \right)^{1/2} \tag{1.11}$$

The lower part of Fig. 1.5 shows four examples of optimal urban growth, in which each area construct (A_1, A_2, ...) has been optimized in two ways: overall shape and angle between each new street and its tributaries. The assumed changes in velocity are listed under each drawing. Comparing examples (a) and (d) we see that when the velocity increase factor V_i/V_{i-1} is large the street pattern spreads fast (in few steps) over the given area, and each area assembly is slender. In the opposite limit, the spreading rate is lower, the assembly steps are more numerous, and each area assembly is less slender. These trends appear together in example (d), where the velocity increase factor decreases as the construction grows.

Comparing Eq. (1.11) with Eq. (1.8), we note that the second degree of freedom (the optimized angle β) plays only a minor role as soon as V_1 is greater than V_0. In other words, the change from V_0 to V_1 does not have to be dramatic for the $\beta = 0$ design (Fig. 1.2) to perform nearly as well as the optimal design. We reach the important conclusion that small internal variations in the organization pattern have almost no effect on the global performance of the organized system ($t_{1,min}$, in this case).

The practical aspect of this observation is that a certain degree of variability (imperfection) is to be expected in the patterns and emerge naturally. These patterns are not identical, nor are they perfectly similar; this accounts for the historic difficulty of attaching a theory to naturally organized systems. Natural patterns are quasi-similar, but only in the same sense in which no two human faces are identical. Their performance, however, is practically the same as that of the best pattern. We call these top performers *equilibrium flow structures* (Bejan and Lorente 2004). The contribution of constructal theory is that the performance and the main geometric features (mechanism, structure) of the organized system can be predicted in purely deterministic fashion.

1.3. The Distribution of Human Settlements*

Every sector of society is a conglomerate of mating flows that morph in time in order to flow more easily: people, goods, money, information, etc. The view that society is a flow system with intertwined morphing (improving) architectures was part of the original disclosure of constructal theory (Bejan 1996, 1997c). This deterministic physics principle is in sharp contrast with the empirical (descriptive, modeling) approaches that have been tried to explain social organization. Society is viewed like the photograph of a turbulent flow. Even though the existence of structure is obvious, the image is so complicated, and so much the result of individual behavior, that description is the norm, not prediction For a review, see Bretagnolle et al. (2000), who argue in favor of introducing a spatial dimension (geography) in modeling, toward the development of an evolutionary theory of settlement systems. Such a theory would provide insights for better policy in the future, and will predict the future evolution of towns, cities, and their heterogeneous distribution on land.

Society may be complicated, but pattern is not. Indeed, pattern is "pattern" because it is not complicated. If it were not simple enough for us to grasp, it would be noise, chaos, turbulence, and randomness. Strikingly clear images such as Fig. 1.7 remain unexplained: the size of a city in Europe is inversely proportional to its rank (Bretagnolle et al. 2000; Bairoch et al. 1988; Moriconi-Ebrard 1994), this throughout history. Why?

Figure 1.7 is derivable from the constructal tree-shaped structures deduced for traffic (Section 1.2), which we now review as an introduction. Consider again the minimization of travel time for traffic between an infinity of points (an area) and one point (Figs. 1.1–1.5). The construction of the tree-shaped architecture of the river basin of people starts with the smallest elemental area A_1, which is fixed by the culture of those who live on A_1. For example, A_1 is the farmland surrounding the smallest road (V_1, L_1) that leads to a single marketplace, M_1. The slow movement covers A_1 and touches every point: the slow movement attaches every single inhabitant to the traffic architecture.

---

* This section is based on Bejan et al. (2006).

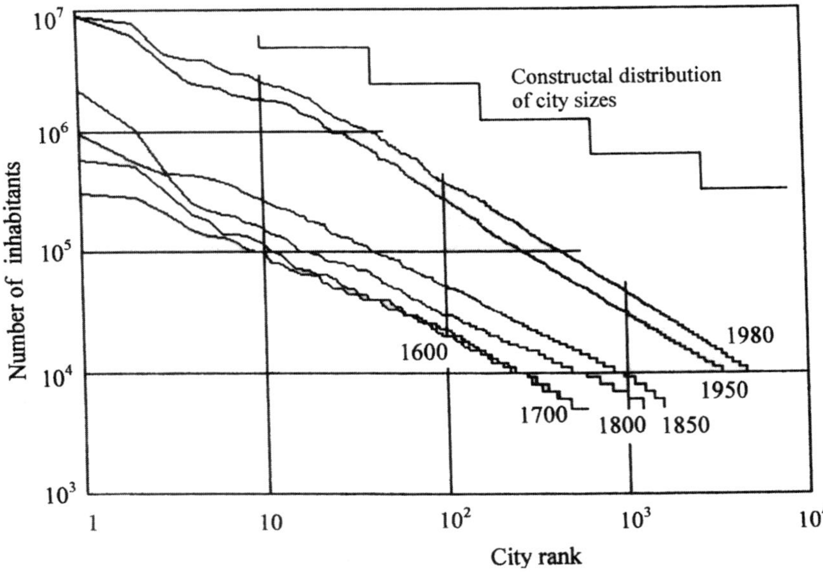

FIGURE 1.7. City size (population) versus city rank, in 1600–1980 Europe (Bejan et al. 2006)

The existence of two modes of movement implies a certain level of civilization: person living not alone (i.e., person with horse) and person with vehicle. Civilization is also the name for the coexistence of farmland (A_1) with markets (M_1). Those who live on A_1 exchange farm products with those who manufacture products and deliver services in compact places such as M_1. It is this balanced counterflow between A_1 and M_1 that justifies this key idea:

The number of those who live on A_1 must be proportional to the number N_1 of inhabitants living at M_1, and both numbers must be proportional to A_1. Both groups are sustained by the agriculture and the "environment" that A_1 provides; therefore, $N_1 = cA_1$, where c is the average number of inhabitants per unit area.

The "culture" factor c accounts for the age and history of the civilization (e.g., technology, commerce, neighbors, natural disasters, plagues, war, peace).

The next larger area that is civilized (A_2) is covered by an assembly of n_1 optimal A_1 rectangles ($A_2 = n_1 A_1$). A central road (speed V_2, such that $V_2 > V_1$) collects or distributes the traffic associated with the elements. This first construct (A_2) can be optimized to provide minimal travel time between A_2 and the new boundary point M_2.

The counterflow of goods between n_1 small markets (M_1) and the largest market (M_2) requires a proportionality between the number (N_2) of inhabitants at M_2 and the total number of inhabitants at the M_1 points. This means that $N_2 = cA_2$. We see here two directions in which hierarchy develops: areas coalesce, from elemental to first construct, and at the same time the population develops concentrations, from farmers on several A_1 plots to traders at several points M_1, and finally to one trading point M_2, which is perhaps a small town.

At the next level of assembly, a number (n_2) of first constructs coalesce into a second construct ($A_3 = n_2A_2$) such that a new central road (speed V_3, where $V_3 \gtrsim V_2$) links the M_2 markets with one new market on the boundary, M_3. In a society, the counterflow of goods and people between the M_3 and the M_2 markets means that the number (N_3) of people living at M_3 must be proportional to n_2N_2. Furthermore, because $n_2 = A_3/A_2$, we conclude that $N_3 = cA_3$.

The analysis of Figs. 1.2–1.5 showed that when V_2 is not much greater than V_1, the optimization of A_3 for minimal travel time yields $n_2 \cong 2$; this feature (pairing, dichotomy) prevails at higher orders of assembly. Dichotomy is a constructal-theory result, not an assumption. The optimized constructs alternate between squares (for i = even) and rectangles with 2:1 aspect ratio (for i = odd), where i is the order of the construct A_i. The balance between the flow of goods from each new construct A_i to its concentration point (M_i) on the boundary requires $N_i = cA_i$, where c is the history-dependent culture factor defined earlier.

The $i \geq 3$ constructs cover the territory with multiscale cities, as shown in the upper part of Fig. 1.5. Because constructs double in size from one construct level to the next ($A_i = 2A_{i-1}$, $i \geq 3$), population sizes also double ($N_i = 2N_{i-1}$) and the number of concentration points decreases to half during each step ($n_i = n_{i-1}/2$). In Fig. 1.8 this is illustrated by disk-shaped cities, the radii of which increase by the factor $2^{1/2}$. The end of this sequence (i = m, odd or even) occurs when the construct area (A_m) matches the size of the available territory (e.g., country or continent). At this uppermost level of assembly, the number of largest cities is

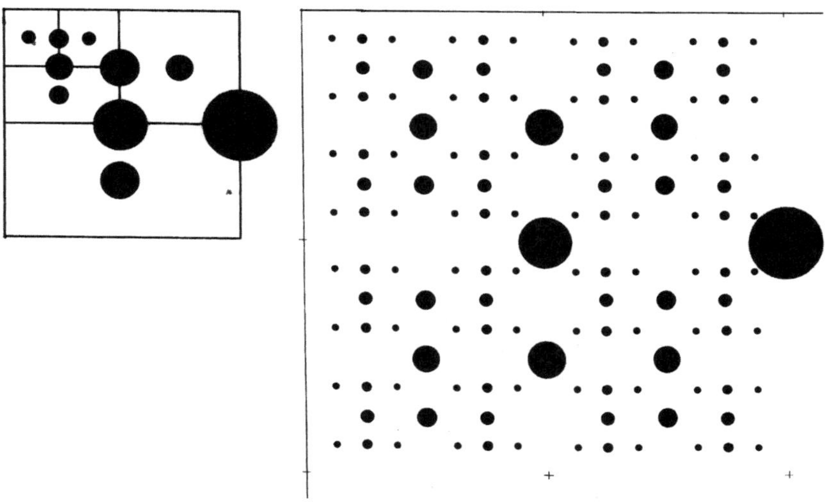

FIGURE 1.8. During the spreading of interlinked city flows, total areas double in size and the size of the largest city doubles (Bejan et al. 2006)

one or two. In conclusion, the existence of one largest city, which is evident in Fig. 1.7 and throughout demography, is a constructal-theory result.

The slope of the distribution of city sizes in Fig. 1.7 is also a constructal feature. The concentration points discovered in Fig. 1.8 have population sizes that decrease by a factor of 2 when the total number of cities increases by a factor of 2. The stepped line in Fig. 1.7 is from constructal theory, and its slope agrees with the trend exhibited by the data. The stepped line is rough in comparison with the data because the construction deduced in Fig. 1.8 does not take into account the complex and the uncertain: geography, geology, and human history. This discrepancy is not the issue. It is the agreement between theory and reality that counts, because it shows the origin of the striking pattern of organization that exists in even the most complex of all flow structures: human society.

The theoretical curve tells us more than the empirical data. In time, the entire curve must shift upward while remaining parallel to itself. This trend is due to the culture factor c (or average population density), which increases in time and has the same value throughout the country (A_m). The constructal pattern of interlinked cities, and all the tree architectures deduced earlier from the constructal law are not fractal objects (Bejan 1997, p. 765). In constructal theory the construction algorithm is deduced, not assumed.

A new theory is a new language, which performs new services. One is to summarize into a simple and predictive statement a large volume of empirical information (e.g., Fig. 1.7). Another is to predict future observations: features that may be present but not evident in the existing volume of empirical data. For example, because dichotomy is replaced by numerous elements when size is small (e.g., A_2 in Fig. 1.5), the curves plotted in Fig. 1.7 should become less steep as the city rank increases (i.e., as towns with less than 5000 inhabitants are counted).

Another theoretical eye opener is provided by Fig. 1.8: if we imagine all cities growing in time, we see that the larger will "eat" the smaller. According to constructal theory, then, urban growth must be stepwise, uneven in direction and time. This prediction is supported by the strange (uneven) fingering exhibited by all large urban areas.

The most special and, potentially, most lasting contribution of this new theory is that it explains the origin of huge volumes of observations in fields far removed from that of the theory and its authors. For example, the constructal distribution of city sizes in Fig. 1.7 happens to have the same trend as a Zipfian distribution of the occurrence of words in language. According to the empirical formula known as *Zipf's law*, the probability of occurrence of words or other items starts high and tapers off: $P_n \sim 1/n^a$, where P_n is the frequency of the nth-ranked item and a is close to 1. In English, for example, the words *and* and *the* occur very frequently, whereas a word such as *juxtaposition* is very rare. This empirical law characterizes the occurrence of words in human and computer languages, operating systems calls, and colors in images, and is the basis of many compression approaches. The fact that in Fig. 1.7 such a trend is predicted from the analysis of a tree-shaped flow system on optimized geography suggests that all the domains in which Zipfian distributions are observed (information, news) are homes to tree-shaped flow systems (point-area, point-volume). These systems inhabit flow architectures (drawings) optimized subject to fixed global size.

Information and news are constructal flows: river basins and deltas akin to those predicted in the next section. In this way, information and news are brought under the great tent of constructal theory: geometry, geography, tree architect-tures, freedom to morph, and optimized finite complexity (hierarchy) are constructal properties of the flow of information.

1.4. Human Constructions and Flow Fossils in General*

Pyramids in Egypt and other ancient sites intrigue us with their size and geometric form. Even by today's standards, size is immense and form is perfect. So impressive are these images that our culture tends to attribute them to an ancient scientific base that was lost, and to presumed links between ancient peoples living on opposite sides of the globe (Fig. 1.9).

Constructal theory offers a considerably more direct explanation: pyramids as results of a universal natural phenomenon, which is so prevalent and permanent that it governs the movement of all materials on Earth. This view does not

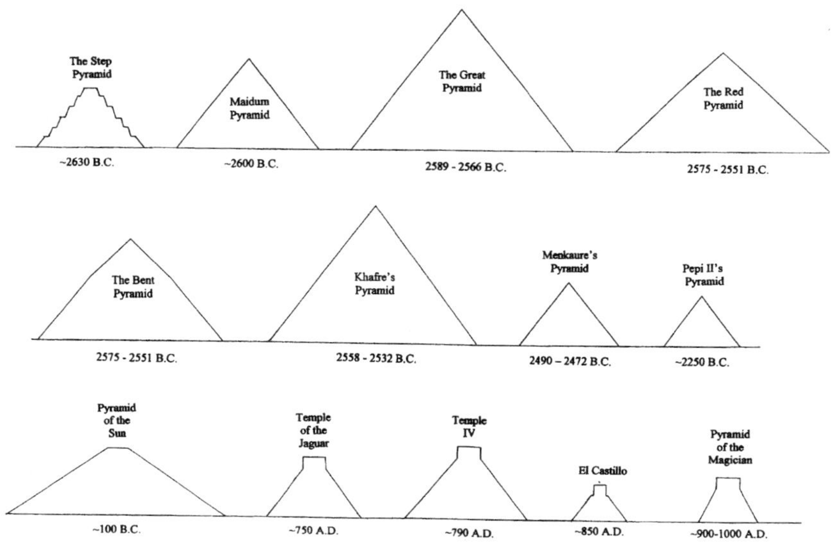

FIGURE 1.9. Pyramids of Egypt and Central America, chronologically (Bejan and Périn 2006)

* This section is based on Bejan and Périn (2006).

take anything away from the known or presumed achievements of the ancient, rather it is a physics argument that what the ancient chose to do is *natural*, that "to engineer is natural," and that the geometry of all material flows (animate, inanimate) can be reasoned based on a single principle. Pyramids and ant hills are like the dried beds of rivers, cracked mud and dendritic crystals (snowflakes). They are traces (fossils) of the optimized flow configurations that once existed.

The universal phenomenon is the generation of flow architecture, and the principle is the constructal law (Section 1.1): for a flow system to persist in time (to survive), its configuration must change such that it provides easier and easier access to its currents. In a flow system, easier access means less thermodynamic imperfection (friction, flow resistances, drops, shocks) for what flows through the river basin or the animal. The optimal distribution of these numerous and highly diverse imperfections is the flow architecture itself (lung, river basin, blood vascularization, atmospheric circulation, etc.). At the global level, the system flows with minimum but finite imperfection, which means that it destroys minimum useful energy (fuel, food, exergy).

In the making of a pyramid, the constructal law calls for the expenditure of minimum work. This principle delivers the *location* and *shape* of the edifice. First, the location is in the middle of the quarry, because minimum work means minimum sliding distance between mining site and construction site (Fig. 1.10). The pyramid and the quarry grow at the same time. If the pyramid is a "positive" architecture ($y > 0$), then the quarry is its negative ($y < 0$), where y points upward. Such positive–negative pairs are everywhere in history and geography, even though modern advances in transportation technology tend to obscure them.

Second, the shape of the pyramid can also be reasoned on the basis of the constructal law. In a pile of stones held together by gravity ("dry-stone construction"), shape means the base angle α. For simplicity, assume a conical pile. If D is the thickness of the stone cut from the quarry, then the conservation of stone volume requires

$$(\pi/3)r^2H = \pi R^2 D \tag{1.12}$$

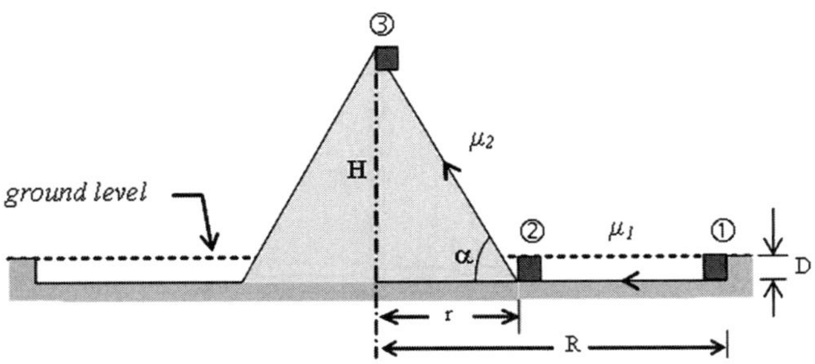

FIGURE 1.10. Constructal growth of a pyramid. Every stone follows an optimal path, which is a refracted ray. Sliding on the horizontal (1–2) is less dissipative than hopping up the incline (2–3). The pyramid shape is the optimal distribution of losses, such that the global movement of stones requires minimal effort (Bejan and Périn 2006)

where r and R are the radii of the base and the quarry, at a time during the construction. The pyramid and the quarry grow together. The newest stones added to the edifice come from the active wall (the perimeter) of the quarry.

The work required to move one stone from the quarry to its place in the pyramid is comparable with the work of moving the farthest stone, from 1 to 2 and 3 as shown in Fig. 1.10. Let N be the weight of one stone, and model the movement from 1 to 2, and from 2 to 3, as Coulomb friction with constant friction coefficients μ_1 and μ_2. The total work spent is

$$W = W_{12} + W_{23} = \mu_1 N (R - r) + (\mu_2 N \cos\alpha + N \sin\alpha) \left(H^2 + r^2\right)^{1/2} \quad (1.13)$$

The edifice has two aspect ratios, r/R and H/R, which are related through Eq. (1.12). Eliminating r/R between Eqs. (1.12) and (1.13), we obtain

$$\frac{W}{NR} = \mu_1 + 3^{1/2} (\mu_2 - \mu_1) \left(\frac{D}{R}\right)^{1/2} \left(\frac{H}{R}\right)^{-1/2} + \frac{H}{R} \quad (1.14)$$

where R/D is a dimensionless parameter that increases with the age of the construction. The total work W can be minimized by selecting H/R:

$$\left(\frac{H}{R}\right)_{opt} = \left(\frac{3}{4}\right)^{1/3} \left(\frac{D}{R}\right)^{1/3} (\mu_2 - \mu_1)^{2/3} \quad (1.15)$$

$$\left(\frac{r}{R}\right)_{opt} = 6^{1/3} \left(\frac{D}{R}\right)^{1/3} (\mu_2 - \mu_1)^{-1/3} \quad (1.16)$$

Remarkable is that the optimal height (H) and base radius (r) are both proportional to $R^{2/3}$, which means that they depend on time in the same way. Consequently, the optimal shape $(H/r)_{opt}$ is *time independent*: it is the same for all constructs, small and large:

$$\alpha_{opt} = \tan^{-1}\left[\frac{1}{2}(\mu_2 - \mu_1)\right] \quad (1.17)$$

An optimal shape exists when $\mu_2 > \mu_1$, i.e., when sliding on the incline is more dissipative than sliding horizontally. This is the case of the pyramids of Egypt: according to the Crozat theory (Crozat 2003; Crozat and Verdel 2002), the incline used for moving the stones upward was the pyramid itself, and the movement from 2 to 3 was effected step by step along the pyramid, by using wood levers and ropes. Each stone was (i) pushed horizontally, (ii) lifted, and (iii) dropped at the next, higher level. Steps (ii) and (iii) and the system of levers and ropes distinguish the movement on the stepped incline from the more common (older, and presumably more perfected) method used on the horizontal, hence $\mu_2 > \mu_1$. Had this not been the case, then the transportation of stones on the horizontal would have been in hops (according to scenario i–iii), not by steady sliding using a wooden sled.

The exact numerical value of the base angle α_{opt} is not the issue, because even if we could estimate the orders of magnitude of μ_1 and μ_2 today, this is not the knowledge that generated the location and shape of the pyramid. That knowledge was the "culture" of the time period—the tendency exhibited by many generations of large numbers of movers and moved (people, objects). Large numbers, time, freedom, and memory are essential in the generation of better flowing configurations. For example, in the construction of a river basin the number of water packets that move is immense: time and freedom allow the configuration to change, and memory is provided by old river beds and seismic faults. *Freedom is good for design*: the relationship between freedom to morph and performance is explored further in Bejan and Lorente (2004, 2005) and Bejan (2006).

The contribution made by this theoretical argument is that an optimal base angle *exists*, and that it is independent of the size of the edifice. The optimized movement (1–3 in Fig. 1.10) is a refracted path in the sense of Fermat. There are two "media" through which the stream of stones flows—two mechanisms—one with low resistivity (μ_1) and the other with high resistivity (μ_2). When the two media are highly dissimilar ($\mu_2 >> \mu_1$), the angle of refraction approaches 90°. Rivers, stones, and animals flow with configurations that come from the same principle (we return to this in Section 1.5).

The law of refraction governs the movement of goods in economics, where it is known as the law of parsimony. The history of the development of trade routes reveals the same tendency. We often hear that a city grew because "it found itself" at the crossroads—at the intersection of trade routes. We believe that it was the other way around: the optimally refracted routes defined their intersection, the city, the port, the loading and unloading site, etc.

More complicated flows are bundles of paths, optimally refracted such that the global flow access is maximized. A river basin under the falling rain is like an area inhabited by people: every point of the area must have maximum access to a common point on the perimeter. There are two media, one with low resistivity (channel flow; vehicles on streets) and the other with high resistivity (Darcy seepage through wet river banks; walking). The shape of the basin comes from the maximization of flow access.

For example, if in the rectangular territory of Fig. 1.11 (also Fig. 1.2) the objective is to have access to point M and if every inhabitant (Q) has two modes of transportation, walking with speed V_0 and riding on a faster vehicle with speed V_1, then the average travel time between all the points of the area and M is minimum when the area shape is $H/L = 2V_0/V_1$. This optimally shaped rectangle is a bundle of an infinite number of optimally refracted paths such as the broken line QRM.

Why is the constructal rectangle of Fig. 1.11 relevant to the discovery that the shape of the pyramid can be optimized? Because optimally shaped rectangular elements are everywhere in city living (e.g., Fig. 1.3), even though the builders of these elemental structures did not rely on the constructal law to design them.

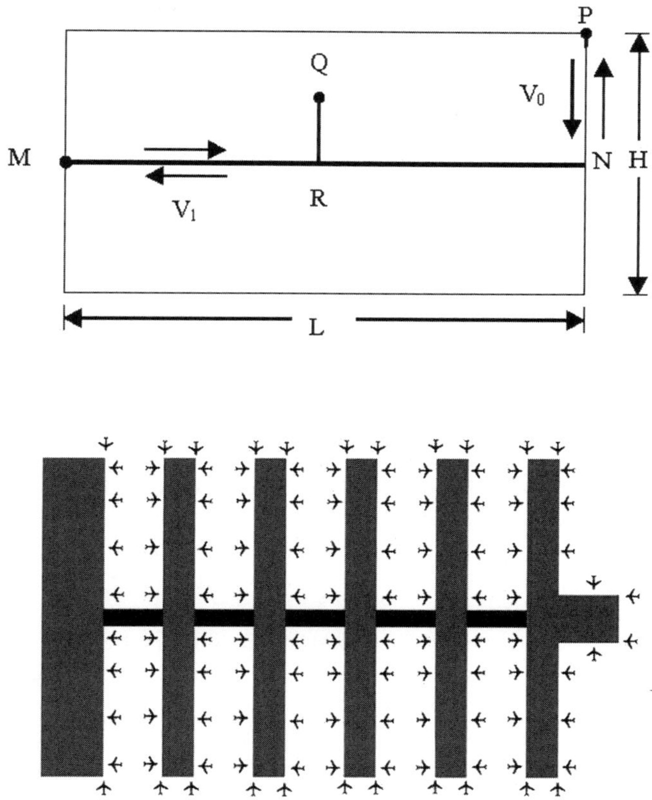

FIGURE 1.11. Optimally shaped rectangle-point access and the Atlanta airport. Because the train speed (V_1) is higher than walking (V_0), the shape of the rectangular territory H × L can be selected such that the inhabitants of the area have maximum access to one point (M). When the shape is optimal, the time of walking is the same as the time of riding on the train. Because of this, the rectangular shape of the Atlanta airport is constructal (Bejan and Lorente 2001)

They behaved in the same way as the builders of the pyramids and their ancestors: they balanced two dissimilar efforts (V_0 versus V_1, or μ_1 versus μ_2) in order to minimize the *global* effort.

As shown in Section 1.2, when the shape H/L is optimal, the travel time from P to N is the same as the travel time from N to M, namely $H/2V_0 = L/V_1$. In constructal theory this is known as the *equipartition of time* (resistance), or *the optimal distribution of imperfection*. This is illustrated by modern edifices such as the Atlanta airport (Fig 1.11). Several objectives were pursued in the development of this tree-shaped flow structure: the minimization of travel time for pedestrians and the minimization of time and transportation cost for the goods flowing between the terminal and each gate. The black line is the high-conductivity stem serviced by a two-way train. The dark bars are the concourses along which travel is much slower (walking, carts). In agreement with constructal theory, the time to walk on a concourse is the same ($\sim$ 5 min) as the time to ride on the train.

1.5. Animal Movement*

The late paleontologist and evolutionary biologist Stephen Jay Gould argued that if the clock of evolution could be rewound to the beginning and allowed to run again to the present day, the resulting animals on Earth would be very different from the ones we know now.

Gould's main point was that a high level of chance has been involved in determining which organisms have survived and evolved over the course of Earth's history. This is likely so but still, perhaps there are some boundaries, some general design rules would always govern the form of any animal life. An assumption behind natural selection is that some designs work better than others. But what are the principles that make some designs work better?

On this topic, biology might be able to take a cue from engineering, specifically from constructal theory. One of the basic goals of any design—whether it is an animal, a building, or a machine—is to get maximum output for minimum fuel or food (useful energy, exergy). This design principle can be seen, for example, in the tree-shaped flows of river basins, lung structure or the cracking pattern of drying mud flats, in the tube shape of pipes, or in the height versus depth proportionality in the cross sections of rivers. All of these designs allow for the maximum throughput of material with the least flow resistance.

The thought that the maximization of flow access could also be a mechanism that is responsible for constructing the configurations of natural flow systems—both animate and inanimate—is the principle of *constructal theory*. Simply stated, constructal theory says that for any finite-sized system to persist over time, it must evolve in such a way that it provides easier and easier access for the currents that flow through it (Section 1.1).

A flow configuration is an equilibrium of areas or volumes with high and low resistivities. This is achieved by an optimal distribution of imperfections, so that the maximum numbers of points of the area are "stressed" as equally as possible. To get this optimal balance of the various resistivities, the material must be distributed in certain ways.

For example, a river basin configures and reconfigures itself so that the water is discharged with less and less resistance through the mouth of the river. The tree shape of a mature river is the easiest-access configuration that connects an infinite number of points (the drainage basin) with one point.

Constructal theory has the potential to influence diverse areas outside of physics, including biology, engineering, and social sciences. For instance, animal

* This section is based on Bejan and Marden (2006a,b).

locomotion can be considered to be a flow of mass from one location to another. Animals move on the surface of Earth in the same way as rivers, winds, and oceanic currents. They seek and find paths and rhythms that allow them to move their mass the greatest distance per expenditure of useful energy while minimizing thermodynamic imperfections such as friction. Animals move in different ways for different purposes, but effective use of useful energy is important over a lifetime, and the basic design of most animals should evolve toward locomotion systems that maximize flow access, i.e., distance per cost.

There have been many analyses of animal locomotion, but most have been based on empirical relations—working backward from observations to find a model that fits the results. Constructal theory is a different approach because it works from a universal (physics) principle to deduce and predict structure and function. It not only predicts maximum-range speeds, but also simultaneously predicts stride/stroke frequencies and net force output.

This theory is not intended to account for all forms of biological variation. It does not maintain that animals must act or be designed in a predictable fashion, only that over large size ranges and diverse species, predictable central tendencies should emerge. Ecological factors will often favor species that move in ways other than that which optimizes distance per cost, such as where energy is abundant and the risk of being captured by active predators is high. Evolutionary history and the chance nature of mutation can also restrict the range of trait variation that has been available for selection. These and other factors should act primarily to increase the variation around predicted central tendencies.

The prevailing point of view is that intrinsic differences exist among the main types of locomotion—running, flying, and swimming. Runners and fliers have weight, whereas swimmers are neutrally buoyant. The wings of birds are structurally different from the limbs of antelopes and the tails of fish. The flapping motion of wings is unlike the hopping legs of a running animal and the undulating body of a swimmer. Birds and fish in cruising mode are at constant altitude and depth, whereas runners are constantly on a hopping (cycloidal) trajectory. Hitting the ground during running is far different from rubbing against air and water. Complicating the picture even further is the great diversity of body sizes, shapes, and speeds found in even a single form of locomotion (e.g., flying birds and insects).

Despite all these differences, numerous investigators have found that there are strong convergences in certain functional characteristics of runners, swimmers, and fliers. The stride frequency of running vertebrates scales with approximately the same relation to body mass (M) as the swimming frequency of fish, $M^{-0.17}$. The velocity of running animals scales with approximately the same relation to mass as the speed of flying birds, $M^{0.17}$. And the force output of the muscles of runners, swimmers, and fliers conforms with surprisingly little variation to a value of about 60 Newtons per kilogram. There are other correlations, such as that between body size, breathing rate, and metabolic rate (Bejan 1997c, 2000, 2001).

In an attempt to explain these consistent features of animal design, biologists have concentrated on potentially common constraining factors, such as muscle contraction—speed or structural-failure limits. However, constructal theory allows us to take a different approach, not of starting with constraints, but beginning with general design goals that can be used to deduce principles for optimized locomotion systems.

1.5.1. Flying

A bird in flight spends useful energy in two ways. One is vertical loss: the body has weight, so it falls incrementally and the bird performs work to lift itself back to cruising altitude. The other is horizontal loss: the bird performs work in order to advance horizontally against air friction. Both losses are needed for flying; neither can be avoided. However, they can be balanced against each other so that their sum is minimal. This optimal distribution of imperfection is flight itself.

Flight is not a steady movement at a constant altitude. Its trajectory is a saw-toothed horizontal line with a tooth size dictated by the flapping stroke (Fig. 1.12). It is an optimized rhythm in which the work of repositioning the body vertically is matched by the work of advancing the body horizontally. The balance is required by two competing trends: the vertical loss decreases and the horizontal loss increases as the flying speed increases. Balance is achieved by flapping such that the flying speed is just right.

With these parameters in mind, constructal theory predicts that flying speeds should be distributed in proportion to the body mass raised to the power 1/6. Flapping frequencies should be proportional to the body mass raised to the power $-1/6$. These predictions agree well with observations over the entire range of flying bodies. The details of this analysis are available in Bejan (2000) and Bejan and Marden (2006a). We derived this formula algebraically, in a process that involved numerous substitutions for equitable terms. In addition, according to the rules of *scale analysis* (Bejan 2004), all factors within one

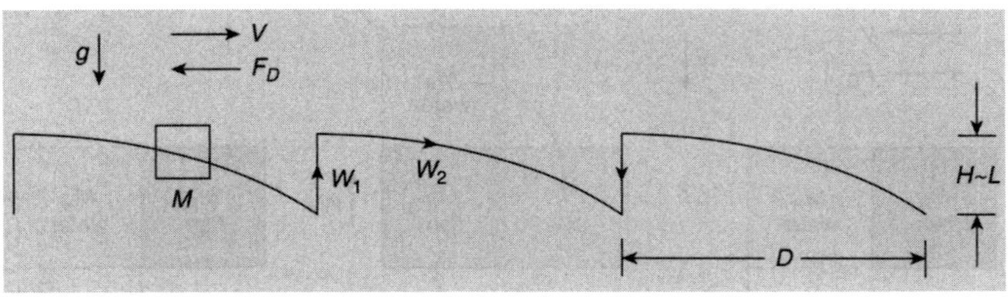

FIGURE 1.12. A simple diagram of the periodic trajectory of a flying animal shows the factors considered in estimating animal locomotion from constructal theory. The sawtooth pattern results because flying velocity (V) is composed of alternating work done to overcome vertical loss (W_1) and work done to overcome horizontal loss (W_2). W_1 is found by multiplying body mass (M), gravity (g), and the height the body falls during the cycle (H), the latter of which scales with body length (L). W_2 is the product of the force of air drag (F_D) and the distance traveled per cycle (Bejan and Marden 2006a,b)

order of magnitude of the value of 1 (0.1 to 10) were dropped for the sake of mathematical simplicity.

1.5.2. Running

If we treat running in the same way as flying—as an optimized intermittence in the Earth's gravitational field—we can also predict the speeds and stride frequencies of all runners. Running is a succession of cycles involving two losses (Fig. 1.13). One loss is the lifting of the body weight to a height that can be generalized as the body length (approximately the length of the limbs). This work is the vertical loss, because when the body lands, its gravitational potential energy is destroyed in the legs and the ground. For simplicity, we ignored elastic storage during landing. The second is horizontal loss: the work performed to overcome friction against the ground, the surrounding air and internal body parts although again for simplicity, we considered all friction to be external.

The vertical and horizontal losses compete, and when they are in balance their sum is minimal. The optimized intermittency called running is again found to be characterized by a speed proportional to $M^{1/6}$ and a stride frequency proportional to $M^{-1/6}$, using a similar derivation as was described in the previous section for the process of flying.

The predictions from this theory of running are quite robust. The horizontal loss may be dominated by dry friction against a hard surface, permanent deformation of a soft surface such as sand, mud or snow, or air drag. All these effects influence the speed and frequency, but they influence them in almost the same way. If air drag is the dominant horizontal loss mechanism, the speed and frequency deviate by only a factor of 10 from what they would be for runners with dry friction and ground deformation.

Another surprise came from the calculation of the work spent on lifting the body off the round. For both runners and fliers, the average force exerted over

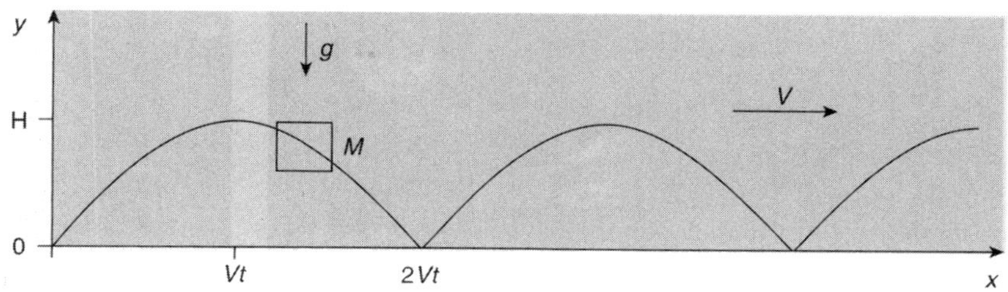

FIGURE 1.13. In the periodic trajectory of a running animal, the distance of each stride is a multiple of the animal's velocity (V) multiplied by the time (t) of frictionless fall from the height of the run (H). Therefore, t is equivalent to H divided by gravity (g), raised to the power 1/2. The stride length and H both scale with the body length, and the body mass (M) is approximated by the body density multiplied by the body length cubed (Bejan and Marden 2006a,b)

the stride or stroke cycle should be twice the body weight. This agrees with the force–weight measurements across all body sizes, for all animals that fly and run.

1.5.3. Swimming

So far we have seen that as an optimized intermittency, running is similar to flying. Is swimming like running and flying? The obvious answer is no, because the movements of the neutrally buoyant bodies of fish seem to have nothing to do with gravity. This view has until now prevented the emergence of a physical theory of locomotion that includes swimming.

The reason why running is no different from swimming or flying (in spite of the fact that swimmers and fliers do not touch the ground) is that the ground supports the weight and feels the movement of every body that exists above it. The same ground serves as a reference against which all moving bodies push, and without it no locomotion is possible.

In swimming, because the bottom of a body of water is immense and stationary, a fish can push and move its body relative to the ground by performing work against gravity and friction, just like a bird or an antelope.

To advance horizontally by one body length, a swimmer's body must do work equivalent to lifting a parcel of water of its own size to a height approximately equal to its body length. This body of water must be lifted because a net vertical displacement is the only way that water can flow around an animal, or any object (Fig. 1.14). The ground under the water does not move, so only the free surface is deformable. This is readily visible as a bow wave lifted in front of a body moving along a water surface, but what has not been appreciated previously is

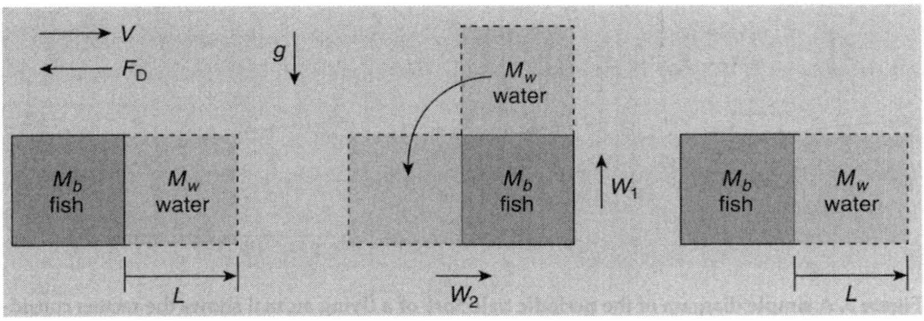

FIGURE 1.14. In order to make forward progress, a fish must move water out of the way, and the net direction the water can go is up. In order to move one body length (L) at a certain velocity (V), a fish with body mass M_b must move an equivalent mass of water M_w. This mass of water can then be thought of as moving downward to occupy space now vacated by the fish. The work required to move the water mass upward (W_1) is approximated by multiplying M_b with L and gravity (g). During the same interval, the fish must do work in order to move horizontally (W_2) that is proportional to the force of water drag (F_D) and the distance traveled per cycle, which in this case is body length (L) (Bejan and Marden 2006a,b)

that this vertical work is non-negligible and is fundamental to the physics of swimming at all depths.

Why don't we see this free-surface deformation caused by every fish that lifts water over itself in order to progress horizontally? Because most fish are small and swim at depths larger than their body length scale. The lifted water is equal to the volume displaced by the free surface as it rises over a very large area—all the larger when the fish is deeper. The lifting of the free surface is visible when the fish is large and near the surface. Elevation of the water surface also has been demonstrated and used in the field of naval warfare, where certain radar systems are able to detect a moving submarine by the change in the surface water height as it passes.

With this, constructal theory accounts for swimming. Its predictions of speed and stroke frequency are the same as those for running on deformable ground, and they agree with much data (Fig. 1.15).

Thus, even though some animals do not touch the ground, they use the ground to propel themselves. They have no choice. The ground is the only "firm spot" in Archimedes' famous dictum: *Give me a firm spot, and I will move the world.* The flapping of the bird's wings produces vortices of air that eventually stagnate against the atmosphere and the ground and increase the pressure that the ground supports. The water lifted by the swimming fish induces a local elevation of the free surface and a greater pressure on the lake bottom. The ground feels, and supports, everything that moves, regardless of the medium in which a particular animal is moving.

In sum, the constructal law predicts complex features of animal design. We have provided evidence that if evolution were rewound and if runners, swimmers and fliers appeared again, the process should consistently produce the same types of speeds, stroke–stride frequencies, and force outputs of these forms of locomotion as exist today. The theory could even be used to predict how these features would evolve on other planets with different gravitational forces and densities of the gaseous and liquid environments.

Humans routinely encounter different terrain and we adjust our speed and stride frequency accordingly. Consciously and unconsciously, we pay attention to the effectiveness of our movement patterns, and we may be wired to select optimal gaits. When astronauts walked on the moon, they encountered a completely different gravitational force, so it would be interesting to see how well their preferred speeds and hopping frequencies match the predictions of our theory.

The predictions of constructal theory are consistent not only for animals, but also for man-made machines, or "man & machine species" (Fig. 1.16). The force–mass relation of engineered motors is the same as that of runners, fliers, and swimmers (Fig. 1.15 c). The constructal theory of animal flight also predicts speeds of machine flight (Fig. 1.17) and unites the animate with the inanimate (Bejan 2000).

The theory can, for instance, help in the design of efficient robots to roam on other planets or in remote environments on Earth. More fancifully, our predictions could be used in animation to choose the speeds and stride patterns of creatures such as dinosaurs, to accurately show how, for instance, a Tyrannosaurus Rex should look while chasing a vehicle in a movie.

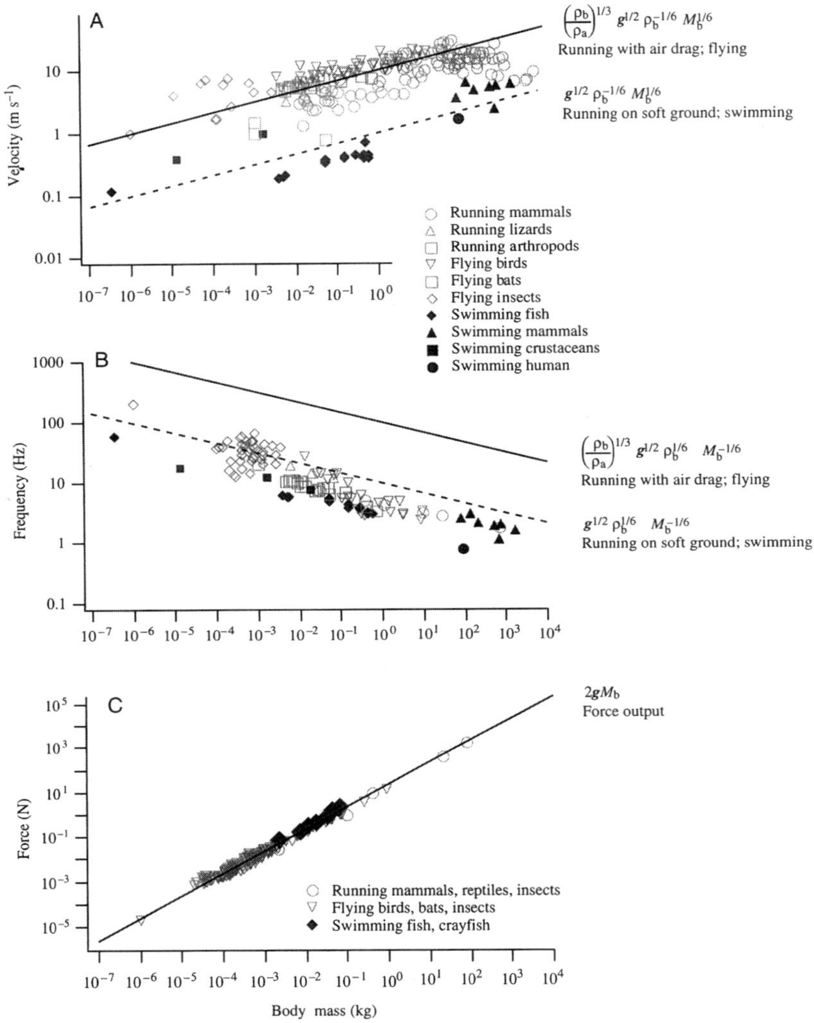

FIGURE 1.15. Theoretical predictions from constructal theory are compared with the velocities, frequencies of strokes or strides, and force outputs of a variety of animals. Solid lines in these log-scale graphs show the predicted velocity (a) or frequency (b) of animals based on body mass for flying animals or running animals where the ground is hard and thus the main frictional loss is due to air drag. Dashed lines show the predicted velocity (a) or frequency (b) of animals based on body mass for swimming animals or running animals where the ground is soft and thus the main frictional loss is due to ground deformation. A dotted line indicates the predicted force output, based on body mass (c). The theoretical predictions ignore constants between 0.1 and 10, and so are expected to be accurate within an order of magnitude (Bejan and Marden 2006a,b)

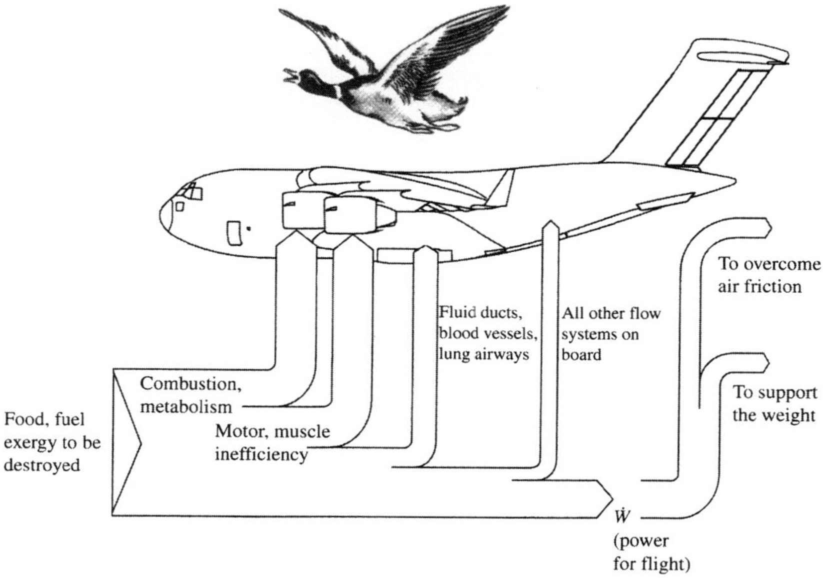

FIGURE 1.16. During flight, useful energy input into a system is destroyed completely in a distributed fashion by components that have equivalent functions in animals and machines. Constructal theory predicts that all flow-system structures result from the clash between two objectives: the need to carry substances from the core to the periphery and the need to avoid direct leakage of these substances and energy (such as heat) into the ambient surroundings. The ultimate purpose of food or fuel taken in by a bird or put into an airplane is to provide power for flight, which overcomes air friction and supports the mass of the animal or machine. In between fuel input and power output, however, useful energy is destroyed by various flow systems (Bejan and Marden 2006b, after Bejan 2005)

1.6. Patterned Movement and Turbulent Flow Structure

A new theory predicts, explains, and organizes a body of knowledge that was growing empirically. This we have seen in this chapter by bringing under the constructal law the movement of goods, people, and animals, and the construction and distribution of human settlements. In particular, animal movement is no different than other flows, animate, and inanimate: they all develop (morph, evolve) architecture in space and time (self-organization, self-optimization, survival), so that they maximize the access of the flow of matter in nature.

All animals, regardless of their habitat (land, sea, air), mix air, water, and soil much more efficiently than in the absence of flow structure. It sounds crude, but when all is said and done, this is what living flow systems accomplish and why their legacy is the same as that of the rivers and the winds. Constructal theory has already predicted the emergence of turbulence by showing that an eddy of length scale L_b, peripheral speed V, and kinematic viscosity ν transports momentum across its body faster than laminar shear flow when the *local Reynolds number*

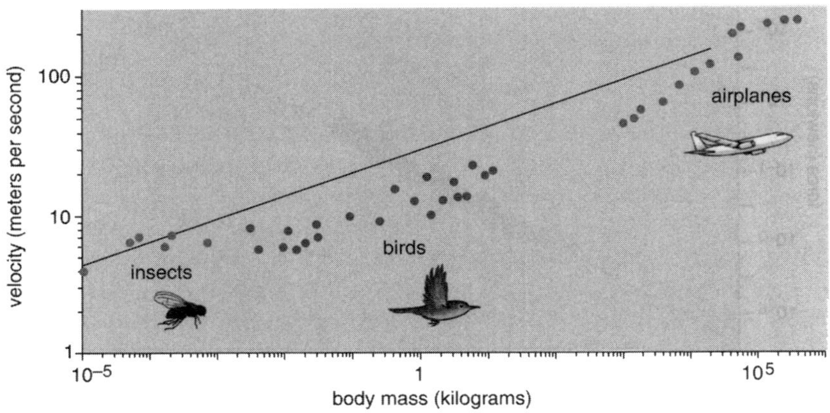

FIGURE 1.17. Constructal theory's predictions of velocity based on body mass extend from animals to man-made machines. The force–mass relation of engineered motors is the same as that for all types of animal locomotion. In this log-scale graph, there is a direct scaling relationship between flying animals (insects and birds) and airplanes. The solid line shows the predicted velocity from constructal theory, whereas dots show measured values of mass versus velocity for insects, birds, and airplanes (Bejan and Marden, 2006b after Bejan 2000)

$L_b V/\nu$ exceeds approximately 30 (cf. Bejan 1997c, 2000). This agrees very well with the zoology literature, which shows that undulating swimming and flapping flight (i.e., locomotion with eddies of size L_b) is possible only if $L_b V/\nu$ is greater than approximately 30.

So we arrive at an unexpected link that this simple physics theory reveals: the generation of optimal distribution of imperfection (optimal intermittence) is governed by the same principle as the generation of turbulent flow structure. The eddy and the animal that produces it are the optimized "construct" that travels through the medium the easiest and mixes Earth's crust most effectively.

The action of the constructal law in the structure of turbulence is evident. Migratory birds fly in large groups organized in precise patterns. The same external organization of movement is visible on the ground, in team bicycle racing, and in the slipstream shed by the lead race car. The principle is the same: aerodynamic drag decreases when individual mass coalesce, and in this way mass travels faster and farther. Strings of racers on the ground (conga lines) and flying carpets (birds, airliners) in the air are all driven by the same principle.

Processions are visible everywhere. Schools of fish are beautiful because they display precise patterns of organization. The synchronized swimming and jumping of dolphins are also due to spatial organization. Ducklings paddle behind their mother arranged in a pattern, like pieces on a chessboard.

Biologists and fluid mechanicists have long recognized the geometric relation between a patterned procession and the regular occurrence of vortices in the wake shed by the body that leads the procession. They are right of course, but not in the general sense that is meant in constructal theory. It is the inanimate fluid in the wake of the leading body that organizes itself. It does so using no brain power whatsoever, so that it may travel and spread itself the fastest through the stationary fluid.

The fish and the ducklings, for their own constructal reasons, act as markers in the flow. They feel the flow, and they follow it. In this way, they visualize the constructal law that accounts not only for the eddies and turbulence in the fluid, but also for the animals' urge (tendency) to advance bodily most easily.

1.7. Science as a Constructal Flow Architecture

Science is our knowledge of how nature works. Nature is everything, including engineering: the biology and medicine of human and machine species. Our knowledge is condensed in simple statements (thoughts, connections), which evolve in time by being replaced by simpler statements. We "know more" because of this evolution in time, not because brains get bigger and neurons smaller and more numerous. Our finite-size brains keep up with the steady inflow of new information through a process of simplification by replacement: in time, and stepwise, bulky catalogs of empirical information (e.g., measurements, data, complex empirical models) are replaced by much simpler summarizing statements (e.g., concepts, formulas, constitutive relations, principles, laws). A hierarchy of statements emerges along the way: it emerges naturally, because it is better (in accordance with the constructal law, Section 1.1).

The simplest and most universal are the laws. The bulky and the laborious are being replaced by the compact and the fast. In time, science optimizes and organizes itself in the same way that a river basin evolves: toward configurations (links, connections) that provide faster access, or easier flowing. The bulky measurements of pressure drop versus flow rate through round pipes and saturated porous media were rendered unnecessary by the formulas of Poiseuille and Darcy. The measurements of how things fall (faster and faster, and always from high to low) were rendered unnecessary by Galilei's principle and the second law of thermodynamics.

The hierarchy that science exhibited at every stage in the history of its development is an expression of its never-ending struggle to optimize and redesign itself. Hierarchy means that measurements, *ad hoc* assumptions, and empirical models come in huge number, a "continuum" above which the compact statements (the laws) rise as needle-shaped peaks. Both are needed, the numerous and the singular. One class of flows (information links) sustains the other. The many and unrelated heat engine builders of Britain fed the imagination of one Sadi Carnot. In turn, Sadi Carnot's mental viewing (thermodynamics today) feeds the minds of contemporary and future builders of all sorts of machines throughout the world.

Civilization with all its constructs (science, religion, language, writing, etc.) is this never-ending physics of generation of new configurations, from the flow of mass, energy, and knowledge to the world migration of the special persons to whom ideas occur (the creative). Good ideas travel. Better-flowing configurations replace existing configurations. Empirical facts are extremely numerous, like the hill slopes of a river basin. The laws are the extremely few big rivers, the Seine and the Danube.

References

Bairoch, P., Batou, J. and Chèvre, A. (1988) *The Population of European Cities from 800–1850*, Droz, Geneva.

Bejan, A. (1996) Street network theory of organization in nature. *J. Adv. Transp.* **30**, 85–107.

Bejan, A. (1997a) Constructal-theory network of conducting paths for cooling a heat generating volume, *Int. J. Heat Mass Transfer* **40**, 799–816.

Bejan, A. (1997b) Theory of organization in nature: pulsating physiological processes. *Int. J. Heat Mass Transfer* **40**, 2097–2104.

Bejan, A. (1997c) *Advanced Engineering Thermodynamics*, 2nd edn., Wiley, New York.

Bejan, A. (2000) *Shape and Structure, from Engineering to Nature*. Cambridge University Press, Cambridge, UK.

Bejan, A. (2001) The tree of convective heat streams: its thermal insulation function and the predicted 3/4-power relation between body heat loss and body size. *Int. J. Heat Mass Transfer* **44**, 699–704.

Bejan, A. (2004) *Convention Heat Transfer*, 3rd edn., Wiley, Hoboken, New Jersey.

Bejan, A. (2005) The constructal law of organization in nature: tree-shaped flows and body-size. *J. Exp. Biol.* **208**, 1677–1686.

Bejan, A. (2006) *Advanced Engineering Thermodynamics*, 3rd edn, Wiley, Hoboken, New Jersey.

Bejan, A. and Lorente, S. (2001) Thermodynamic optimization of flow geometry in mechanical and civil engineering. *J. Non-Equilib. Thermodyn.* **26**, 305–354.

Bejan, A. and Lorente, S. (2004) The constructal law and the thermodynamics of flow systems with configuration. *Int. J. Heat Mass Transfer* **47**, 3203–3214.

Bejan, A. and Lorente, S. (2005) *La loi constructale*, L'Harmattan, Paris.

Bejan, A. and Lorente, S. (2006) Constructual theory of generation of configuration in nature and engineering, *J. Appl. Phys.* **100**, 041301.

Bejan, A. and Marden, J. H. (2006a) Unifying constructal theory for scale effects in running, swimming and flying. *J. Exp. Biol.* **209**, 238–248.

Bejan, A. and Marden, J. H. (2006b) Constructing animal locomotion from new thermodynamics theory. *American Scientist*, July–August, 343–349.

Bejan, A. and Périn, S. (2006) Constructal theory of Egyptian pyramids and flow fossils in general. Section 13.6 in Bejan (2006).

Bejan, A., Lorente, S., Miguel, A. F. and Reis, A. H. (2006) Constructal theory of distribution of city sizes. Section 13.4 in Bejan (2006).

Bretagnolle, A., Mathian, H., Pumain, D. and Rozenblat, C. (2000) Long-term dynamics of European towns and cities: towards a spatial model of urban growth, *Cybergeo* **131**, March 29.

Carone, M. J. (2003) Applying constructal theory for product platform design in the context of group decision-making and uncertainty, M.S. thesis, Georgia Institute of Technology, Atlanta, GA.

Carone, M. J., Williams, C. B., Allen, J. K. and Mistree, F. (2003) An application of constructal theory in the multi-objective design of product platforms, ASME Paper

DETC2003/DTM-48667, *Proceedings of DETC'03, ASME 2003 Design Engineering Technical Conferences and Computers and Information in Engineering Conference*, Chicago, September 2–6.

Crozat, P. (2003) Le Génie des Pyramides, Doctoral dissertation, National Polytechnical Institute of Lorraine, Nancy, France.

Crozat, P. and Verdel, T. (2002) Système constructif des pyramides: de la géologie à l'édification, presented at Journées Nationales de Géotechnique et de Géologie de l'Ingénieur, JNGG 2002, Nancy, France, Oct. 8–9.

Errera, M. R. and Bejan, A. (1998) Deterministic tree networks for river drainage basins, *Fractals*. **8**, 245–261.

Hernandez, G. (2001) Design of platforms for customizable products as a problem of access in a geometric space, Ph.D. thesis, Georgia Institute of Technology, Atlanta, GA.

Hernandez, G., Allen, J. K. and Mistree, F. (2003) Platform design for customizable products as a problem of access in a geometric space, *Eng. Opt.* **35**, 229–254.

Ledezma, G. A. and Bejan, A. (1998) Streets tree networks and urban growth: optimal geometry for quickest access between a finite-size volume and one point, *Physica A* **255**, 211–217.

Ledezma, G. A., Bejan, A. and Errera, M. R. (1997) Constructal tree networks for heat transfer, *J. Appl. Phys.* **82**, 89–100.

Lewins, J. (2003) Bejan's constructal theory of equal potential distribution, *Int. J. Heat Mass Transfer* **46**, 1541–1543.

Moriconi-Ebrard, F. (1994) *Geopolis*, Anthropos-Economica, Paris.

Poirier, H. (2003) Une théorie explique l'intelligence de la nature, *Sci. Vie* **1034**, 44–63.

Reis, A. H. (2006) Constructal theory: from engineering to physics, and how flow systems develop shape and structure, *Appl. Mech. Rev*. **59**, 269–282.

Rosa, R. N., Reis, A. H. and Miguel, A. F. (2004) *Bejan's Constructal Theory of Shape and Structure*. Évora Geophysics Center, University of Évora, Portugal.

Torre, N. (2004) La Natura, vi svelo le formule della perfezione *Macch. Tempo* **5**, 36–46.

Upham, G. and Wolo, J. (2004) Coordination of dynamic rehabilitation experiments of people living with post-polio syndrome and call for research into PPS as a forgotten disease, presented at the Global Forum for Health Research, Forum 8, Mexico, D.F., Nov.

Chapter 2
Constructal Models in Social Processes

Gilbert W. Merkx

2.1. Introduction

This chapter argues that the patterns predicted by constructal theory[1] for natural phenomena are also found in social phenomena involving large number of actors who are linked through networks involving exchanges. Among the types of social phenomena that can be expected to reflect the same tree patterns that are predicted by constructal theory for natural flows are social networks, social stratification systems, episodes of collective behavior such as panics and crazes, commodity chains, capital investment flows, and migration streams. Examples of the last two phenomena, capital investments and migration, are discussed at greater length.

In recent years, a considerable literature has developed around the concepts of social networks and social capital, the former being a source of the latter.[2] This literature makes several general points. First, social networks are neither random nor uniform (fishnet-like) but consist of strong ties of interactions between many units and a center node. Second, the network will have weak ties or bridges to other networks leading to flows among networks. Third, networks are more efficient systems than alternative forms of linkage. Fourth, the greater efficiency of nodal networks as compared to other systems provides greater benefits (also called social capital) in various forms (e.g., improved information, trust) than found in alternative systems. Fifth, networks grow as the node attracts new satellites. Sixth, larger networks grow at the expense of smaller networks. Finally, the dimensions (the relative size and number of branches) of social networks

---

[1] Adrian Bejan, *Shape and Structure from Engineering to Nature* (Cambridge University Press, Cambridge, 2000).

[2] Yochai Benkler, *The Wealth of Networks* (Yale University, New Haven, CT, 2006); Pamela Walker Laird, *Pull: Networking and Success Since Benjamin Franklin* (Harvard, Cambridge, MA, 2006); Maria Forsman, *Development of Research Networks; the Case of Social Capital* (Abo Academi, 2005). Mark Buchanan offers a broad popular account of the literature in *Nexus: Small World and the Groundbreaking theory of Networks* (Norton, New York, 2002).

reflect fractal-like properties similar to those of tree-shaped networks in nature, such as those found for river basins by Rodriguez-Iturbe and Rinaldo.[3]

Maps or illustrations of social networks, such as the Internet or an air traffic grid, usually portray them as two-dimensional systems with linkages radiating out from the nodes to the periphery (or from the periphery to the nodes). However, this type of network is merely a version of the tree structures predicted by constructal theory. Thus, a real three-dimensional tree with leaves if viewed from above or below would appear to be a node with radial arms, or a network. Rivers are also three-dimensional tree structures (with one dimension somewhat flattened that would appear to be networks if they could be viewed from the source to the mouth or vice versa).

If it is the case that social networks are in fact three-dimensional tree structures rather than two-dimensional radial networks, then what is characterized as a node in social network theory is not actually a point, but rather a trunk or a channel along which social exchanges flow once they are collected from the periphery, or before they are distributed to the periphery. This chapter presents two cases in which networks are connected by channels rather than nodes. These cases suggest that the networks conform to the predictions of constructal theory. They also suggest that the origins and characteristics of the main trunk (which are historically contingent on exogenous variables) are other important determinants of the rise of the social network.

2.2. Natural Versus Social Phenomena: An Important Distinction?

The idea that constructal theory, which describes properties of physical or natural systems, should also describe properties of social systems may appear questionable. There is, after all, a long intellectual tradition dating back to Kant based on the conception that there are two different realms of human knowledge, the natural sciences and the studies of culture and social behavior. Perhaps the most famous expression of this perspective is found in the German sociologist Max Weber's concept of "verstehen."[4] Verstehen, or roughly "sympathetic understanding," is the notion that the behavior of social actors is motivated by thought and culture, allowing an understanding of the reasons for that behavior that is very different in character from explanations that describe changes in inanimate or physical units.

If it is granted that people are not like drops of water, why would there be by any reason to think that social networks should resemble river networks or trees? At least four possible explanations come to mind. The first is that the unique

[3] Ignacio Rodriguez-Iturbe and Andrea Rinaldo, *Fractal River Basins* (Cambridge University Press, Cambridge, 1997).

[4] Max Weber, *The Theory of Economic and Social Organization* (Free Press, New York, 1964), p. 87.

characteristics of each of the individuals that compose a network are irrelevant to the character of the network itself. Thus, no two leaves on an oak tree are identical, but they perform similar functions as members of the same tree system. Weber's concept of bureaucracy is premised on a similar assumption, which is that the rules of the bureaucratic system determine behavior, not the unique characteristics of the individuals in the bureaucracy. Thus, the individuals can be replaced without changing the nature of the bureaucratic organization.

A second explanation is that individual motivations are cancelled out in situations involving large numbers of people, or masses, a topic studied by the field of collective behavior. The behavior of individuals in a mob may be radically different from the behavior of those individuals when alone. Similar effects may be noted when large numbers of people are motivated by shared beliefs, even when they are not physically together, as in crazes (e.g., the tulip mania in 17th century Holland) and panics (the stock market crash of 1929). If the process of sharing beliefs, either in a physical crowd or in a virtual crowd, constitutes the formation of a social network, then collective behavior episodes may also fall under the general rubric of being system-determined behavior.

A third explanation could be that while people's motivations can be understood, such an understanding leads to the discovery that most people most of the time are rational actors who aspire to minimize the costs and maximize the benefits of their behavior. This is of course the basis of rational choice theory, which underlies modern economics and is gaining ground in the other social sciences. In any case, to the extent that people behave to maximize their benefits, they will select, or rediscover, or invent more efficient systems of interaction. Constructal theory explains that tree networks exist because they require the least amounts of useful energy, and hence are the most efficient. Hence, rational actors will tend to gravitate toward social networks that exhibit the same properties of efficiency shown by systems in nature.

Fourth, it may simply be the case that any sustained series of human interactions becomes by its very nature a network, and that all networks, other things being equal, evolve toward efficiency. If they do not, they are replaced by other networks. The more that such a network evolves in the direction of efficiency, the more it will tend to look like other efficient networks, i.e., the more it will resemble the tree systems found in nature. One of the contributions of constructal theory is to show that the trees in nature are not fractal, because they have a finite, predictable smallest scale (e.g., the alveolus of the lung). In any case, the individual motivations in social contexts, or the differences among constituent parts in systems of nature, are simply not relevant to understanding the evolution of networks, which is driven by emergent properties common to tree networks in both nature and social life.

It is obvious, of course, that these four explanations are not mutually exclusive. While they start from slightly different premises, they lead to the same conclusion, which is that the emergent properties of the system are what that counts, not the irreducible or unique properties of the individual units that populate and move through the system.

2.3. Case Studies: Two Social Networks

2.3.1. The Argentine Railway Network: 1870–1914

Between 1870 and 1914 Argentina grew more rapidly than any other economy in the world. In 1870 it was one of the poorest countries in the world. By 1914 it is estimated to have had the seventh highest standard of living per capita in the world. The extraordinary growth of Argentina was a by-product of the rapid industrialization of Europe and North America, which created an unprecedented demand for beef, grains, and wool. It was made possible by a timely series of technological breakthroughs. Argentina had vast riches of arable soil and, in the pampas, the best pastureland in the world. The invention of barbed wire and the windmill allowed that land to be exploited. The invention of refrigerated shipping provided a means for fresh Argentine beef to reach Europe. Italian and French migrants provided the necessary work force both in the countryside and in the cities. And last, but not the least, British capital and engineering expertise provided the basis for the development of a railroad network that could link the pampas with the coast.[5]

The exponential growth of British investment, all foreign investment, immigration, land, and railroad mileage in Argentina from 1870 and 1940, is presented in Fig. 2.1. Note logarithmic scale on the ordinate. The growth of these factors accelerates up to the start of World War in 1914, at which time growth essentially terminates. Table 2.1 shows the operational mileage of the Argentine railways between 1880 and 1900. Figures 2.2–2.5 present the outline of the Argentine railroad network in 1866, 1882, 1896, and 1914.[6]

The articulation of the Argentine railway structure resulted in a tree-like structure of the type predicted by constructal theory, with the port of Buenos Aires being the point of access to the trunk. The trunk line of the network consisted of the shipping route from Buenos Aires to the northeastern horn of Brazil and thence to Europe, where routes diverged from the different trunk ports. Those ports were linked by railroads to markets, creating a second tree structure or delta.

There are some other interesting phenomena from this case that are worth exploring. One is the flow of capital toward Argentina. The railroads were financed not by direct foreign investment, but by the floating of bonds on the London market. The bond issues were packaged by consolidators, primarily the Baring Brothers firm, and sold to the British public. The interest rates offered were high and the sales of these bonds created a broad network of investors, with Baring Brothers serving as the node or trunk.

The success of Argentine railway bonds sales triggered not one but two episodes of collective behavior evincing constructal patterns. The first episode

[5] Gilbert W. Merkx, "Political and Economic Change in Argentina, 1870–1966", Ph.D. Dissertation (Yale University, New Haven, CT, 1968).

[6] The railroad network maps are taken from Colin M. Lewis, *British Railways in Argentina 1857–1914* (Athlone Press, New York 1983).

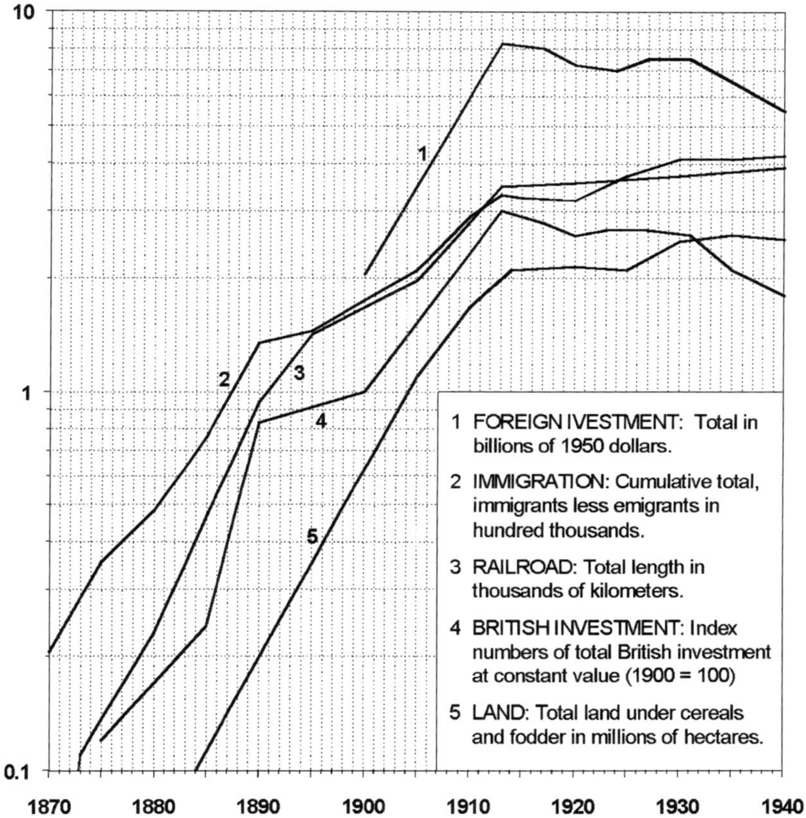

FIGURE 2.1. The growth of economic factors in Argentina between 1870 and 1940

was an explosion of investor interest, or a craze, in which investors avidly sought Argentine railway bonds, resulting in a geometric growth of bond issues and a rapidly expanding investor network thanks to word-of-mouth dissemination. As Colin Lewis notes, "Argentine railway issues were the long-awaited *eldorado*. Quotations on the London Stock Exchange rose to hitherto unimagined heights. Within the context of a boom mentality, new subscribers displayed an amazing ability for self-delusion." Thus, the Argentine railroads were made possible by tree network in England in which the money that flowed from thousands of individual investors to Baring Brothers was transferred to Argentina, and then flowed out as new mileage for the Argentine railway network.

Unfortunately, this virtuous cycle was not to last. The process spun out of control. In 1890, Argentina's inability to provide sufficient foreign currency to cover gold-denominated debt-service payments led to a suspension of debt-payments. This led to the second collective behavior episode, a panic in which the public frantically tried to unload Argentine securities at the very time that new

TABLE 2.1. Argentine railways: expansion of the network

Year	Operational mileage	Increase
1880	1563	–
1881	1563	0
1882	1636	73 (4.7%)
1883	1966	330 (20.2%)
1884	2261	295 (15.0%)
1885	2798	537 (23.8%)
1886	3627	829 (29.6%)
1887	4157	530 (14.6%)
1888	4705	549 (13.2%)
1889	5069	364 (7.7%)
1890	5861	792 (15.6%)
1891	7752	1891 (32.3%)
1892	8502	750 (9.7%)
1893	8607	105 (1.2%)
1894	8718	111 (1.3%)
1895	8772	54 (0.6%)
1896	8986	214 (2.4%)
1897	9169	183 (2.0%)
1898	9601	432 (4.7%)
1899	10199	598 (6.2%)
1900	10292	93 (0.9%)

Source: Compiled and calculated from República Argentina, Ministerio de Obras Públicas, Dirección General de Ferrocarriles *Estadística de los ferrocarriles en explotación durante el año 1913* (Bueons Aires 1916), pp. 396–398; A. E. Bunge *Ferrocarriles argentinos: contribucion al estudio del patrimonio nacional* (Bueons Aires 1918), pp 120–121

Argentine securities were flooding the market. Indeed, in 1890 Baring Brothers had a portfolio of unsold Argentine securities that exceeded the entire reserves of the Bank of England. Although the British financial establishment was able to weather the storm and England's rate of investment recovered somewhat toward the end of the decade, the rate of growth was lower, and ceased entirely with the advent of World War I. The loss of British capital in that war was such that British investment in Argentina was never to resume.

The flow of investment was not the only factor of production to dry up in 1914. The flow of immigration also stopped. Expansion of the railroad network ceased, and the size of the network began a long slow decline. In the absence of further expansion of the railway network, little new land was brought under cultivation. More fundamentally, World War I left Europe close to bankruptcy, and European demand for Argentine stopped growing. When the Great Depression of 1930 hit, trade between Argentina and Europe collapsed and Argentina defaulted on its debt payments. The flows of commodities to Europe and capital to Argentina dried up. The Argentine railway system became as frozen in time as a dried river delta in the desert.

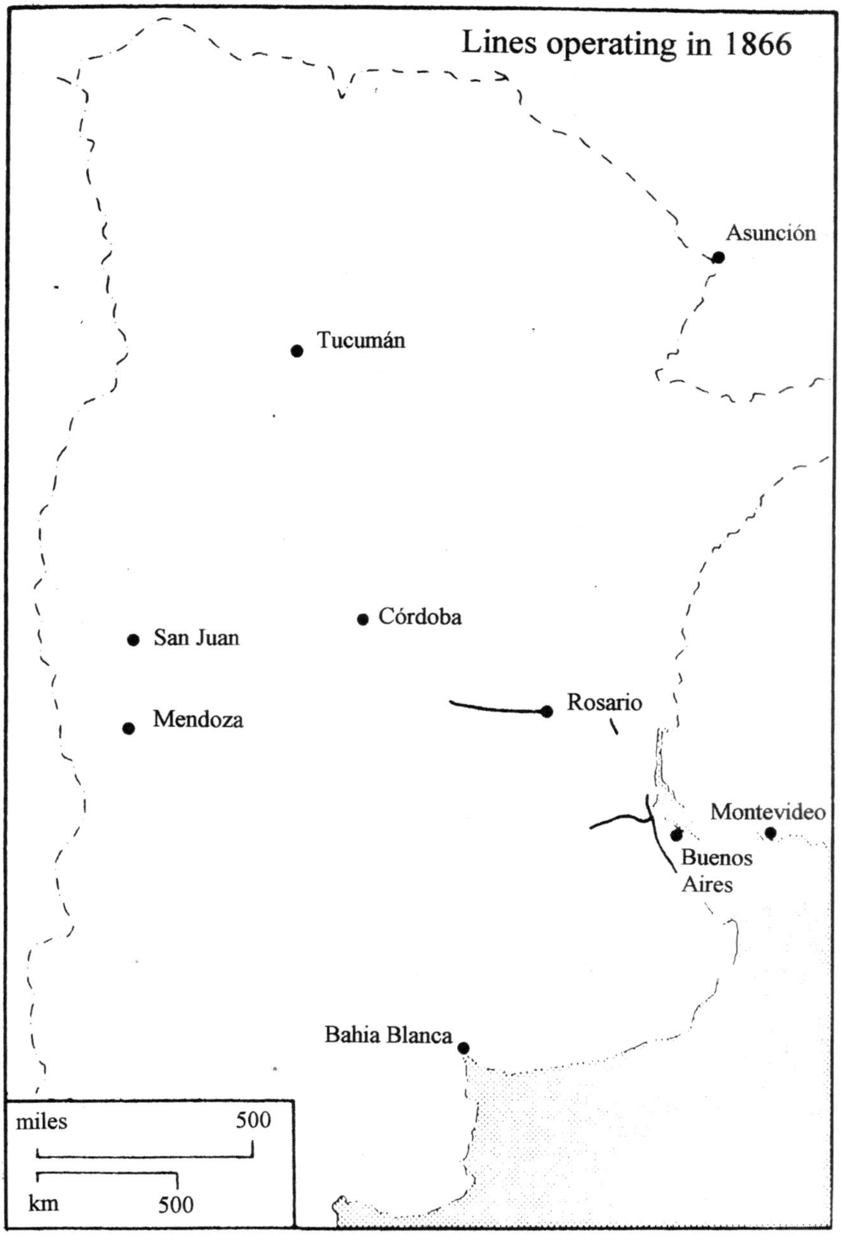

FIGURE 2.2. Argentine rail grid, 1866

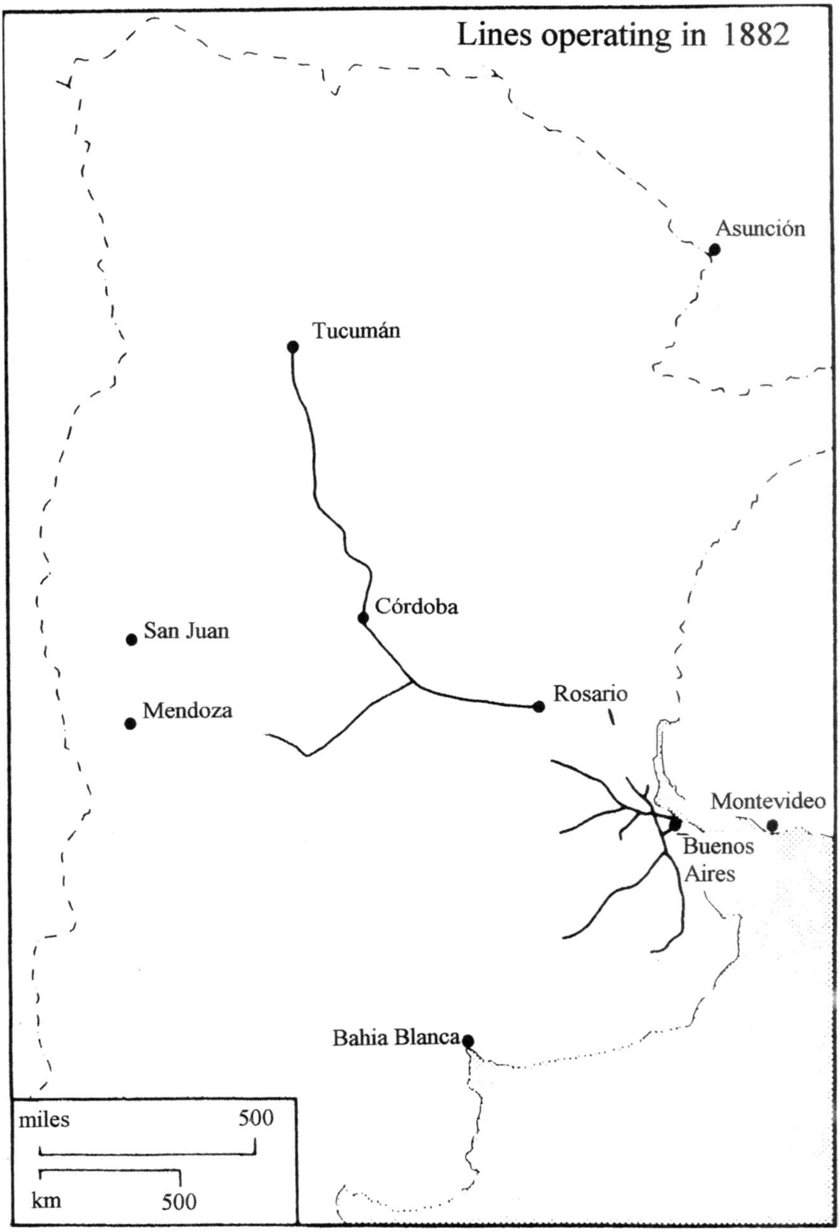

FIGURE 2.3. Argentine rail grid, 1882

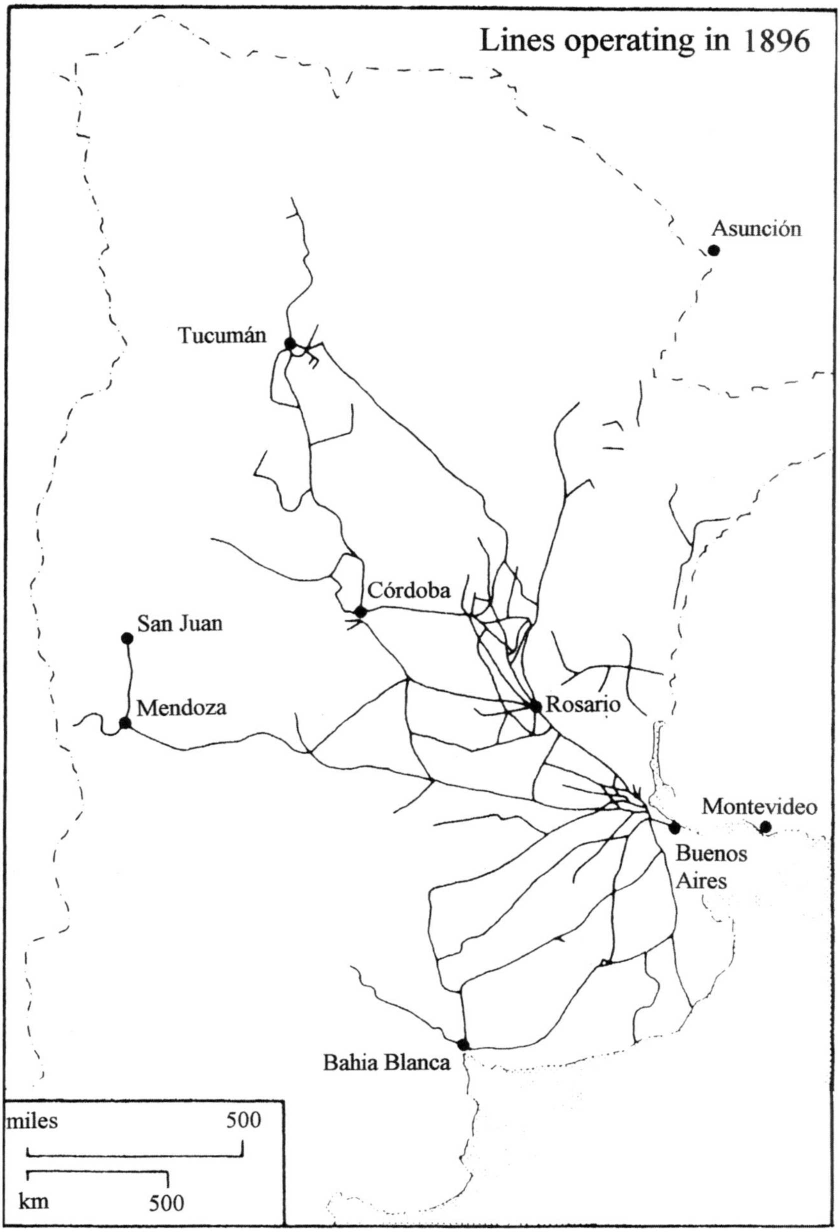

FIGURE 2.4. Argentine rail grid, 1896

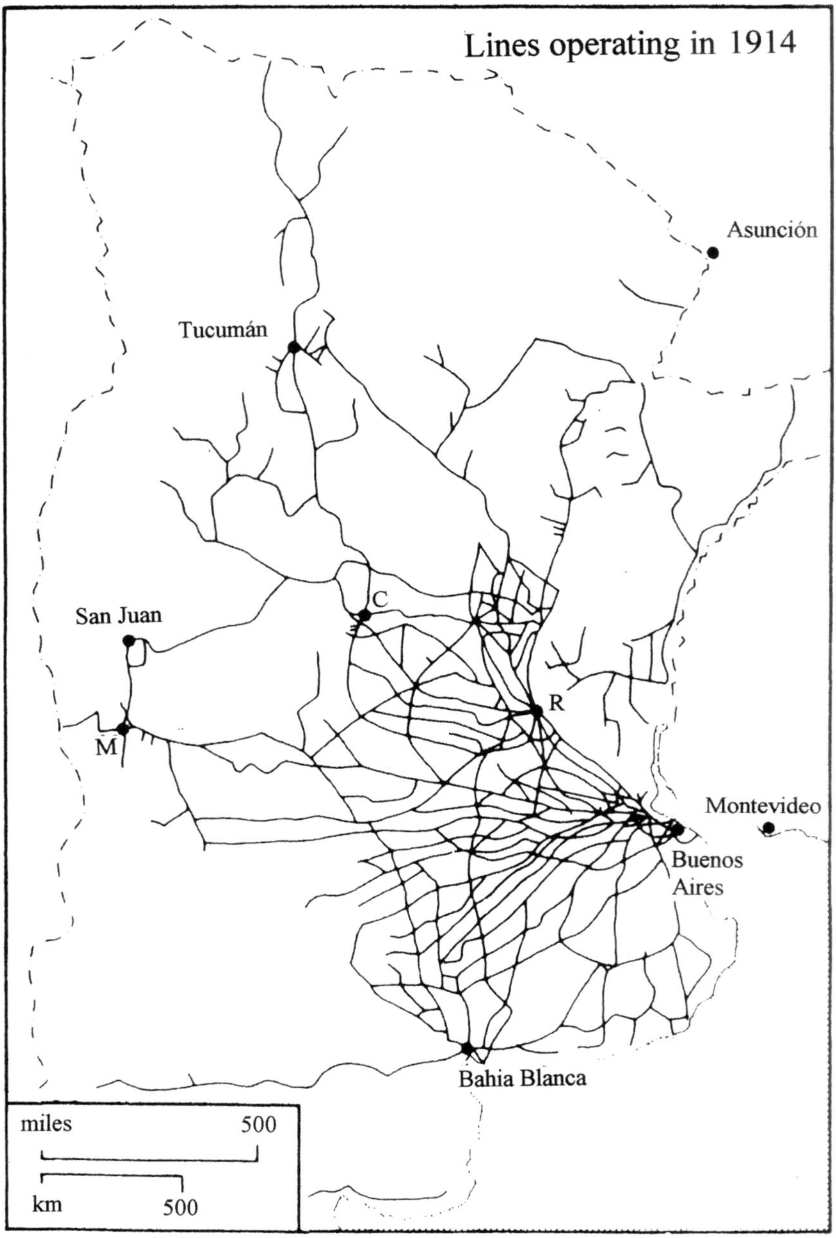

FIGURE 2.5. Argentine rail grid, 1914

2.3.2. *Mexican Migration to the United States, 1980–2006*

The history of migration between the US and Mexico is well documented.[7] Throughout most of the 19th century, there was substantial US migration to northern Mexico, resulting in the annexation of most of northern Mexico by the United States in 1848. Mexican migration to the United States was negligible until the last two decades of the century. Until that time Chinese laborers were an easy source of inexpensive immigrant manpower in the US's new western territories. Two developments set the stage for change. One was the exclusion of migration from China in 1882. The other was completion of the railway lines linking El Paso to Kansas City (1881) and to Mexico City (1884).

The railroad links between Kansas City and Mexico City served as a trunk line for what was eventually to become a vast migration network involving hundreds of sending communities in central Mexico and hundreds of work destinations in the United States. The opportunity cost of migration from the central Mexican plateau through El Paso had been substantially lowered. A trickle of migration began northward, stimulated by the demand for labor to build the Southern Pacific railway through the Gadsden Purchase toward California.

Because the Mexican railways did not reach California from central Mexico until 1923, the El Paso route remained the trunk line, and El Paso served as the primary port of entry for Mexican laborers (even those going to California), for nearly 40 years. After 1923, San Diego become a second important port of entry, establishing a new trunk that would grow for another 40 years. A third, but less significant point of entry was the lower Rio Grande valley in South Texas. That pattern was still visible in a map (Fig. 2.6) of in-migration fields for Mexican undocumented workers in the 1970s prepared by Richard C. Jones.[8]

The volume of Mexican migration to the US took a quantum leap with the introduction of the Bracero Program of 1942 in response to wartime manpower shortages in the US labor market. While the Bracero Program continued after the war, additional and undocumented migration also began to rise. In the 1950s "Operation Wetback," a paramilitary campaign crackdown called on undocumented workers, was instituted. Operation Wetback proved disastrous for US agricultural harvest. The result was a backlash from agricultural interests, followed by bureaucratic backpedaling on enforcement. The *status quo ante* was re-established along with a doubling of Bracero permits to about 400,000 per year.

The Bracero Program was for agricultural laborers whose work was seasonal. As a result, the flows along the migration network were tidal-like, a wave effect in which labor would flow north over the border for planting and harvest, and

---

[7] Gilbert W. Merkx, "A review of the issues and groups involved in undocumented migration to the United States," in Robert S. Landman, ed., *The Problem of the Undocumented Worker* (Washington, DC: Community Services Administration, 1980), pp. 7–14.

[8] Richard C. Jones, *Patterns of Undocumented Migration, Mexico and the United States* (Totowa, N.J., Rowman and Allanheld, 1984), p. 3, Fig. 1.1.

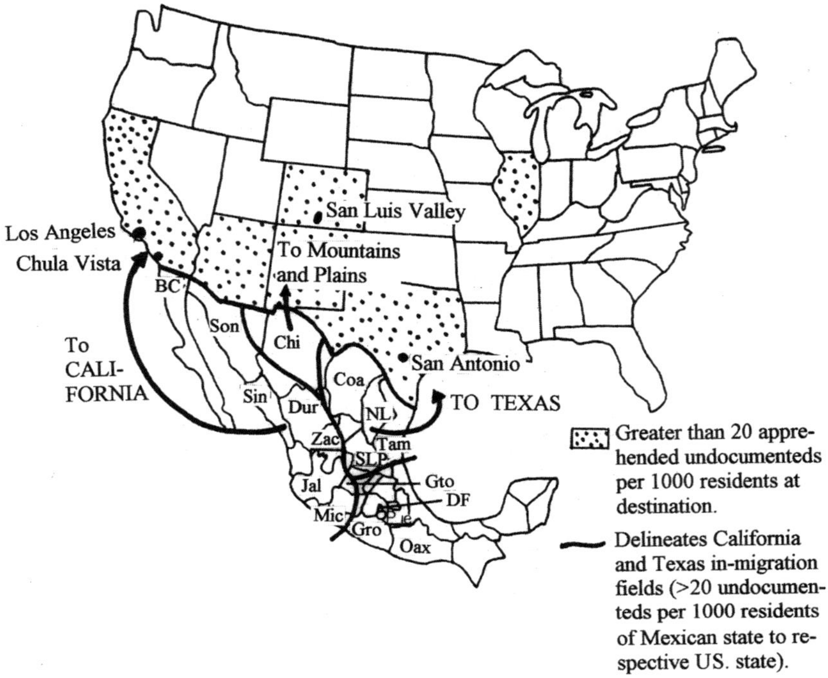

FIGURE 2.6. In-migration fields for Mexican undocumenteds to the US, mid-1970s (State abbreviations: BC = Baja California; Chi = Chihuahua; Coa = Coahuila; DF = Distrito Federal; Dur = Durango; Gro = Guerrero; Gto = Guanajuato; Jal = Jalisco; Mic = Michoacan; NL = Nuevo Leon; Oax = Oaxaca; Pue = Puebla; Sin = Sinaloa; SLP = San Luis Potosi; Son = Sonora; Tam = Tamaulipas; Zac = Zacatec

flow back over the border in the off seasons. For those possessing the work permits, the transaction costs of the back and forth migration stream were low.

In 1964 a coalition of labor unions, civil rights groups, and religious organizations succeeded in having the Bracero Program terminated. Shortly afterwards, the 1965 Hart-Celler Act scrapped the nationality-based immigration quotas established in the 1920s and capped total immigration from the Western Hemisphere at 120,000 per year. This set the stage for restrictions on migration from Mexico, which began in 1968. In 1976 the Congress placed the Western Hemisphere nations under a quota of 20,000 migrants per country, which meant that Mexico had the same quota, like Antigua a small Caribbean island. This and a stream of following legislation progressively tightened legal migration from Mexico while increasing border security. The result did not stem the flow of migrant, but it did make it increasingly illegal and costly. As the transaction costs of crossing the border grew, the seasonal flow of migrants back to Mexico shrank. The network was increasingly becoming a one-way flow of people north, while the flow south began to be one of money transfers instead of people.

The effects of tightening both border controls and legal access can be seen in the following two flow-density charts prepared by Richard Jones. The first chart (Fig. 2.7) shows the flow density of deportable Mexican aliens per million population of INS districts in 1973. As expected, the highest densities are in the southwest region, centering on El Paso and extending to California and Texas, and the lowest densities in New England.[9] The second wave flow chart (Fig. 2.8) shows the *percentage increase* in the flow density of deportable aliens between 1973 and 1978, a period in which border controls were being tightened and the legal quota was reduced. What this table shows is densities dropping in the dense areas along the border and increasing in the low-density areas are farthest from the border. Thus, with the flow back across the border being reduced, the wave effect was lengthened in the opposite direction or, to put it differently, the delta was widening.

In 1986 the Immigration and Reform and Control Act (IRCA) set the stage for a long battle over the US–Mexico border policy. Under IRCA, the Immigration and Naturalization Service received a 50% budget increase to hire additional Border Patrol officers. This led to a series of crackdowns such as "Operation Hold-the-Line" in El Paso and "Operation Gatekeeper" in San Diego. The same dynamics have intensified since IRCA, with massive increases in personnel and funding for border security.

The border crackdowns have been shutting down the traditional flow points at or near El Paso and San Diego. With each new tightening of border security, the

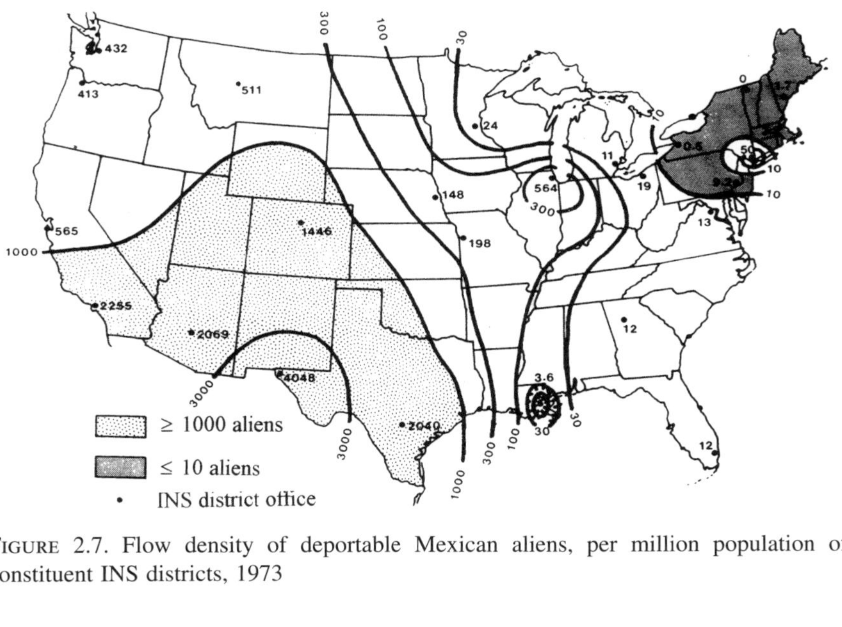

FIGURE 2.7. Flow density of deportable Mexican aliens, per million population of constituent INS districts, 1973

[9] Jones (1984), p. 39, Fig. 3.1.

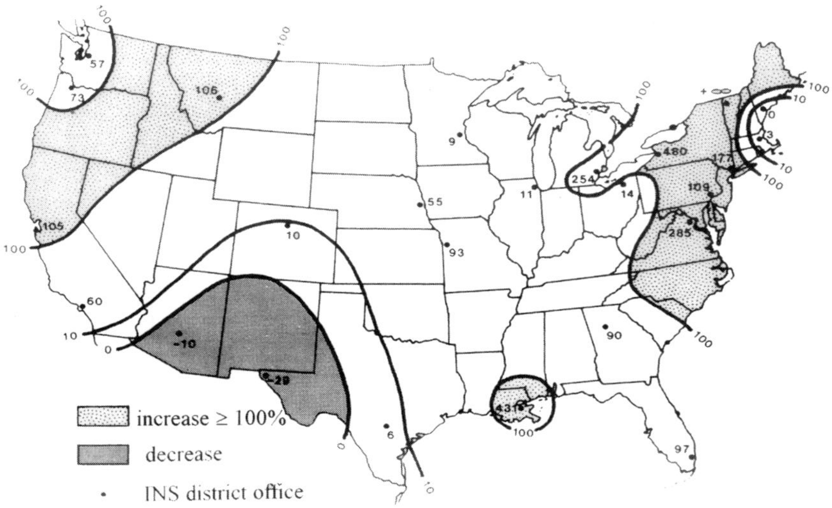

FIGURE 2.8. Percentage increase in flow density of deportable Mexican aliens, 1973–1978

transaction costs for the migrants have increased, reducing back-migration and, in consequence, leading to an increase in the migration of females who previously would have remained in Mexico waiting for their partners to return. Also, with the closing of the traditional points of entry, migration has shifted to other, more dangerous points in remote and hazardous desert regions, with a concomitant rise in fatalities. Efforts to close the channels simply lead to leakages in other places. In short, current policies, including those in the new immigration bills passed by the Congress, have not reduced the flow or eliminated the network, but they have distorted it, creating huge inefficiencies and high costs for all parties.

The efforts to shut down the Mexican migration network have had a social effect contrary to the intent of the policymakers. The rising costs of crossing the border of Mexico have transformed a seasonal migration flow into permanent immigration and into permanent rather than seasonal employment. The migration flow has actually increased as the new immigrants send for their wives and children. As a result, a network that was once composed largely of males traveling back and forth to agricultural sites now consists of both sexes distributed widely across economic and geographic sectors of the United States. The obvious policy lesson has not been learned, namely that *it is difficult, if not impossible, to shut down a social network of large size and high motivation.*

2.4. Conclusions

Several inferences are suggested by the cases of Argentina's trade network with Europe and Mexican migration network to the United States. The first inference is the importance of the trunk or channel in making possible the larger network. In both cases, the trunk flow is made possible by the introduction of new technology: the central railroad in the Mexican case, and the refrigerated steamship in the Argentine case. Also in both cases, new financial technology in the form of interest-bearing bonds played a role in making the railroads possible. A difference between the two cases is that in Mexico the railroad line was the main channel, while in Argentina the railroads were part of the tree delta, not the primary channel (which was the shipping lane to Europe).

These two cases suggest that the introduction of any new technology that creates a flow mechanism of greater efficiency and lower cost than prior alternatives will become the basis for a new social network. It can be suggested that the introduction of technologies such as the newspaper, telephone, air travel, and Internet have indeed had such an effect.

These case studies also suggest that while the introduction of the technology may be a necessary condition for the development of a tree network, they are not sufficient. Other conditions must create the demand or pressure for peripheral units or actors to link to the trunk. In the Anglo-Argentine case, an excess of British capital was seeking outlets for investment, while at the same time there was an unsatisfied demand for better foodstuffs from Britain's rapidly growing industrial labor force. On the Argentine side there was an excess of agricultural production seeking new markets, as well as vast territories of rich grassland awaiting exploitation.

In the Mexican–US case, the initial creation of the trunk line to El Paso did not lead to a migration stream in and of itself. The migration stream began to develop rapidly with the advent of World War II, when the United States responded to labor shortages by recruiting Mexicans into the Bracero Program. In this case, demand for labor in the US had as its counterpart in Mexico a growing demand for employment, triggered by rapid population growth.

In both cases, there are demands or needs that create incentives to utilize the trunk, analogous to pressure in a hydraulic system. Once these incentives are present, then the efficiencies offered by the trunk lead to the rapid growth of tree networks at both ends of the connection.

Another point that can be deduced from the Anglo-Argentine network is that the tree system will cease to grow or perhaps even to function if the pressures or incentives are removed from the system. In the US–Mexican network, the incentives have remained due to labor shortages in the US and wage disparities between the US and Mexico. However, efforts to dam the flow have distorted the natural tree patterns and wave flow, leading to various forms of leakage, and widening ripples on the US side of the border. The border enforcement policies have no chance of success, but the network would cease to function if the incentives and pressures were removed. This would be the case if a depression in the US removed the demand for labor, or if Mexican population growth dropped significantly and led to labor shortages in that country. Both developments are not beyond the realm of possibility.

In sum, the application of constructal theory to social networks appears to be highly promising. Next steps could include (1) systematizing the predictions of constructal theory for the size, shape, and evolution of social networks; (2) refining the concepts that cause social pressure (which here have been referred to as need, demand, or incentives); (3) developing a general definition of a social trunk or main channel (can we specify when one exists?); (4) operationalizing the measurement of the dimensions of the tree; and then (5) testing the accuracy of the predictions of constructal theory as applied to social phenomena, using empirical data on social networks.

The persistence or emergence of obstacles that interfere with the maximization of efficiency through the development of constructal patterns suggests a final line of investigation. I would like to suggest that obstacles or inefficiencies are the result of competing networks that try to occupy the same property space as an existing network. Thus, a tree growing into a river, is like the river itself, a constructal system seeking efficiency, but in so doing its roots slow the water and create inefficiencies in the river system. Likewise, a crime syndicate engaging in corrupt practices may be an emerging constructal system, but it is one which interferes with the normal flows of the unfettered market. If this suggestion is correct, inefficiencies or obstacles are not anomalies, but predictable consequences of the intersection of competing constructal systems.

References

Bejan, A. (2000) *Shape and Structure from Engineering to Nature*, (Cambridge University Press, Cambridge, UK).

Benkler, Y. (2006) *The Wealth of Networks* (Yale, New Haven, CT).

Buchanan, M. (2002) *Nexus: Small World and the Groundbreaking theory of Networks* (University Press, Abo. Norton, New York).

Forsman, M. (2005) *Development of Research Networks: The Case of Social Capital* (Abo Academi).

Jones, R. C. (1984) *Patterns of Undocumented Migration, Mexico and the United States* (Totowa, N.J., Rowman and Allanheld), p. 3, Fig. 1.1; p. 39, Fig. 3.1.

Laird, P. W. (2006) *Pull: Networking and Success Since Benjamin Franklin* (Harvard, Cambridge, MA).

Merkx, G. W. (1968) Political and Economic Change in Argentina, 1870–1966, Ph.D. Dissertation (Yale University, New Haven, CT).

Merkx, G. W. (1980) "A review of the issues and groups involved in undocumented migration to the United States," in Robert S. Landman, ed., *The Problem of the Undocumented Worker* (Washington, DC: Community Services Administration), pp. 7–14.

Rodriguez-Iturbe, I. and Rinaldo, A. (1997) *Fractal River Basins* (Cambridge University Press, Cambridge, UK).

The railroad network maps are taken from Colin M. Lewis, *British Railways in Argentina 1857–1914* (Athlone Press: 1983).

Weber, M. (1964) *The Theory of Economic and Social Organization* (Free Press, New York), p. 87.

Chapter 3
Tree Flow Networks in Urban Design

Sylvie Lorente

3.1. Introduction

Configuration (geometry, topology) is the chief unknown and major challenge in design. Better designers make choices that work, i.e., configurations that have been tested. Experience and longevity are useful when the needed system is envisioned as an assembly of already existing parts. Is this the way to proceed in every application? It would be useful to the designer to have access from the beginning to the infinity of configurations that exist. Access means freedom to contemplate them all, without constraints based on past experience. It would also be useful to have a strategy (road map, guide) for discovering architectural features that lead to more promising configurations, and ultimately to better-optimized designs.

The objective of this chapter is to illustrate this approach to urban flow system design, specifically the view that configuration itself is the unknown that is to be discovered (Bejan 2000). We do this by considering the fundamental problem of distributing a supply of hot water as uniformly as possible over a given territory. This is a classical problem of urban design and urban heating, with related subfields in piping networks, sewage and water runoff, irrigation, steam piping, etc. (Barreau and Moret-Bailly 1977; Bonnin 1977; Dupont 1977; Falempe 1993; Falempe and Baudoin 1993; Nonclercq 1982; Plaige 1999). Finally, a possible contribution of the constructal theory of flow networks to the French system of access to elite schools in superior education is proposed.

3.2. How to Distribute Hot Water over an Area

The distribution of hot water to users on a specified territory presents two problems to the thermal designer: the fluid mechanics problem of minimizing the pumping power, and the heat transfer problem of minimizing the loss of heat from the piping network. The water flow is from one point (the source) to an area—the large number of users spread uniformly over the area.

The area A is supplied with hot water by a stream of flow rate $\dot{m}$ and initial temperature T_i. The stream enters the area from the outside, by crossing one of its boundaries. The area is inhabited by a large number of users, $n = A/A_0$, where

A_0 is the area element allocated to a single user. Let A_0 be a square with the side length L_0, so that A constitutes a patchwork of n squares of size A_0. Each elemental square must receive an equal share of the original stream of hot water, $\dot{m}/n$. As in all the point-area tree flows considered previously (Bejan 1997; 2000), the fundamental question is how to connect the elements so that the ensemble (A) performs best. The constraint is the total amount of insulation wrapped around the pipe:

$$V = \pi \int_0^L (r_o^2 - r_i^2)\, dx \tag{3.1}$$

The geometry of the pipe and its insulation is described completely by r_o and r_i (or r_o/r_i and r_i) as functions of x, where r_i is the inner radius wetted by the flow and r_o is the outer radius (including the insulation). The optimal distribution of pipe size can be derived from the minimization of the pumping power $\dot{W}$ required to drive $\dot{m}$ through the entire system. The pumping power varies as the product $\dot{m}(dP/dx)$, while in fully rough turbulent flow the pressure gradient is proportional to $\dot{m}^2/r_i^5$.

One design alternative is to introduce branches in the path of the stream, and to distribute area elements to each branch (Wechsatol et al. 2001). We explore this alternative by starting with the smallest (and therefore simplest) area element, and continuing toward larger areas by assembling elements into larger and larger constructs. One simple rule of assembly is to use four constructs into the next, larger assembly, Fig. 3.1. In this case, each construct covers a square area,

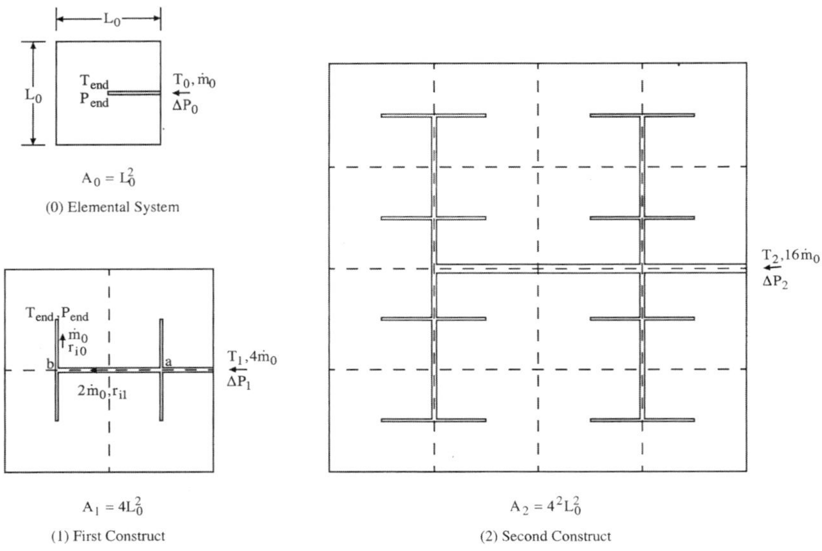

FIGURE 3.1. Sequence of square-shaped constructs containing tree-shaped streams of hot water (Wechsatol et al. 2001)

in the sequence $A_0 = L_0^2, A_1 = 4L_0^2, A_2 = 4^2L_0^2, \ldots, A_i = 4^iL_0^2$. This assembly rule is "simple" because the shape (square) of each construct is assumed, not optimized.

The objective is to supply hot water to the users distributed uniformly over A_i, and to accomplish this task with minimal pumping power and a finite amount of thermal insulation. The geometry of each pipe is described by its length (a fraction or multiple of L_0), inner radius wetted by the flow (r_i), and ratio of insulation radii ($R = r_o/r_i$). The pipe wall thickness is neglected for the sake of simplicity. The subscripts 0, 1, and 2 indicate the elemental area, first construct, and second construct, in accordance with the notation shown in Fig. 3.1.

To minimize the pumping power requirement at the elemental level ($\dot{W}_0 = \dot{m}_0 \Delta P_0/\rho$) is to minimize the pressure drop along the elemental duct of the length $L_0/2$. Assuming that the flow is a fully developed turbulent in the fully rough regime (f = constant), we find that the pressure drop derived from the definition of friction factor is (Padet 1991)

$$\Delta P_0 = \frac{f}{\pi^2} \frac{\dot{m}_0^2 L_0/2}{\rho \, r_{i0}^5} \tag{3.2}$$

We assume that the dominant thermal resistance between the stream and the ambient is posed by a cylindrical shell of thermal insulation installed on the outside of the pipe that carries the stream. In this limit the rate of heat loss per unit of pipe length is (Bejan et al. 1996)

$$q' = \frac{2\pi k}{\ln\,(r_o/r_i)} \; [T(x) - T_\infty] \tag{3.3}$$

where r_o, r_i, k and T_∞ are the outer and inner radii of the insulation, the thermal conductivity of the insulating material, and the ambient temperature respectively. Written for a pipe length dx, the first law of thermodynamics requires

$$\dot{m}\, c_p\, dT = \frac{2\pi k}{\ln\,(r_o/r_i)} \; (T - T_\infty)\, dx \tag{3.4}$$

Equations (3.2) and (3.3) can be integrated to obtain the temperature drop from the inlet to the element (T_0) to the user (T_{end}),

$$\frac{T_{end} - T_\infty}{T_0 - T_\infty} = \exp\left(-\frac{N_0}{\ln R_0}\right) \tag{3.5}$$

where the number of heat loss units is based on elemental quantities

$$N_0 = \frac{\pi k L_0}{\dot{m}_0 c_p} \tag{3.6}$$

At the first-construct level there are two pipe sizes, one central pipe of length $(3/2)L_0$ and radii r_{i1} and $R_1 = (r_o/r_i)_1$, and four elemental branches of length

$(1/2)L_0$ and radii r_{i0} and $R_0 = (r_o/r_i)_0$. The flow rate is $4\,\dot{m}_0$ through the root of the tree, and $\dot{m}_0$ through each branch. By writing the equivalent of Eq. (3.2) for each segment of pipe without branches, we find that the drop in pressure from the root to the most distant user (the center of the farthest element) is

$$\Delta P_1 = \frac{f}{\pi^2}\,\frac{\dot{m}_0^2\,L_0}{\rho}\left(\frac{12}{r_{i1}^5} + \frac{1}{2r_{i0}^5}\right) \tag{3.7}$$

The pressure drop ΔP_1 can be minimized by selecting the ratio of pipe sizes r_{i1}/r_{i0} when the total volume occupied by the ducts is fixed:

$$\frac{3}{2}\,L_0\,r_{i1}^2 + 2L_0\,r_{i0}^2 = \text{constant} \tag{3.8}$$

The minimization of ΔP_1 subject to constraint (3.8) yields the optimal step in pipe size,

$$(r_{i1}/r_{i0})_{opt} = 2^{5/7} \tag{3.9}$$

The optimization of the geometry of the thermal insulation shells wrapped around each pipe proceeds in the same steps as the pressure-drop minimization. We write a temperature drop expression of type (3.5) for each segment of pipe without branches. We omit the algebra and report only the overall temperature drop from the root of the tree (T_1) to the temperature (T_{end}) of the water stream delivered to the most distant user:

$$\theta_1 = \frac{T_{end} - T_\infty}{T_1 - T_\infty} = \exp\left(-\frac{N_0}{\ln R_0} - \frac{5\,N_0}{4\,\ln R_1}\right) \tag{3.10}$$

The dimensionless end temperature θ_1 depends on three parameters, R_0, R_1, and N_0. The geometric parameters R_0 and R_1 are related through the thermal insulation volume constraint

$$V_1 = \pi\,L_0\,r_{i0}^2\left[\frac{3}{2}\left(\frac{r_{i1}}{r_{i0}}\right)^2 (R_1^2 - 1) + 2\,(R_0^2 - 1)\right] \tag{3.11}$$

for which (r_{i1}/r_{i0}) is $2^{5/7}$, cf. Eq. (3.9). Constraint (3.11) may be put into the dimensionless form $\tilde{V}_1\,(R_0, R_1)$ by recognizing r_{i0} as the smallest pipe size and defining the dimensionless insulation volume $\tilde{V}_1 = V_1/(\pi r_{i0}^2\,L_0)$.

The maximization of expression (3.10) with respect to R_0 and R_1, and subject to constraint (3.11) yields the optimal step change in radii ratio:

$$\left(\frac{R_1\,\ln R_1}{R_0\,\ln R_0}\right)_{opt} = \left(\frac{5}{6}\right)^{1/2}\left(\frac{r_{i0}}{r_{i1}}\right)_{opt} \tag{3.12}$$

In view of the $(r_{i1}/r_{i0})_{opt}$ values, we conclude that $R_{1,opt} < R_{0,opt}$, i.e., the shell of the central duct is relatively thin in comparison with the shells of the elemental ducts. Relatively 'thin' means that the shell thickness is small in comparison with the radius of the same tube.

The optimization of the internal architecture of the second construct (A_2, Fig. 3.1) is performed by executing the same steps as in the optimization of the first construct. New are the larger size of the construct ($A_2 = 4A_1$, $\dot{m}_2 = 16\dot{m}_0$) and the new central duct of length $3L_0$, inner radius r_{i2}, and insulation radii ratio $R_2 = r_{o2}/r_{i2}$. The optimized geometric features of the first construct are retained.

In the fluid flow part of the problem, we minimize the overall pressure drop from the root of the fluid tree (P_2) to the farthest user (P_{end}), namely $\Delta P_2 = P_2 - P_{end}$. After some algebra we obtain

$$\Delta P_2 = \frac{f}{\pi^2} \frac{\dot{m}_2^2 L_0}{\rho} \left(\frac{3/2}{r_{i2}^5} + \frac{12 + \beta/2}{16^2 r_{i1}^5} \right) \tag{3.13}$$

where $\beta = (r_{i1}/r_{i0})^2$ is $(2^{5/7})^2$. By varying r_{i2} and r_{i1} subject to this constraint, we minimize ΔP_2 and find the optimal relative size of the central duct of the A_2 construct:

$$\left(\frac{r_{i2}}{r_{i1}} \right)_{opt} = \left(\frac{3 + 4/\beta}{24 + \beta^{5/2}} \cdot 2^9 \right)^{1/7} = 2^{6/7} \tag{3.14}$$

Noteworthy is the inequality $(r_{i2}/r_{i1})_{opt} > (r_{i1}/r_{i0})_{opt}$, which states that the step change in duct size is more abrupt at the second-construct level than at the first-construct level.

The second part of the analysis of A_2 is concerned with the temperature (T_{end}) of the water stream received by the farthest elemental user, and the maximization of this temperature subject to the constrained total volume of thermal insulation allocated to the A_2 construct. The analysis yields an expression for the dimensionless overall temperature drop

$$\theta_2 = \frac{T_{end} - T_\infty}{T_2 - T_\infty} = \theta_1 \exp\left(-\frac{5N_0/8}{\ln R_2} \right) \tag{3.15}$$

where $\theta_1(N_0, \tilde{V}_1)$ is a function that is available numerically based on the optimization of the A_1 construct (cf. Fig. 3.7b).

3.3. Tree Network Generated by Repetitive Pairing

The sequence of square-shaped constructs used beginning with Fig. 3.1 is an assumption, not a result of optimization. To see whether a better rule of assembling small constructs into larger constructs can be found, consider the area-doubling sequence shown in Fig. 3.2. Each area construct is obtained by putting together two constructs of the immediately smaller size. The area supplied with

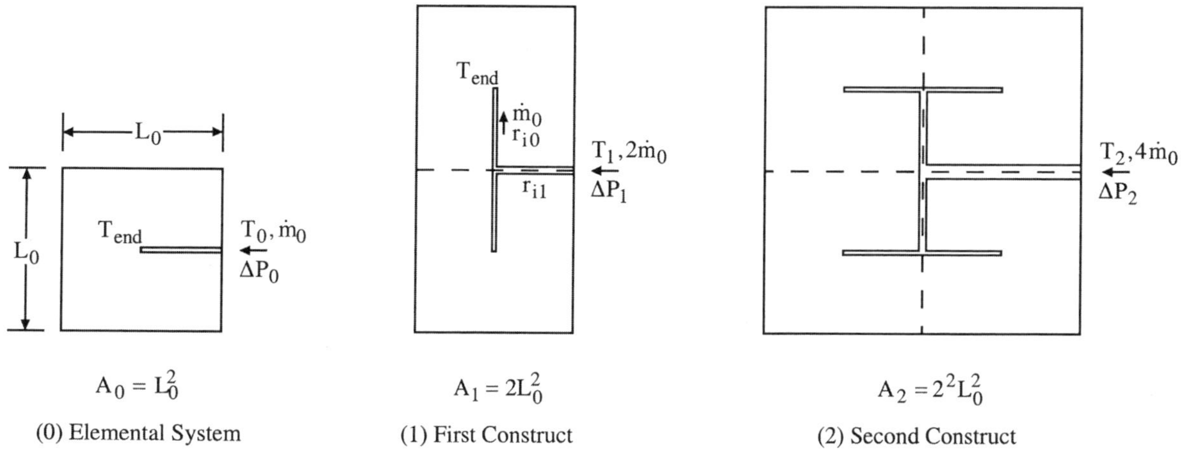

FIGURE 3.2. Sequence of area constructs obtained by pairing smaller constructs (Wechsatol et al. 2001)

hot water increases in the sequence $A_0 = L_0^2$, $A_1 = 2L_0^2$, $A_2 = 2^2L_0^2, \ldots, A_i = 2^iL_0^2$, and the shape of the area alternates between square and rectangular. The elemental area that starts the sequence in Fig. 3.2 is the same as in Fig. 3.1, namely A_0. The second construct of Fig. 3.2 covers the same area ($4L_0^2$) as the first construct of Fig. 3.1. One objective of the optimization work reported in this section is to see which area construction sequence serves the farthest (end) user of the $4L_0^2$ territory better, Fig. 3.1 or Fig. 3.2?

For the sake of brevity, we report only the results of the minimization of water flow resistance and loss of heat to the ambient. The analysis follows step by step the analysis detailed in the preceding section. For the first construct of Fig. 3.2, in place of Eq. (3.9) we obtain

$$(r_{i1}/r_{i0})_{opt} = 2^{3/7} \tag{3.16}$$

Equations (3.10) and (3.11) are replaced by

$$\theta_1 = \frac{T_{end} - T_\infty}{T_1 - T_\infty} = \exp\left(-\frac{N_0}{\ln R_0} - \frac{N_0}{2\ln R_1}\right) \tag{3.17}$$

$$\tilde{V}_1 = \frac{V_1}{\pi L_0 r_{i0}^2} = \frac{1}{2}\left(\frac{r_{i1}}{r_{i0}}\right)^2 (R_1^2 - 1) + (R_0^2 - 1) \tag{3.18}$$

At the second-construct level, in place of Eq. (3.14) we obtain $(r_{i2}/r_{i1})_{opt} = 2^{3/7}$. The user temperature and total insulation volume are

$$\theta_2 = \frac{T_{end} - T_\infty}{T_2 - T_\infty} = \theta_1 \exp\left(-\frac{N_0}{2\ln R_2}\right) \tag{3.19}$$

$$\tilde{V}_2 = \frac{V_2}{\pi L_0 r_{i0}^2} = \left(\frac{r_{i2}}{r_{i0}}\right)^2 (R_2^2 - 1) + 2\tilde{V}_1 \tag{3.20}$$

The numerical values and trends are similar to what we saw earlier. More to the point, we can compare the relative goodness of the doubling sequence (Fig. 3.2) versus the sequence of square areas (Fig. 3.1). In Fig. 3.3 we show the ratio of the maximized end temperatures, with the observation that both cases refer to the same construct size ($4L_0^2$), and that the insulation volume is the same in both cases. The total duct volume is the same in the two designs sketched above the graph. Also note that r_{i0} of Fig. 3.1 is not the same as the r_{i0} of Fig. 3.2 and, consequently, the dimensionless volumes $\tilde{V}_{1,\text{Fig. 3.1}}$ and $\tilde{V}_{2,\text{Fig. 3.2}}$ are not the same. The comparison shown in Fig. 3.3 allows us to conclude that the tree structure generated by repeated pairing (A_2, Fig. 3.2) is superior to the square structure (A_1, Fig. 3.1). The temperature of the hot water received by the end user in Fig. 3.2 (A_2) is consistently higher.

We pursued this comparison to an even higher level of assembly—the construct of size 16 L_0^2—on the basis of the same mount of insulation material and total duct volume. The two structures are illustrated in the upper part of Fig. 3.4, with the observation that the tree generated by repeated pairing (A_4)

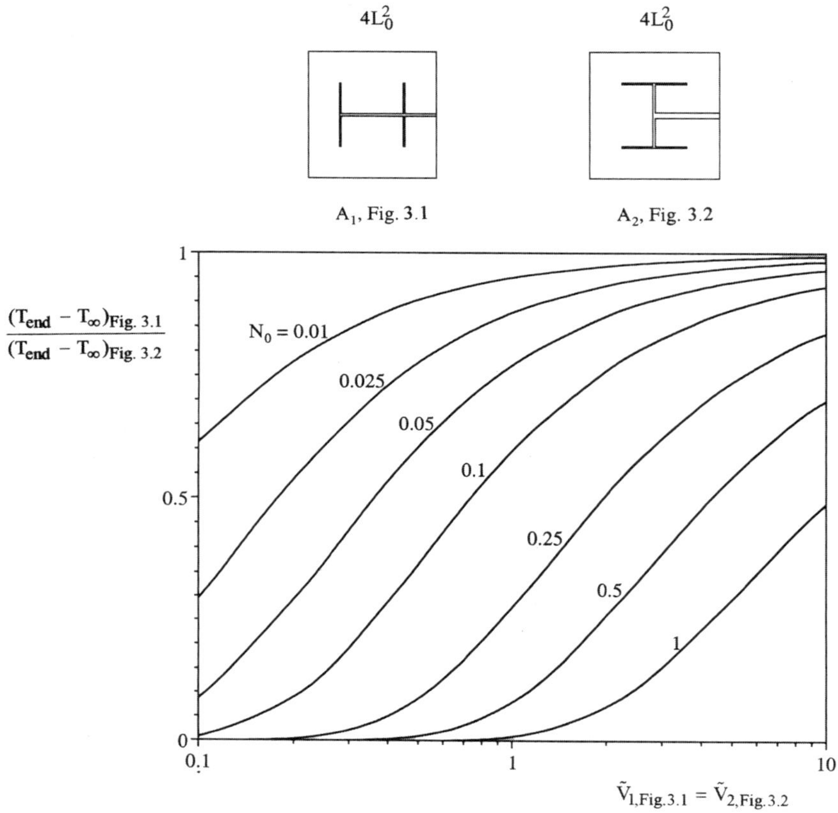

FIGURE 3.3. Comparison between the maximized end-user water temperature on the $4L_0^2$ construct, according to the construction sequences of Figs. 3.1 and 3.2 (Wechsatol et al. 2001)

is not shown in Fig. 3.2. Once again, the tree design based on the sequence of Fig. 3.2 outperforms the design based on Fig. 3.1. By comparing Figs. 3.4 and 3.3 on the same basis (the same L_0, total pipe volume, total amount of insulation), it can be shown that the difference in the global performance of the two types of trees (Figs. 3.1 and 3.2) diminishes as each optimized tree structure becomes more complex. The global performance becomes progressively less sensitive to the actual layout of the tubes, provided that the distributions of tube step sizes and shells of insulation material have been optimized. When the optimized tree structure becomes more complex it also becomes more *robust* with respect to changes in the tree pattern.

Another useful property of the pairing sequence (Fig. 3.2) is this: each user receives hot water at the same temperature. The $\dot{m}_0$ stream received by each user passes through the same sequence of insulated tubes. This is a very useful

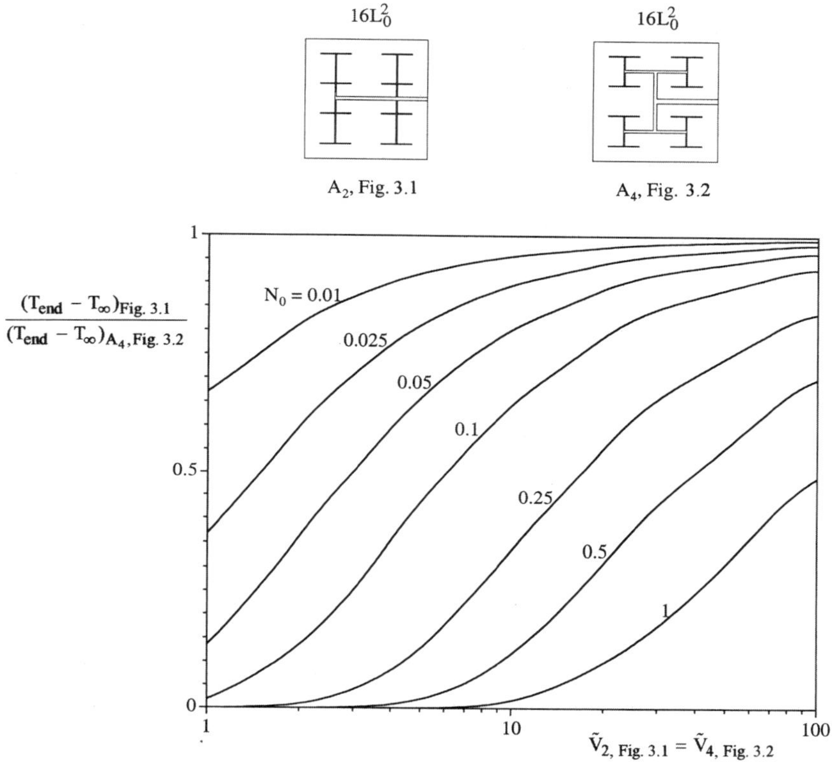

FIGURE 3.4. Comparison between the maximized end-user water temperature on the $16L_0^2$ construct, according to the construction sequences of Figs. 3.1 and 3.2 (Wechsatol et al. 2001)

property, because in other tree structures such as in Fig. 3.1 the users located closer to the root of the tree receive warmer streams than the farther users. The tree designs of Fig. 3.2 deliver hot water to the territory uniformly—uniformly in space and in temperature.

3.4. Robustness and Complexity

To summarize, in Figs. 3.1–3.4 we treated in some detail the basic problem of distributing hot water to users spread over a territory, when the amount of insulation and other constraints are in place. The optimal distribution of insulation over the many pipes of the network was a necessary step, but not the main focus of this example. The main focus was on the *configuration* of the water distribution network—the layout of the pipes over the territory. Main questions were how the layout can be deduced from the objective of maximizing the global performance of the hot water distribution system, and to what extent the choice of layout influences global performance.

The difference in performance between one type of tree structure over another decreases as the complexity of the structure increases. It is as if "any tree will do" if it is large and complex enough, and if its link dimensions and insulation have been optimized. Tree-shaped distribution systems perform consistently and substantially better than string-shaped or coil-shaped systems (Wechsatol et al. 2001). The robustness of the tree-flow performance to differences in internal layout (Fig. 3.1 vs Fig. 3.2) is important because it simplifies the search for a nearly optimal layout, and because a constructed system will function at near-optimal levels when its operating conditions drift from the values for which the system was optimized.

The chief conclusion so far is that the use of geometric form (shape, structure) is an effective route to achieving high levels of global performance under constraints. The brute force approach of delivering hot water by using large amounts of insulation and flow rates (small N) is not economical. Much faster progress toward the goal of global performance maximization can be made by recognizing and treating the geography of the flow system as the main unknown of the problem.

3.5. Development of Configuration by Adding New Users to Existing Networks

The new approach to the design of tree-shaped flows that forms the subject of this section is inspired by the spontaneous generation of tree-shaped patterns of traffic in 'urban growth' (Martinand 1986; Bastié and Dézert 1980; Pelletier and Delfante 1989). New neighborhoods are added to existing tree structures. In other words, the new (larger) tree-shaped flow is not optimized again, globally, as a new assembly. Instead, the newcomer is simply grafted onto the tree in the spot that is the most advantageous to the newcomer. We refer to this approach as the "one-by-one tree growth." We follow the growth and evolution of the tree based on this principle, and show that the somewhat irregular tree structure that results has many features in common with the organized, constructal tree design. One such feature is the global performance of the tree flow structure: the one-by-one grown structure performs nearly as well as the constructal structure if the structure is sufficiently large and complex.

We illustrate this new approach by considering the fundamental problem of distributing a supply of hot water as uniformly as possible over a given territory. We add new users to an existing structure, while not having the means to reoptimize the structure after the new users have been added. Furthermore, the addition of each new user is decided from the point of view of the user, not from the point of view of maximizing the global performance. Each new user is connected to the tree network in the place that maximizes the user's benefit, namely the temperature of the water stream that the new user receives.

In this new approach we must have a start, and for this we choose the optimized A_2 construct of Fig. 3.2, which is reproduced in Fig. 3.4. The A_2 construct has the area $4L_0^2$, and the property that its four users receive water at the same temperature, $T_1 = T_2 = T_3 = T_4$. The overall pressure drop has been minimized according to the pipe size ratios r_{i2}/r_{i1} and r_{i1}/r_{i0}. The total amount of insulation has been distributed optimally according to the insulation radii ratios R_0, R_1, and R_2.

Consider now the decision to add a new elemental user ($\dot{m}_0$, L_0^2) to the A_2 construct. Because of symmetry, we recognize only the three possible positions (a, b, c) shown in Fig. 3.5. For the pipe that connects the new user to the existing A_2 construct, we use the pipe size and insulation design of the elemental system of A_2, namely r_{i0} and R_0. In going from the A_2 construct of Fig. 3.2 to the constructs of Fig. 3.5, the overall mass flow rate has increased from $4\dot{m}_0$ to $5\dot{m}_0$. The addition of the new user disturbs the optimally balanced distribution of resistances and temperatures, which was reached based on the analysis presented in the preceding section.

To deliver the hottest water from the network to the new user we must use the shortest pipe possible. By analyzing the user water temperature in the three configurations identified in Fig. 3.5, we find that $T_{5,b} > T_{5,a} > T_{5,c}$. For example, for the configuration of Fig. 3.5a, it can be shown that the temperature of the hot water received by the new user (T_5) is

$$\frac{T_5 - T_\infty}{T_0 - T_\infty} = \exp\left(-\frac{5N_0}{2\ \ln R_0} - \frac{N_0}{3\ \ln R_1} - \frac{2N_0}{5\ \ln R_2}\right) \tag{3.21}$$

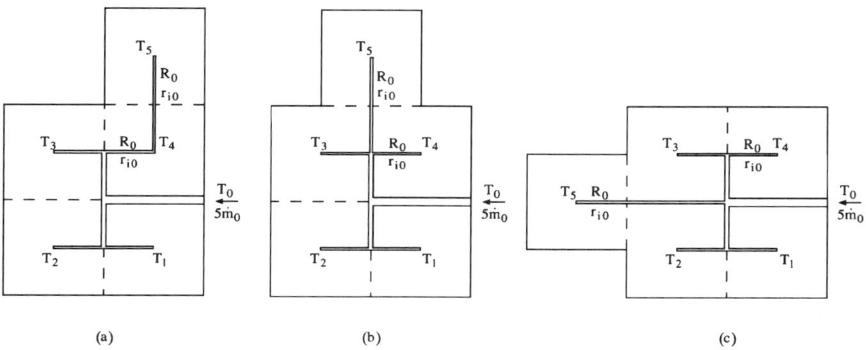

FIGURE 3.5. Three possible ways of attaching a new user to the optimized A_2 construct of hot water pipes of Fig. 3.2 (Wechsatol et al. 2002)

The corresponding expressions for the hot water temperature of the new user in the configurations of Figs. 3.5b and c are

$$\frac{T_5 - T_\infty}{T_0 - T_\infty} = \exp\left(-\frac{2N_0}{\ln R_0} - \frac{N_0}{3 \ln R_1} - \frac{2N_0}{5 \ln R_2}\right) \tag{3.22}$$

$$\frac{T_5 - T_\infty}{T_0 - T_\infty} = \exp\left(-\frac{3N_0}{\ln R_0} - \frac{2N_0}{5 \ln R_2}\right) \tag{3.23}$$

These equations for T_5 and the corresponding expressions for the temperatures of the hot water streams received by the existing four users ($T_1, \ldots, T_4$) are plotted in Fig. 3.6 for the same elemental number N_0. At first sight, these figures show an expected trend: the temperature of the delivered water increases as the total mount of insulation ($\tilde{V}_2$) increases. These figures, however, tell a more important story: they show not only how the choice of grafting (a, b, c) affects the temperature T_5, but also how the new (fifth) water stream affects the temperatures felt by the previous users.

Before the fifth user was added, the constructal design of A_2 delivered water of the same temperature to the four users. Note the single curve drawn for $T_{\text{end},A_2,\ \text{Fig. 3.2}}$ in each of Fig. 3.6. This curve serves as reference in each of the designs of Fig. 3.5, where the temperatures of the existing users are altered by the insertion of a fifth user. In the configuration of Figs. 3.5a and 3.6a, the fifth-user temperature drops below the original temperature of hot water delivery ($T_{\text{end},A_2,\ \text{Fig. 3.2}}$). In the same design, the temperatures of the four older users increases, the largest increases being registered by the users that are situated the closest to the newly added user. It is as if the new user insulates (or shields) its closest neighbors from cold water. It does so by drawing a larger stream of hot water in its direction and in the direction of its immediate neighbors.

These features are also visible in the performance of the configurations of Figs. 3.5b and c (cf. Figs. 3.6b and c), but the differences between the water temperatures of the five users are not as great as in the case of Fig. 3.5a (Fig. 3.6a). It is interesting that the best configuration (Figs. 3.5b and 3.6b), where the water received by the new user is the hottest, is also the configuration in which all the users receive water at nearly the same temperature. This finding is in line with the seemingly general conclusion that the optimal design of a complex flow system is the one where the imperfections are distributed as uniformly as possible (Bejan 2000). In the class of problems treated here, by imperfections we mean the decrease of the hot water temperature, which is due to the loss of heat to the ambient across the insufficient (finite) amount of insulation. This point is stressed further by Fig. 3.6c, which shows that the water temperatures of the four existing users are affected in equal measure by the insertion of the fifth user. The symmetry in this case is explained by the fact that in Fig. 3.5c the new user is attached to the hub of the previously optimized A_2 construct, at equal distance away from the existing four users.

The relative lack of success of design (c) in Fig. 3.5 may be attributed at first sight to the fact that the new user requires a longer pipe ($1.5\, L_0$) for its connection

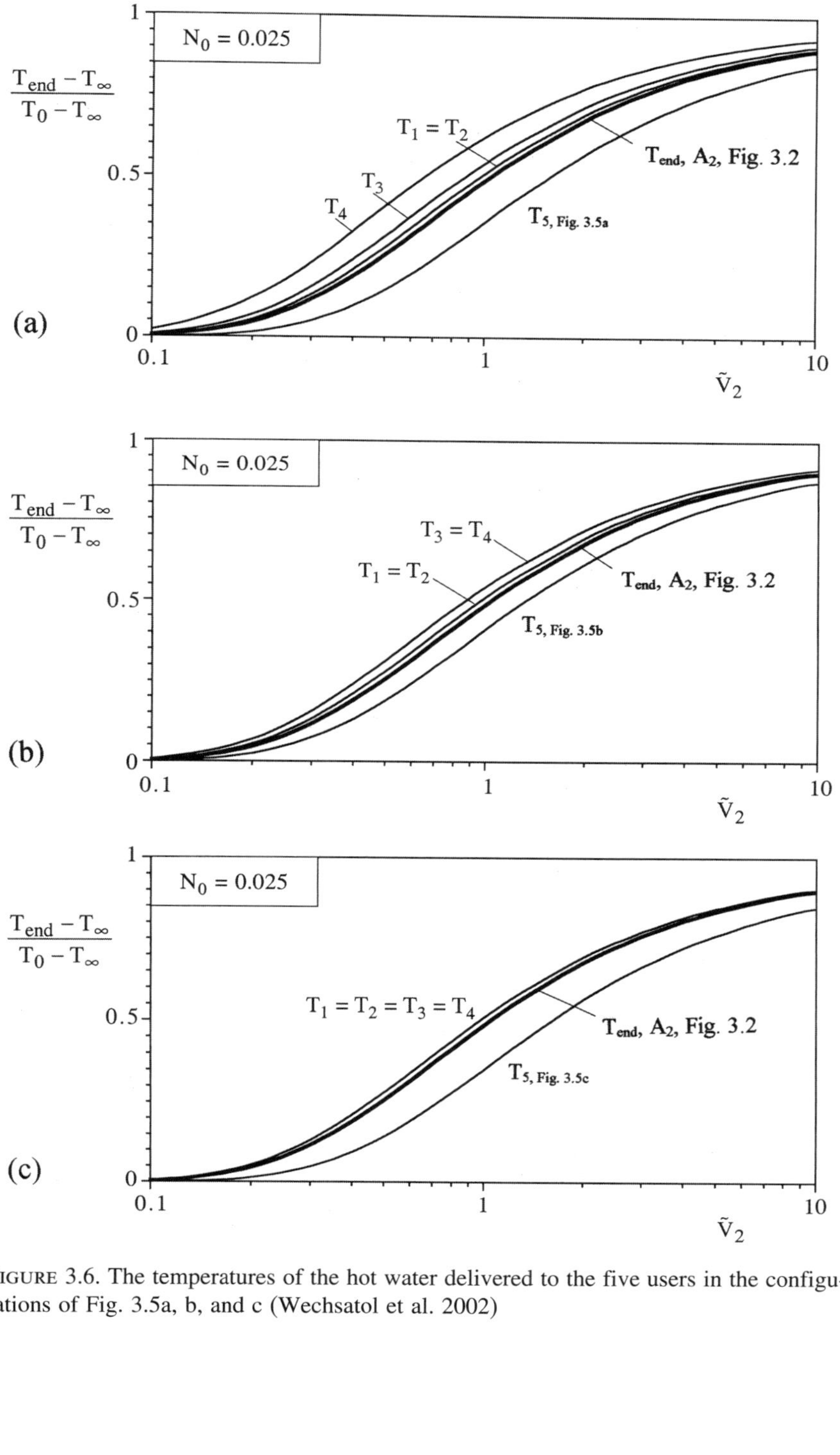

FIGURE 3.6. The temperatures of the hot water delivered to the five users in the configurations of Fig. 3.5a, b, and c (Wechsatol et al. 2002)

to the original construct, whereas in designs (a) and (b) the required connection is shorter (L_0). This explanation is not as straightforward as this, because design (c) has the advantage that the new user is connected to the heart (hot region) of the construct, while in designs (b) and (c) the new user is connected to peripheral locations. This competition between the length of the link and the temperature of the point of attachment to the original structure make the entire procedure subtle, i.e., not transparent. It is necessary to try all the possible configurations before deciding which design step or rule is beneficial.

In conclusion, we retain the design of Fig. 3.5b, and proceed to the next problem, which is the placing of a new (sixth) user in the best spot on the periphery of the five-user construct. The two most promising choices are shown in Fig. 3.7. They are the most promising because in each case the new user is attached to a source with relatively high temperature. In other words, unlike in Fig. 3.5 where we considered without bias all the possible ways of attaching the fifth user, now and in subsequent steps of growing the structure we expedite the place-selection process by using conclusions and trends learned earlier.

The performance curves plotted in Fig. 3.8 reinforce some of the earlier trends. Figure 3.8a shows that when the new user is placed symmetrically relative to the fifth user (Fig. 3.7a), symmetry is preserved in the temperature distribution over the entire tree. Note that in this configuration T_6 equals T_5. Symmetry works in favor of making the distribution of temperature (and heat loss, imperfection) more uniform.

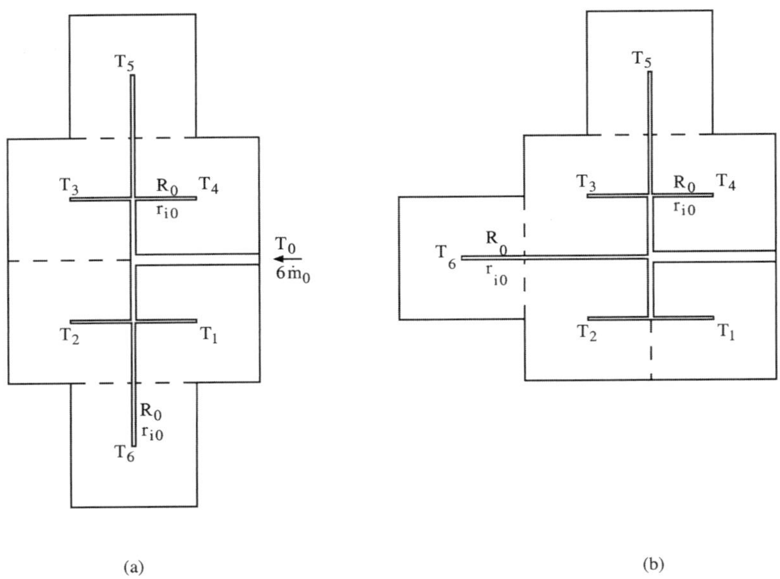

FIGURE 3.7. Two ways of attaching a sixth user to the construct selected in Fig. 3.5b (Wechsatol et al. 2002)

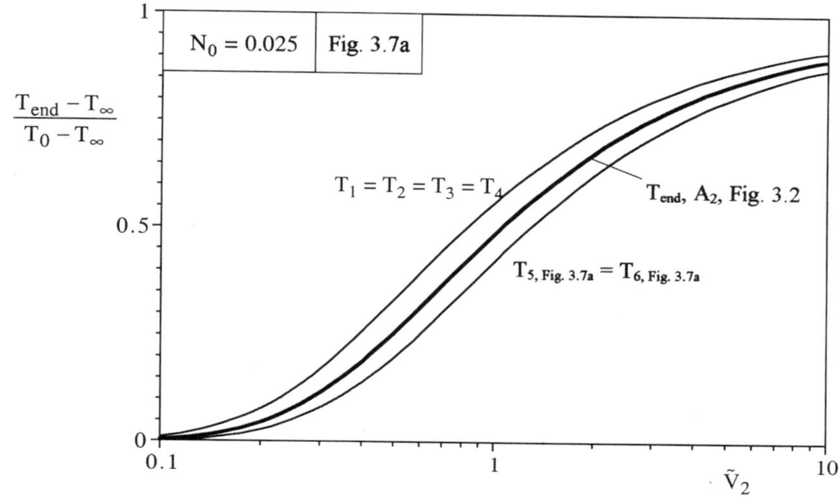

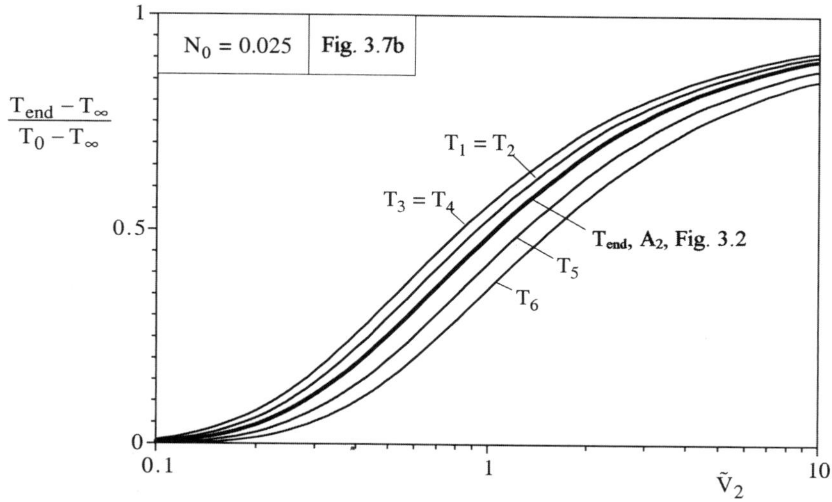

FIGURE 3.8. The temperatures of the hot water delivered to the fifth and sixth users in the configurations of Figs. 3.7a and b (Wechsatol et al. 2002)

The competing configuration (Fig. 3.7b) has the new user attached to the center of the original A_2 construct, in the same place as in Fig. 3.5c. Its performance is documented in Fig. 3.8b. The configuration of Fig. 3.7b is inferior to that of Fig. 3.7a because the temperature T_6 in Fig. 3.8b is consistently lower than the corresponding temperature in Fig. 3.8a. Furthermore, the configuration of Fig. 3.7b is inferior because it enhances the nonuniformity in the temperature of the water received by all the users.

In sum, we retain the configuration of Fig. 3.7a for the construct with six users. It is important to note that this is the first instance where the symmetrical placement of the even-numbered user emerges as the best choice. This conclusion was tested (one new user at a time) up to the 12th new user, beyond which it was adopted as a rule for expediting the optimized growth of structure. In other words, each even-numbered new user was placed symmetrically relative to the preceding odd-numbered user (the placement of which was optimized).

Figure 3.8 shows the three most promising positions that we tried for a new seventh user. Temperature distribution charts (such as Fig. 3.8) were developed for each configuration, but are not shown. On that basis we found that the seventh-user temperature T_7 decreases, in order, from Fig. 3.9b to Fig. 3.9a, and finally to Fig. 3.9c. We retained the configuration of Fig. 3.9b, and proceeded to the selection of the best place for an eight user. The best position for the eight user is the symmetric arrangement shown in Fig. 3.10a. The rest of Fig. 3.10 shows the optimal positioning of subsequent users up to the sixteenth. The rule that is reinforced by these choices is that the better position for a new user is that one that requires a short connection to that side of the existing construct where the water delivery temperatures are higher.

We are now in a position to compare the performance of designs based on one-by-one growth (Fig. 3.10) with the constructal designs obtained by repeated doubling (Fig. 3.2). The most complex design of the one-by-one sequence

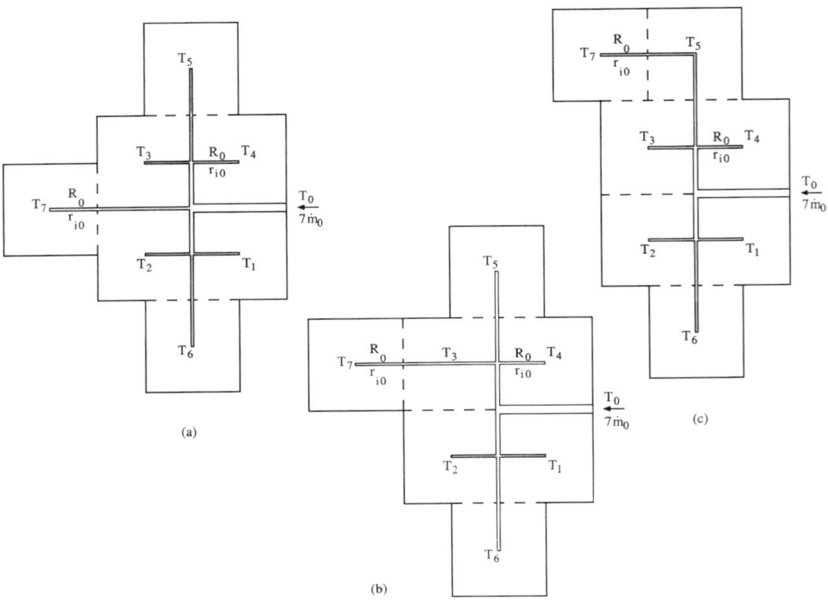

FIGURE 3.9. Three ways of placing a seventh user in the construct selected in Fig. 3.7a (Wechsatol et al. 2002)

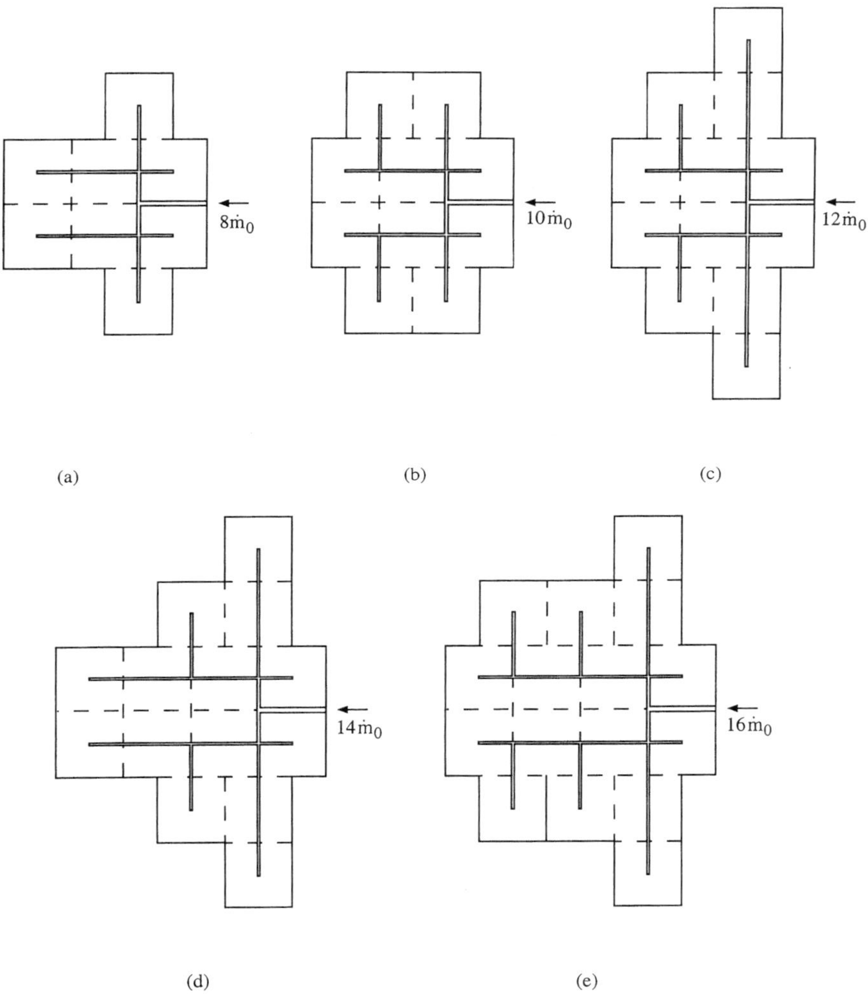

FIGURE 3.10. The best configurations for the constructs with 8 to 16 users (Wechsatol et al. 2002)

(Fig. 3.10e) is compared with its constructal counterpart in Figure 3.11a. The comparison is done on the same basis, i.e., the same total amount of insulation material, and the same serviced territory ($16L_0^2$). We see that the excess temperature of the latest (16th) user in Fig. 3.10e is consistently less than the excess temperature felt by each of the users in A_4 construct generated by doubling (recall that in Fig. 3.2 each user receives water at the same temperature).

The discrepancy between the performances of the two designs compared in Fig. 3.11(a) diminishes as the total amount of insulation ($\tilde{V}_{A_{16},\ \mathrm{Fig.\ 3.10(e)}}$) increases, and as N_0 decreases. The smaller N_0 corresponds to better insulation materials (smaller k) and denser populations of users (smaller L_0).

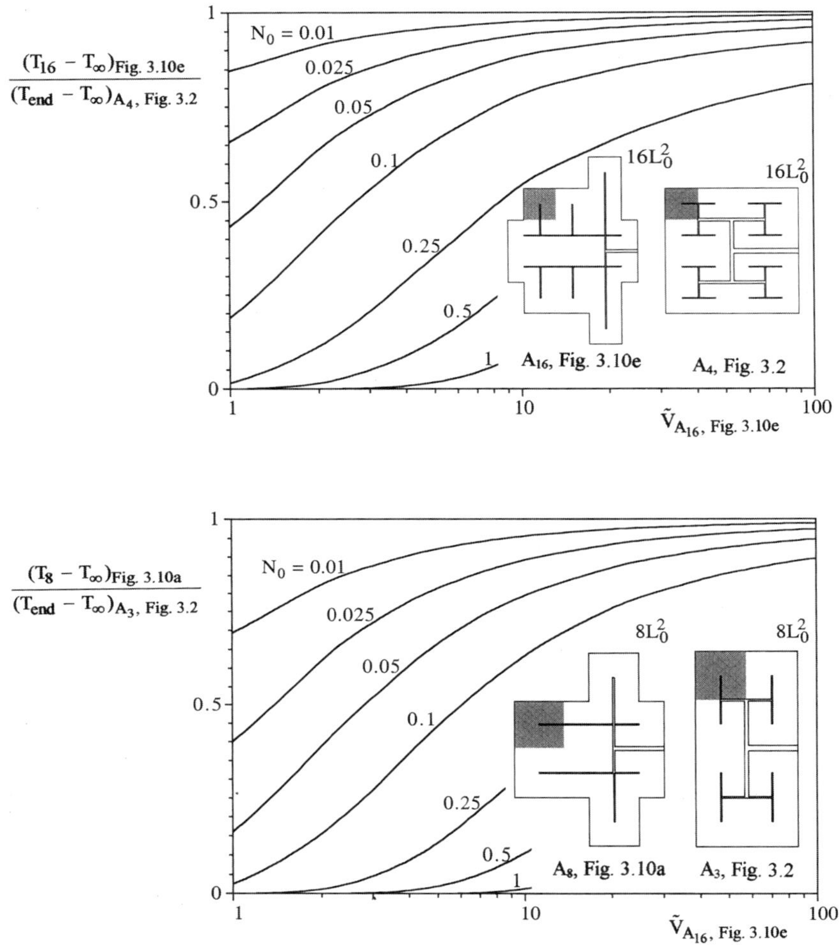

FIGURE 3.11. **a** Comparison between the excess temperature of the water received by the last user added to the structure of Fig. 3.10e and the corresponding temperature in the design based on the sequence of Fig. 3.2. **b** Comparison between the excess temperature of the water received by the last user added to the structure of Fig. 3.10a, and the corresponding temperature in the design based on the sequence of Fig. 3.2 (Wechsatol et al. 2002)

3.6. Social Determinism and Constructal Theory

To illustrate the connection between constructal theory and social determinism, consider the example of the access to elite superior education in France. In France, the system of superior education is divided into two kinds of structures. Historically, the university was dedicated to the transmission of knowledge and culture and to research, while "les grandes écoles," a French specificity, were created to educate the engineers and the executive of the nation and of private companies. If the first entity does not select its students (any pupil who graduates from high school can in principle enter the university), "les grandes écoles," are highly selective. The latter have been accused for a long time to be responsible for *elite reproduction*, instead of what they are intended for, namely *elite production*. Indeed, today the probability to enter "une grande école," is less then 2% when the student's father belongs to the category of workers, while the probability is greater then 20% when the father is an engineer, an executive, or a professor (Albouy and Wanecq 2003). Note that the mother's profession is not mentioned, not a single bit of information being available at the time of the study.

Such figures are typical of social determinism: the major part of the students in "les grandes écoles" are children of the upper middle class, upper class, or children of teachers and professors. We have a main stream with very low flow resistances conducting these privileged children to the "les grandes écoles," and we have other channels with extremely high flow resistances through which, from time to time, a brilliant pupil can enter one of these schools of excellence. The flow through these last channels is so badly connected that sometimes we end up with dead arms!

The consequences of this report are depressive: the social lift does not work anymore, and this feeds the despair of thousands of young people. It may be one of the explanations of desperate acts such as the riots at the end of October 2005 in certain French suburbs.

In constructal theory, we are aware that we cannot eliminate the flow resistances, but that we can distribute them optimally over the network. Here, the network is made of paths leading to places of excellence. Instead of the existing radial paths with direct access for privileged social categories to elite school, tree branching is a clever alternative because it allows children from other and less favored origins to join the main stream.

As an additional contribution, a few years ago a group consisting of people conscious of the need for *Social Opening* in the French elite schools was created. This initiative is called "L'Ouverture Sociale." Its objective is to decrease the resistance along the paths (e.g. strengthen the knowledge of the pupils, overcome self-censorship), eliminate the dead arms (wrong orientation choices), enlarge the cross-section of these ducts, and create special access to reduce the local (social) pressure losses. Bridges are created that help the students from difficult social backgrounds to access the main stream: these bridges are the loops that we can see when we observe carefully a leaf. The loops are here as add-ons to the tree-shaped network. In constructal theory, we have studied the effect of superimposing loops on our dendritic fluid networks (Wechsatol et al. 2005). We know that they increase the required pumping power (the work requirement), but they are the solution that prevents damages. This is how the optimal distribution of imperfections can help us achieve an overall optimal functioning.

References

Albouy, V. and Wanecq, T. (2003) Les inégalités sociales d'accès aux grandes écoles. *Economie et Statistique* **361**, 27–52.

Barreau, A. and Moret-Bailly, J. (1977) Présentation de deux méthodes d'optimisation de réseaux de transport d'eau chaude à grande distance. *Entropie* **75**, 21–28.

Bastié, J. and Dézert, B. (1980) *L'Espace Urbain*, Masson, Paris.

Bejan, A. (1997) Constructal-theory network of conducting paths for cooling a heat generating volume. *Int. J. Heat Mass Transfer* **40**, 799–816.

Bejan, A. (2000) *Shape and Structure, from Engineering to Nature*, Cambridge University Press, Cambridge, UK.

Bejan, A., Tsatsaronis, G. and Moran, M. (1996) *Thermal Design and Optimization*, Wiley, New York.

Bonnin, J. (1977) *Hydraulique Urbaine Appliquée aux Agglomérations de Petite et Moyenne Importance*, Eyrolles, Paris.

Dupont, A. (1977) *Hydraulique Urbaine, Ouvrages de Transport*, Elévation et Distribution, Eyrolles, Paris.

Falempe, M. (1993) Une plate-forme d'enseignement et de recherche du procédé de cogénération chaleur-force par voie de vapeur d'eau. *Revue Générale de Thermique* **383**, 642–651.

Falempe, M. and Baudoin, B. (1993) Comparaison des dix méthodes de résolution des réseaux de fluides à usages énergétiques. *Revue Générale de Thermique* **384**, 669–684.

Martinand, C. (1986) *Le Génie Urbain Rapport au Ministre de l'Equipement, du Logement, de l'Aménagement du Territoire et des Transports*, La Documentation Française, Paris.

Nonclercq, P. (1982) *Hydraulique Urbaine Appliquée*, CEBEDOC, Liège.

Padet, J. (1991) *Fluids en Écoulement. Méthodes et Modèles*, Masson, Paris.

Pelletier, J. and Delfante, C. (1989) *Villes et Urbanisme dans le Monde*, Masson, Paris.

Plaige, B. (1999) Le Chauffage Urbain en Pologne. *Chauffage, Ventilation, Conditionnement d'Air* **9**, 19–23.

Wechsatol, W., Lorente, S. and Bejan, A. (2001) Tree-shaped insulated designs for the uniform distribution of hot water over an area, *Int. J. Heat Mass Transfer* **44**, 3111–3123.

Wechsatol, W., Lorente, S. and Bejan, A. (2002) Development of tree-shaped flows by adding new users to existing networks of hot water pipes. *Int. J. Heat Mass Transfer* **45**, 723–733.

Wechsatol, W., Lorente, S. and Bejan, A. (2005) Tree-shaped networks with loops. *Int. J. Heat Mass Transfer* **48**, 573–583.

Chapter 4
Natural Flow Patterns and Structured People Dynamics: A Constructal View

A. Heitor Reis

4.1. Introduction

We show here that flow patterns similar to those that emerge in nature can also be observed in flows of people and commodities and can be understood in light of the same principle—the Constructal Law. River basins are examples of naturally organized flow architectures whose scaling properties have been noticed long ago. We show that these scaling laws can be anticipated based on constructal theory, which views the pathways by which drainage networks develop in a basin not as the result of chance but as flow architectures that originate naturally as the result of minimization of the overall resistance to flow. Next we show that the planetary circulations and the main global climate zones may also be anticipated based on the same principle. Finally, we speculate that the same principle governs "rivers of people," whose architecture develops in time to match the purpose of optimizing access within a territory.

4.2. Patterns in Natural Flows: The River Basins Case

Flow architectures are everywhere in nature, from the planetary circulation to the smallest scales. In the inanimate world, we can observe an array of motions that exhibit organized flow architectures: general atmospheric circulation, oceanic currents, eddies at the synoptic scale, river drainage basins, etc. In living structures, fluids circulate in special flow structures such as lungs, kidneys, arteries, and veins in animals and roots, stems, and leaves in plants.

Rivers are large-scale natural flows that play a major role in the shaping of the Earth's surface. River morphology exhibits similarities that are documented extensively in geophysics treatises. For example, Rodríguez-Iturbe and Rinaldo (1997) provided a broad list of allometric and scaling laws involving the geometric parameters of the river channels and of the river basins.

In living structures, heat and mass flow architectures develop with the purpose of dissipating minimum energy, thereby reducing the food or fuel requirement, and making all such systems more "fit" for survival (Reis et al., 2004, 2006b).

Fractal geometry has been used to describe river basin morphology (e.g., Cieplak et al., 1998; Rodríguez-Iturbe and Rinaldo 1997). Fractals do not account for dynamics, hence are descriptive rather than predictive.

Different from the fractal description, constructal theory views the naturally occurring flow structures (their geometric form) as the end result of a process of area to point flow access maximization with the objective of providing minimal resistance to flow (see Bejan 2000; Bejan and Lorente 2004). The Constructal Law first put forward by Bejan in 1996 states that "for a finite-size system to persist in time (to live), it must evolve in such a way that it provides easier access to the imposed (global) currents that flow through it."

What is new with Constructal theory is that it unites geometry with dynamics in such a way that geometry is not assumed in advance but is the end result of a tendency in time. Constructal theory is predictive in the sense that it can anticipate the equilibrium flow architecture that develops under existing constraints. In contrast with fractal geometry, self-similarity need not be alleged previously, but appears as a result of the constructal optimization of river networks.

4.2.1. Scaling Laws of River Basins

River basins are examples of area-to-point flows. Water is collected from an area and conducted through a network of channels of increasing width up to the river mouth. River networks have long been recognized as being self-similar structures over a range of scales. In general, small streams are tributaries of the next bigger stream in such a way that flow architecture develops from the smallest scale to the largest scale, ω (Fig. 4.1).

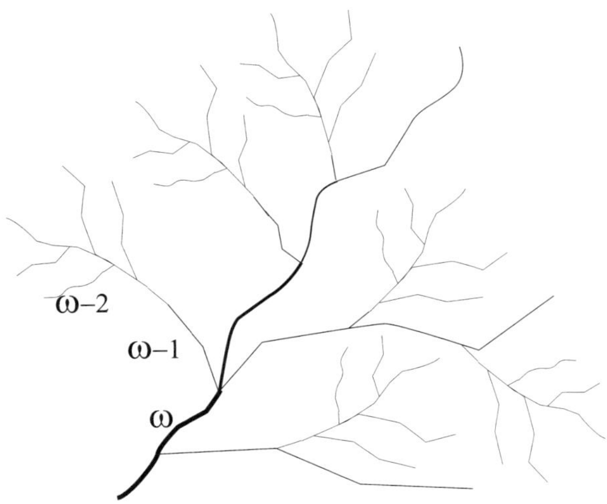

FIGURE 4.1. Hierarchical flow architecture in a river basin. Streams of order ω_{i-1} are tributaries of streams of order ω

The scaling properties of river networks are summarized in well-known laws (Beven, 1993). If L_i denotes the average of the length of the streams of order i, Horton's law of stream lengths states that the ratio

$$L_i / L_{i-1} = R_L \tag{4.1}$$

is a constant (Horton 1932; see also Raft et al. 2003; Rodríguez-Iturbe and Rinaldo 1997). Here, the constant R_L is Horton's ratio of channel lengths. On the other hand, if N_i is the number of streams of order i, Horton's law of stream numbers asserts constancy of the ratio

$$N_{i-1} / N_i = R_B \tag{4.2}$$

where R_B is Horton's bifurcation ratio. In river basins, R_L ranges between 1.5 and 3.5 and is typically 2, while R_B ranges between 3 and 5, typically 4 (Rodríguez-Iturbe and Rinaldo, 1997).

The mainstream length L_ω and the area A_ω of a river basin with streams up to order ω are related through Hack's law (Hack 1957; see also Rodríguez-Iturbe and Rinaldo 1997; Schuller et al. 2001):

$$L_\omega = \alpha (A_\omega)^\beta \tag{4.3}$$

where $\alpha \sim 1.4$ and $\beta \sim 0.568$ are constants.

If we define a drainage density $D_\omega = L_T / A$ (where L_T is the total length of streams of all orders and A the total drainage area) and a stream frequency $F_s = N_s / A$ (where N_s is the number of streams of all orders) then Melton's law (Melton 1958; see also Raft et al. 2003; Rodríguez-Iturbe and Rinaldo 1997) indicates that the following relation holds:

$$F_s = 0.694 (D_\omega)^2 \tag{4.4}$$

Other scaling laws relating discharge rate with river width, depth, and slope may be found in the book by Rodríguez-Iturbe and Rinaldo (1997).

The scaling laws of geometric features of river basins have been predicted based on Constructal theory, which views the pathways by which drainage networks develop in a basin not as the result of chance but as flow architectures that originate naturally as the result of minimization of the overall resistance to flow (Bejan 2006; Reis 2006a).

The ratios of constructal lengths of consecutive streams match Horton's law for the same ratio, while the same holds for the number of consecutive streams that match Horton's law of ratios of consecutive stream numbers.

Hack's law is also correctly anticipated by the constructal relations that provide Hack's exponent accurately.

Melton's law is anticipated approximately by the constructal relationships that indicate 2.45 instead of 2 for Melton's exponent. However, the difficulty to calculate correctly the drainage density and the stream frequency from field data indicates that some uncertainty must be assigned to Melton's exponent (Reis 2006a).

4.3. Patterns of Global Circulations

Here, constructal theory is extended to the problem of atmospheric and oceanic circulation driven by heating from the sun. Climate means the average thermo-hydrodynamic conditions that prevail over a significant period (generally 30 years) at a particular region of Earth's surface. Due to nonuniform heating, flows develop on the Earth's surface carrying heat from hot to cold regions. Atmospheric and oceanic circulations of a wide range of magnitudes participate in this transfer. Coupling between different scales of heat and mass flows is highly non-linear, thereby making prediction of the thermo-hydrodynamic state of the atmosphere a very hard task.

Thermodynamically, the Earth as a whole is a nonequilibrium closed system—a flow system with heat input and heat rejection, and with internal flows. It is not an isolated thermodynamic system. Furthermore, the Earth's thermodynamic system is not in a steady state, as can be concluded from the observational evidence that temperature and pressure have temporal variations. However, if we consider local values averaged over a long period, temperature and pressure become time independent even if their spatial variation is preserved. This procedure implies loss of information of short-period phenomena, but it allows the construction of useful steady-state models of the whole system. These models capture the long-term performance of the Earth system and allow useful predictions about climate.

Atmospheric and oceanic circulations are the largest flow structures on Earth. Modeling of such flow structures relies on deterministic equations, e.g., conservation of mass, energy, angular momentum, and momentum, some of which are nonlinear and give rise to additional terms that result from the averaging procedure. The closure of such a system of equations is not easy and is achieved with the help of empirical information. In spite of the large number of parameters used, such models have shed light on global circulation and climate, and have served as reminders that theory is needed.

A different approach is made possible by constructal theory (see Bejan and Reis 2005; Reis and Bejan 2006). Although constructal theory also relies on mass and energy conservation, it derives the actual flow field (the flow architecture) from the maximization of the flow access performance of the whole system under the existing constraints. The nonuniform heating of the Earth's surface and atmosphere drives the Earth circulation. According to constructal theory, the purpose of the circulation (the objective of any flow with configuration) is to provide maximum access to the currents that flow, in this case to the transfer of heat from the equatorial zone to the polar caps. The zones and caps are organized in such a way that they perform this transport in the most efficient way, which is the one that maximizes the heat flow or, alternatively, by the flow structure that minimizes the resistance to the global heat flow. The theory showed that the poleward heat flow is maximized if the Earth's surface is partitioned into a heat source between 25°40′ N and 25°40′ S, and two heat sinks located in the polar caps bounded by the latitudes 53°10′ N and S. Between the heat source and sinks, in each hemisphere, there is a third surface that participates in the heat transfer process and has an average temperature of 281.5 K. These intermediate surfaces correspond to vertical circulating loops, as was seen when the average temperatures defining this partitioning of surfaces are included in the calculation of the flow variables of the model.

It is worth noting that the latitude 25°40′ corresponds very accurately to the boundary between the Hadley and Ferrel cells. Furthermore, the latitude 53°10′ is close to the latitude 60°, which is recognized as the boundary between the Ferrel and the polar cells that represent the mean meridional atmospheric circulation on Earth. In Fig. 4.2 we can observe the cells corresponding to the long-term latitudinal circulation: the Hadley cells, which start at the Equator and go close to 25°, the Polar cells that develop between each pole and latitude 60°, and the Ferrel cells located between the Hadley and the Polar cells. These cells are considered to drive the Ferrel cells, which develop between them.

It is a remarkable coincidence that the invocation of the Constructal Law predicts these latitudes as the optimal partitioning of the Earth's surface with respect to the heat flow along the meridian. The predicted average temperature of the Earth's surface, the convective conductance in the horizontal direction as well as other parameters defining the latitudinal circulation also match the observed values.

In a second part of the study (Reis and Bejan 2006), the Constructal Law was invoked in the analysis of atmospheric circulation at the diurnal scale. Here the heat transport is optimized against the Ekman number. Even though this second optimization is based on very different variables than in the first part of the paper, it produces practically the same results for the Earth's surface temperature and the other variables. The Earth's average temperature difference between day and night was found to be approximately 7 K, which matches the observed value. The accumulation of coincidences between theoretical predictions and natural facts adds weight to the claim that the Constructal Law is a law of nature (Reis and Bejan 2006).

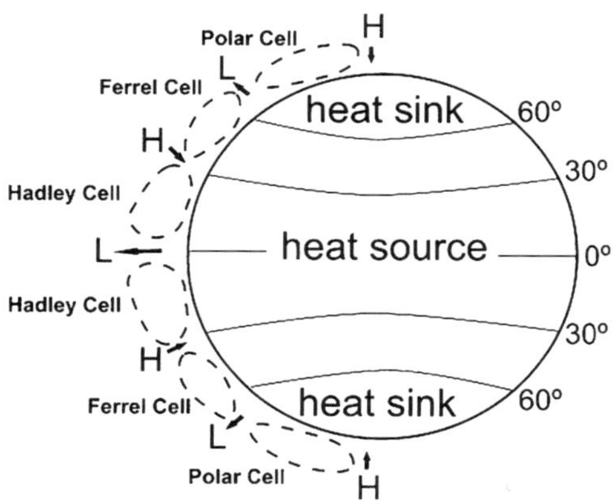

FIGURE 4.2. The main cells of global circulation that determine Earth's climate

4.4. Flows of People

Individuals living in large cities usually have to travel long distances to carry out daily activities. Walking is the first mode of movement for people when they depart from home. Then they can use cars, buses, trains, or planes in the way to their destination. Each of these means of transportation uses a proper channel to move on. Public transportation exists because individuals agree in moving together in some direction. It is also a way of saving exergy, i.e., useful energy, and time. The movement of people is in the beginning disorganized (erratic, if we consider a group of individuals) showing the characteristics of diffusive flow, then becomes progressively organized (more and more people moving in the same way) as people move into larger streams. The speed of transportation also increases as individuals proceed from home to the larger avenues (Fig. 4.3). The global movement of people can be regarded as an area-to-point flow as people move from their homes to the largest way and then as a point-to-area flow as they disperse onto the area of destination. These patterns have been encountered before in many naturally organized flows (river basins, lungs, etc.) and occur in nature because they offer minimum global resistance to flow (Bejan 2000).

Modeling flows of people may be carried out in a framework similar to flows of inanimate matter. In that case, we know the forces that drive the movement. Humans are not only subjected to external (physical) forces and constraints but are also subjected to "internal" driving forces. However, whatever the forces that drive individuals along a path i are (see Fig. 4.3), they always result in internal energy dissipation (entropy generation) that can be represented by (see Reis 2006b):

$$\dot{E}_i = R_i I_i^2 \tag{4.5}$$

where $\dot{E}_i$ means energy dissipation rate, R_i is resistance to flow and $I_i = \Delta m_i / \Delta t_i$ is current, i.e., the ratio of mass Δm_i (bodies, cars) that crosses the channel section to passage time Δt_i.

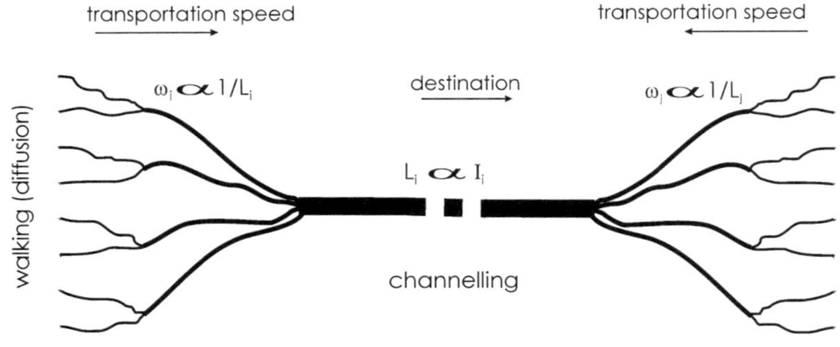

FIGURE 4.3. Area-to-point and point to area flows of people form a double tree. Constructal theory provides the optimal flow tree configuration

The physical significance of resistance in flows of people can be inferred from Eq. (4.5) and corresponds to the energy that is destroyed by a current flowing at the rate of 1 kg/s. The resistance to flow depends on technological factors such as the efficiency and capacity of the available means of transportation as well as on the morphology of street and road networks and the topography of the territory.

The current I may vary during the day and usually matches permanent transportation needs corresponding to the existing economic and social activities of the communities. In this way, most of the times I must be considered as a constant while being also a constraint to the transportation system.

Finally, the energy is the physical prime-mover of the economic and social systems (Reis 2006c). Everything that flows in these systems is driven by the available energies. Thus, energy is essential for keeping systems alive, and because it is scarce it must be used adequately. This is why a "living" system must optimize the use of energy in every particular activity or, what is the same, it must minimize the resistance to the respective internal flows [see Eq. (4.5)]. Therefore, flows of people are also a field of application of the Constructal Law.

Easier access means lesser global resistance to flow, i.e., better utilization of the available energy. In this way, "living" systems must develop internal flow structures that match this purpose. Modern societies have developed networks with the purpose of making the flows of people and goods easier and easier. People and goods flow through these networks in organized ways. These organized movements configure tree patterns as people move from home to destination (see Fig. 4.3). These tree architectures may be optimized under the Constructal Law. However, this does not mean that the existing flows of people and goods are optimized. From the point of view of the Constructal Law these are evolving flow structures that morph in time as a result of the struggle of the global system for better performance. What Constructal Theory provides is the optimum flow structure under the existing constraints and elects it as the equilibrium flow structure in the direction of which the actual configuration will move as the result from the struggle for better performance.

4.4.1. Optimal Flow Tree

As discussed before, individuals moving daily from one area to another follow paths in the transportation network that form flow trees. When starting from different points spread over an area people's movement (walking) is little organized, while converging to access points to the first means of transportation, which usually carry little numbers of people (cars, buses). As people move together, the resistance per unit length (resistivity) diminishes. The next step for diminishing flow resistance might be converging to railway stations in order to use trains that present lower flow resistivity and then, if it is the case, to accede airports to proceed by plane. The distribution of people over the destination area proceeds inversely, i.e., people do use means of successively higher resistivity while moving to their destinations. Each pattern of area-to-point and point-to-area flow shapes as a tree (Fig. 4.3).

The form of the flow tree that offers the lowest resistance to flow of people under the existing constraints may be derived analytically as follows. Let r_i, L_i and ω_i be the resistivity, the average distance traveled, and the number of pathways traveled by means of transportation of kind i, respectively. Then, the average resistance of each branch i of the conceptual flow tree is given by

$$R_i = \frac{r_i L_i}{\omega_i} \tag{4.6}$$

The total distance L to be traveled using n means of transportation—the branching level of the tree—is fixed,

$$\sum_{i=1}^{n} L_i = L \tag{4.7}$$

as it is the total current I flowing in the tree

$$\sum_{i=1}^{n} \omega_i I_i = nI \tag{4.8}$$

where I_i is people's current flowing in pathway i. Let $I_i = N_i / \Delta t_i$, where N_i is the number of people transported at the average velocity v_i through all the i pathways during the interval $\Delta t_i = L_i / v_i$. The energy dissipated in branch i per unit distance and per unit of current is $\dot{e}_i = \dot{E}_i / I_i L_i$, and is fixed because it depends on the performance of the means of transportation that move in that branch. Then by using Eqs. (4.5) and (4.6) we can write Eq. (4.8) as:

$$\sum_{i=1}^{n} \omega_i^2 \frac{\dot{e}_i}{r_i} = nI \tag{4.9}$$

By minimizing the flow resistance (Eq. 4.6) under the existing constraints (Eqs. 4.7 and 4.9) we obtain

$$\omega_i = \lambda_1 r_i \tag{4.10}$$

and

$$L_i = 2\lambda_1^2 \lambda_2 I_i \tag{4.11}$$

where λ_1 and λ_2 are constants that may be calculated from Eqs. (4.7) and (4.8). The distances L_i define the lengths of the branches of the tree shown in Fig. 4.3. Given that $I_i = I / \omega_i$, Eq. (4.11) may be rewritten as

$$L_i = 2\lambda_1^2 \lambda_2 \frac{I}{\omega_i} \tag{4.12}$$

or, in view of Eq. (4.10), as

$$L_i = 2\lambda_1 \lambda_2 \frac{I}{r_i} \tag{4.13}$$

Equations (4.10–4.13) indicate that the best conceptual configuration—i.e., the best combination of all available pathways for area-to-point and point-to-area access—is a flow tree in which the number of pathways traveled by means of transportation of kind i is proportional to the respective resistivity (see Eq. 4.10), while the average length to be traveled varies inversely with resistivity (see Eq. 4.13). In this way, the "best flow tree" is composed of a large number of pathways of high resistivity and progressively smaller number of low-resistivity pathways. On the other hand, pathway length increases with decreasing resistivity. This means that the largest distances must be covered by low-resistivity means of transportation. Furthermore, as pathway length varies inversely with resistivity (Eq. 4.13), the largest current of people must flow along the longest pathway (Eq. 4.11).

Finally, Eq. (4.12) shows the geometric form of such a conceptual flow tree: the number of pathways must vary inversely with pathway length. A representation of such a tree is shown in Fig. 4.3.

The double flow tree shown in Fig. 4.3 matches two constrains only, which are represented by Eqs. (4.7) and (4.8). However, real trees occur as the result of a much higher number of constraints, though their importance is of second order. The freedom to morph is limited by the existing street and road networks, terrain configuration, and other particularities that must also be considered as constraints. However being the result of millions of individual movements, flow patterns appear as a consequence of a collective perception that movement is easier when people move in some special pathways. This kind of "crowd intelligence" designs flow patterns in city and intercity networks of communication. Networks pre-exist to flow patterns but these define which parts of the existing networks must be used for easier flow access. Morphing of transportation networks in time occurs as a response to pressure of people in search for flow trees that allow best area-to-point and point-to-are flow access.

4.4.2. Fossils of Flows of People

The Constructal Law indicates that the pattern that develops in time is the one that provides maximum flow access (minimum resistance) under the existing constraints. This universal behavior (law) has been observed both in inanimate and animate systems (e.g. Bejan 2000, 2006; Reis, 2006a,b)

The first human communities that established in fixed places had to establish their own "farmland," i.e., the area that fed the community. Even in our time we can observe these "farmlands" that feed small rural communities, and the pathways that serve for transportation of goods and people (Fig. 4.4). Also in ancient times, movement of people between the settlements and the surrounding area designed flow pathways some of which can be seen still in the present time. They are fossils of ancient flows of people that remind us of how our ancestors solved area-to-point flow access problems.

In many of our cities that are of ancient origin, we can observe such fossil patterns which are reminiscent of ancient eras. Figure 4.5 represents an aerial

FIGURE 4.4. Rural community and the surrounding farmland. Pathway network serves the purpose of flow access of people and commodities (Courtesy of IGP, Portugal)

view of the city of Évora in the south of Portugal, which was named as world heritage by UNESCO. We can observe an inner nucleus that matches the ancient perimeter of the city under the Roman Empire, with a network of small streets which were suited for flows of people in those times. The street network reflects the randomness characteristic of walking, which was the main way of locomotion. The pattern reveals a great number of small streets with high resistivity to flow,

FIGURE 4.5. Historical center of Évora. Note that the gradient of people flow resistivity points to the city center, i.e., to the past (Courtesy of IGP, Portugal)

in accordance with Eqs. (4.10) and (4.13). Next to the center a second pattern develops with broader and longer streets, which corresponds to the medieval part of the city, a time when horse and cart transportation were widespread. Note that reduction in flow resistivity entails increasing length of flow channels as stated by Eq. (4.13). Next we find the part of the city that corresponds to the times when automotive transportation (cars, buses, etc.) appeared. During the twentieth century the new streets became progressively wider and longer in response to increase in the automotive traffic.

In Fig. 4.6 we can also observe an aerial view of the downtown of Lisbon. The part to the right-hand side shows a circular street pattern developing around the medieval castle. Here we see again a great number of small streets corresponding to the high resistivity city network of the medieval era. In contrast with this part of the city, on the left-hand side we observe a very regular network of a smaller number of streets of lower resistivity. This part is from the middle of the eighteenth century after the big earthquake (1755), and was erected in order to match people flows of those times. In fact, during reconstruction, big ways were opened for people and goods transportation from the Lisbon harbor to the city center. Here we see again patterns that match Eqs. (4.10) and (4.13).

FIGURE 4.6. Downtown of Lisbon. Note the high resistivity area close to the medieval castle and the lower resistivity area of the eigtheenth century (Courtesy of IGP, Portugal)

In this way, street networks tell us the story of development of people's flow patterns in history, which is also the story of the continuous search for easier flow access.

4.5. Conclusions

Constructal theory that has been successfully applied to planetary circulations and climate and to river basin morphology is shown to provide a useful framework for describing flows of people. We showed here, with simple examples, that intuitive rules of traffic organization can be anticipated based on principle, i.e., based on the Constructal Law. In addition, and similarly to the case of flows of inanimate matter, in the case of flows of people, flow patterns emerge as a necessary consequence of reduction of global flow resistances. These flow patterns point to decreasing resistivity to flows of people and commodities. Pathway length varies inversely with resistivity while pathway number increases with resistivity.

In summary, constructal theory provides a broad framework for the analysis of all types of flow including flows of people, which despite their complexity share common features with flows of inanimate matter.

References

Bejan, A. (2000) *Shape and Structure, from Engineering to Nature*, Cambridge University Press, Cambridge, UK.

Bejan, A. (2006) *Advanced Engineering Thermodynamics*, Chapter 13, 3rd Edn, Wiley, New York.

Bejan, A. and Lorente, S. (2004) The Constructal law and the thermodynamics of flow systems with configuration, *Int. J. Heat Mass Transfer* **47**, 3203–3214.

Bejan, A. and Reis, A. H. (2005) Thermodynamic optimization of global circulation and climate, *Int. J. Energy Res.*, **29**(4), 303–316.

Bejan, A., Dincer, I., Lorente, S., Miguel, A. F. and Reis, A. H. (2004) *Porous and Complex Flow Structures in Modern Technologies*, Springer-Verlag, New York.

Beven K. (1993) Prophesy, reality and uncertainty in distributed hydrological modelling, *Adv. Water Resources* **16**, 41–51.

Cieplak, M., Giacometti, A., Maritan, A., Rinaldo, A., Rodriguez-Iturbe, I. and Banavar, J. R. (1998) Models of Fractal River Basins, *J. Stat. Phys.* **91**, 1–15.

Hack, J. T. (1957) Studies of longitudinal profiles in Virginia and Maryland. USGS Professional Papers 294-B, Washington DC, pp. 46–97.

Horton, R. E. (1932) Drainage basin characteristics, *EOS Trans.* AGU **13**, 350–361.

Melton, M. A. (1958) Correlation structure of morphometric properties of drainage systems and their controlling agents, *J. Geology* **66**, 35–56.

Raft, D. A., Smith, J. L. and Trlica, M. J. (2003) Statistical descriptions of channel networks and their shapes on non-vegetated hillslopes in Kemmerer, Wyoming, *Hydrol. Processes* **17**, 1887–1897.

Reis, A. H. (2006a) Constructal view of scaling laws of river basins, *Geomorphology* **78**, 201–206.

Reis, A. H. (2006b) Constructal theory: from engineering to Physics, and how flow systems develop shape and structure, *Appl. Mech. Rev.* **59**, 269–282.

Reis, A. H. (ed.) (2006c) Energy Based Analysis of Economic Sustainability, *Perspectives in Econophysics*, University of Évora, Portugal, pp. 147–159.
Reis, A. H. and Bejan, A. (2006) Constructal theory of global circulation and climate, *Int. J. Heat Mass Transfer* **49**, 1857–1875.
Reis, A.H., Miguel, A.F. and Aydin, M. (2004) Constructal theory of flow architectures of the lungs, *Med. Phys.* **31**(5), 1135–1140.
Reis, A. H., Miguel, A. F. and Bejan, A. (2006) Constructal particle agglomeration and design of air-cleaning devices, *J. Phys. D: Appl. Phys.* **39**, 2311–2318.
Rodríguez-Iturbe, I. and Rinaldo A. (1997) *Fractal River Basins*, Cambridge University Press, New York.
Schuller, D. J., Rao, A. R. and Jeong, G. D. (2001) Fractal characteristics of dense stream networks, *J. Hydrol.* **243**, 1–16.

Chapter 5
Constructal Pattern Formation in Nature, Pedestrian Motion, and Epidemics Propagation

Antonio F. Miguel

5.1. Introduction

The emergence of shape and structure in animate and inanimate systems is among the most fascinating phenomena in our planet. Rivers carry water and sediment supplied by the hydrological cycle and erosional mechanisms, which play a major role in the shaping of the Earth's surface. River basins result, for instance, from these naturally organized flow architectures (Bejan 1999; Reis 2006a). Flow architectures also underline the phenomenon by which wet soil exposed to the sun and wind loses moisture, shrinks superficially, and develops a network of cracks (Bejan 2000). Similarly, in complex cellular systems such as the vertebrates, the requirement of large amount of oxygen for the metabolic needs of the cells constitutes the basis of the development of specialized and hierarchically organized flow systems, such as the respiratory tract and the circulatory system (Bejan et al. 2004; Reis et al. 2004; Bejan 2005; Reis and Miguel 2006).

The formation of dissimilar patterns inside similar systems under different environmental conditions is especially intriguing. Examples of these phenomena can be found in almost every field, ranging from physics to the behavior of social groups. For instance, in a horizontal fluid layer that is confined between two plates, and which is heated from below to produce a fixed temperature difference, there is a critical Rayleigh number for which the fluid breaks off from its macroscopically motionless form and starts to present a roll configuration—the Rayleigh–Bénard convection (Gettling 1998; Bejan 2000). It is equally well known that stony corals collected from exposed growth sites, where higher water currents are found, present more spherical and compact shape than corals of the same species growing in sheltered sites, which display a thin-branched morphology (Kaandorp and Sloot 2001; Merks et al. 2003). Furthermore, bacterial colonies that have to cope with hostile environmental conditions have more branched growth forms than colonies of the same species from nutrient-rich environments (Ben-Jacob et al. 1995; Thar and Kühl 2005). In a similar manner, plant roots seem to be able to respond to localized regions of high nutrient supply by proliferating or elongating root branches into the nutrient-rich patches (Robinson 1994; Hodge et al. 1999). Root systems in soil are more open and more thinly branched than roots which are growing in hydroponics regime.

Pedestrian crowd motion also exhibits a variety of conjunction patterns, from more-or-less ‘chaotic’ appearance (Schweitzer 1997) to spontaneous organization in lanes of uniform walking direction looking like river-like streams (Navin and Wheeler 1969; Helbing et al. 2001). What determines the condition for the existence of so dissimilar patterns? Understanding crowd motion is essential on a wide range of applications including crowd safety (Donald and Canter 1990; Langston et al. 2006). From what principle can pedestrian facilities be deduced?

There has been a renewed impetus for the study of geotemporal spread of epidemics, following concerns over the increasing potential outbreak of infectious diseases (Koopmans et al. 2004; Hsu et al. 2006; Suwandono et al. 2006). Throughout history several pandemics have occurred in many areas of our planet, infecting and wiping out a large number of people. One of the most catastrophic pandemics was the bubonic plague (Black Death) in the fourteenth century. It is estimated that a third of the European population at the time died as a consequence of this outbreak. Moreover, historic records of the progress of this outbreak suggest a wave-like propagation of the disease (Langer 1964). What are the conditions for the existence of such a traveling wave? Would pandemics nowadays follow a similar propagation pattern as those that occurred in the fourteenth century?

In this chapter, we will examine the formation of dissimilar patterns inside similar systems from the viewpoint of the constructal theory of organization in nature (Bejan 1997, 2000; Bejan et al. 2004, 2006; Rosa et al. 2004; Bejan and Lorente 2005; Reis 2006b). Based on this view, common features between systems in very different fields are evidenced, and the importance of an optimum balance of competing trends (flow regimes, resistances, etc.) on the generation of patterns (architecture) is stressed. In particular, we aim to provide an answer to a very fundamental question: Have their patterns (architecture) in nature been developed by chance, or do they represent the optimum structure serving an ultimate purpose?

5.2. Constructal Law and the Generation of Configuration

Constructal theory is about the physics principle from which geometric form in flow systems can be deduced (Bejan 2000). Consider, e.g., the drainage of fluid through a nonhomogeneous surface (e.g., a surface with a central strip having a low resistance to fluid flow and lateral strips with high resistances to fluid flow) depicted in Fig. 5.1. According to the constructal law put forward by Bejan (1997, 2000), in order “for a flow system to persist (to survive) it must morph over time (evolve) in such a way that it provides easier access to the imposed currents that flow through it.” Thus, the shape and flow architecture of

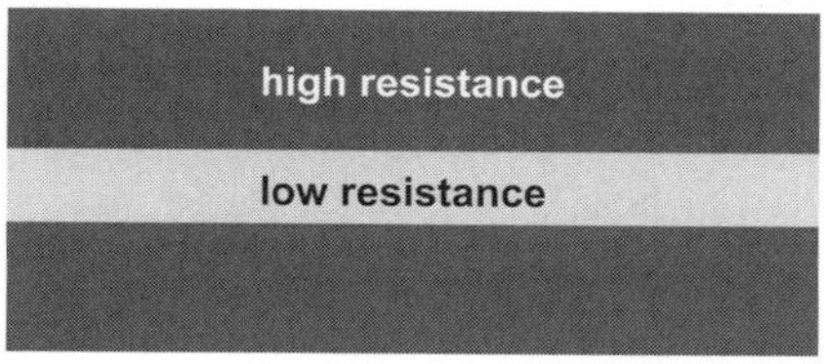

FIGURE 5.1. A surface with a central strip having a low resistance to fluid flow surrounded by strips with high resistance to fluid flow

the system in Fig. 5.1 do not develop by chance. In fact, the constructal theory proposes that every flow system exists with a purpose and that flows are free to morph their configuration in search of the best architectural solution within a framework of existing constraints (area or volume allocated, material properties, etc.). Therefore, the shape and flow architecture of the system are the result of the optimum balance between two competing trends—slow and fast (a surface with different flow resistances in this example, although it could be different flow regimes or other)—that ensures the maximization of fluid drainage.

In summary, the optimum balance between competing trends—slow (high resistivity) and fast (low resistivity)—is at the origin of shape and flow architecture.

5.3. Constructal Pattern Formation in Nature

5.3.1. Formation of Dissimilar Patterns Inside Flow Systems

The spreading of a tracer or a solute, and the transport of heat or fluid can be analyzed within the framework of diffusive–convective phenomena (Fig. 5.2). For example, the one-dimensional tracer transport within a fluid is governed by the macroscopic equation:

$$\frac{\partial n}{\partial t} + u\frac{\partial n}{\partial z} = D\frac{\partial^2 n}{\partial z^2} \tag{5.1}$$

where u is the average fluid velocity, D is the tracer diffusion coefficient, n is the tracer concentration, and t is the time. The time scales can be obtained by applying scale analysis (Bejan 2000) to the above equation:

$$\frac{\Delta n}{t} \sim D\frac{\Delta n}{L^2}, u\frac{\Delta n}{L} \tag{5.2}$$

Therefore, the characteristic times corresponding to the diffusive and convective driven transport are

$$t_{dif} \sim \frac{L^2}{D} \tag{5.3}$$

$$t_{cv} \sim \frac{L}{u} \tag{5.4}$$

FIGURE 5.2. The spreading of ink within water (slow or high resistivity) and boiling water (fast or low resistivity)

while the velocities are

$$v_{dif} = \frac{dL_{dif}}{dt} \sim \frac{1}{2}\left(\frac{D}{t}\right)^{1/2} \tag{5.5}$$

$$v_{cv} = \frac{dL_{cov}}{dt} \sim u \tag{5.6}$$

where t_{dif} and t_{cv} are the characteristic times corresponding to diffusive and convective driven transport, and v_{dif} and v_{cv} are the velocities corresponding to diffusive and convective driven transport, respectively.

The transition time from diffusive to convective driven transport, t^*, is obtained from the intersection of Eqs. (5.3) and (5.4):

$$t^* = \frac{D}{u^2} \tag{5.7}$$

If $t < t^*$ diffusion overcomes convection. Conversely, when $t > t^*$ tracer transport is mainly driven by convection.

Diffusion coefficients are usually much smaller than $1\,m^2s^{-1}$ (e.g., the diffusion coefficient for oxygen in air is approximately $2 \times 10^{-5}\,m^2s^{-1}$). When $u << D$ (e.g., u is close to zero), the time of transition t^* becomes very large ($t^* \to \infty$). In this situation, transition from diffusive to convective driven transport is not very likely to occur and the tracer transport is linked only to a diffusive phenomenon. However, if fluid velocity is much larger than the tracer diffusivity, the transition time (Eq. 5.7) is very small. What are the consequences of this? The initial diffusive velocity (Eq. 5.5) is larger than any convective velocity

(Eq. 5.6), but decreases as $t^{-1/2}$. Diffusion is the main driving mechanism at very beginning of the transport process but for times slightly greater than t*, convective transport takes its place as the main mechanism. Similar results were obtained by Bégué and Lorente (2006) for the ionic transport through saturated porous media.

In summary, the architecture of the flow is the result of the trade-off between two competing trends—diffusion (slow or high resistivity) and convection or channeling (fast or low resistivity)—which ensure the maximization of the transport process.

Other flow systems exhibit a similar tendency. Bejan (2000) showed that the onset of a roll configuration in fluid layers heated from below (Rayleigh-Bénard convection) can be predicted based on the constructal theory. Conduction or thermal diffusion (high resistivity) prevails as the main mechanism, as long as it provides the shortest time in transporting heat across a surface layer. Conversely, rolls/channeling or convection cells (low resistivity) start to occur when the Rayleigh number reaches a critical value (>1700), so as to maximize the heat transport process (Fig. 5.3). Therefore, flow architecture is the result of the trade-off between two competing trends, and rolls are the optimized access for internal currents (e.g., the optimal architecture). Bejan (2000) also showed that a turbulent flow is a combination of the same two mechanisms—viscous diffusion (high resistivity) and eddies (low resistivity)—and can therefore be covered by the constructal law.

5.3.2. The Shapes of Stony Coral Colonies and Plant Roots

In stony corals and other organisms, which have a relatively weakly developed transport system, the amount of nutrients arriving at a certain site in the tissue, as well as the local deposition velocity of the skeleton material, are limited both by the locally available suspended material and the local amount of contact with the environment (Sebens et al. 1997; Anthony 1999). In plants, the water and dissolved minerals necessary for their survival are provided by the

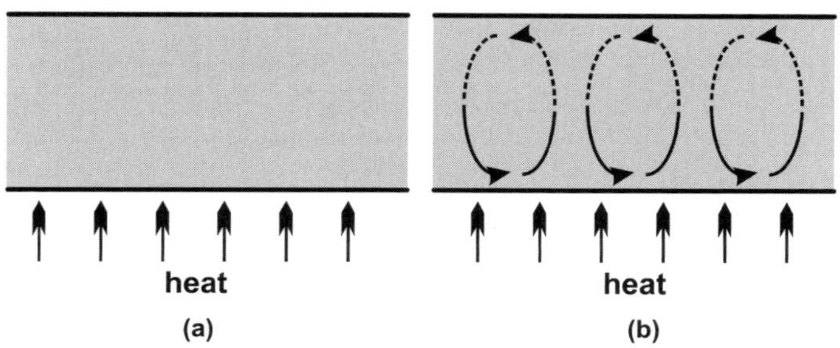

FIGURE 5.3. Rayleigh–Bénard convection: the fluid layer remains macroscopically motionless (a) but after the critical Rayleigh number the fluid presents a roll configuration (b)

root systems. Consequently, plants produce new roots to maximize nutrient absorption and continue to grow.

Both coral colonies and plant roots show an intraspecific variability of the shape. They may develop a branched or a more round shape apparently in a differentiated response to the variability of environmental conditions (Thaler and Pages 1998; Merks et al. 2003). For example, it is known that stony corals collected from exposed growth sites, where higher water currents are found, present more spherical and compact shape, than corals of same species from sheltered sites, which display more thin-branched morphologies (Merks et al. 2003).

Branched and circular (round) shapes are quite different regarding their ability to fill space (Fig. 5.4). Consider that l is the characteristic length/radius of the biological system and w is the width of the branch/needle (e.g., a very small quantity). The surface area of the biological system is $\sim$ lw for branches/needles and $\sim l^2$ for a circular shape. Undoubtedly, $l^2 >> lw$ which means that the circular shape is the most effective arrangement for filling the space in the shortest time and thus, according to the constructal law, it constitutes the optimal architecture. But sometimes stony corals and roots develop a branched shape. How can one reconcile such an obvious contradiction with the maximization of flow access?

The answer was provided by Miguel (2004, 2006), based on the constructal description made by Bejan (2000) of the structure of a dendritic crystal formed during rapid solidification. Consider, e.g., stony corals growing in exposed (open) sites. In this case, nutrient transport is driven mainly by convection. The velocity of the nutrient-rich water which surrounds the coral colony is much larger than the growth velocity of corals (which for a specie named *Porites* spp. is, for instance, about 12 mm per year). This implies that the coral grows always inside a region where nutrients are readily available. Consequently, it is able to spread (diffuse) in all directions and develop the most effective arrangement for filling the space in the shortest time—a round shape.

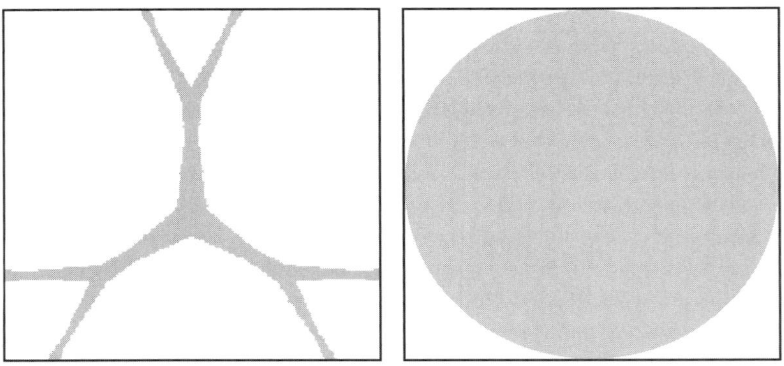

FIGURE 5.4. Branched and circular shape geometries and their ability to fill the space

Assume that the coral is growing in a sheltered site. When the velocity of the water containing nutrients becomes close to zero, the transport of nutrients is mainly due to diffusion and nutrients close to the coral colony are quickly depleted. Thus, the decrease of nutrient concentration around the coral triggers a wave of nutrients with a velocity of propagation $v_{dif} = \frac{dL_{dif}}{dt} \sim \frac{1}{2}\left(\frac{D}{t}\right)^{1/2}$, Eq. (5.5), and a characteristic length $L_{dif} \sim \frac{1}{2}(Dt)^{1/2}$. The initial velocity of propagation $v_{dif} \rightarrow \infty$ is much larger than the growth velocity of corals (~ 12 mm per year), but decreases as $t^{-1/2}$. Thus, there is a moment when the growth speed of the biological system exceeds the speed of nutrient propagation.

The temporal evolution of the characteristic length of the coral and nutrient propagation are presented in Fig. 5.5. This plot shows that at the critical time, t_{ct}, the characteristic length of the coral system overtakes the characteristic length corresponding to the diffusive transport. From this moment on the circular shape is no longer the most effective arrangement to fill the space. At times slightly larger than t_{ct}, the biological system starts to grow outside of the nutrient diffusion region. Guaranteed survival, branches are then generated to promote the easiest possible access to nutrients. This "biological channeling" enables the system to experience again growth inside the nutrient diffusion region from t_{ct} until $2t_{ct}$. At times slightly greater than $2t_{ct}$, the coral once again sticks out the nutrient diffusion region. New branches are consequently sent forward in order to promote the growth inside the nutrient diffusion region until a new critical time is reached. This means that each branch generates a new group of branches, and the result of this process is a dendritic-shaped system. Thus, in these circumstances "biological channeling" clearly becomes the most competitive shape configuration.

To conclude, the coral system, in its struggle for survival, must morph toward the configuration that provides an easier access to nutrients. The generation of branches is the response of the system when growth takes it out of the nutrient-rich region. These branches provide thus the paths that maximize the access to nutrients (e.g., the branches constitute the optimal shape under these circumstances).

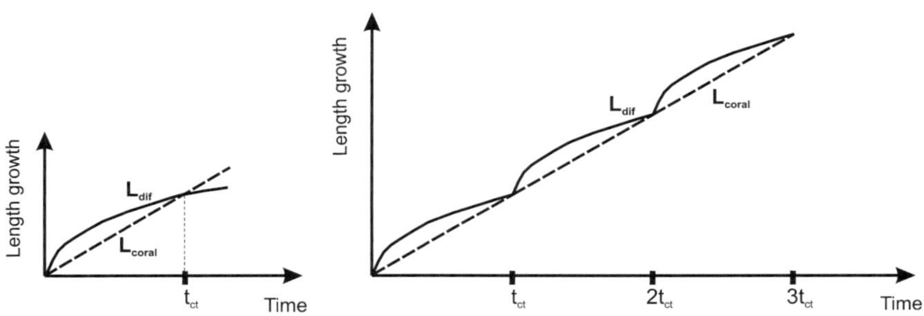

FIGURE 5.5. Simultaneous growth of characteristic lengths L_{coral} and L_{dif} and the occurrence of critical times t_{ct}

It is interesting to note that the optimal architecture of the system composed by nutrients and coral colony once more results from two competing trends: a convective (or channeling) transport of nutrients (e.g., low resistivity) implies a round (diffusive) coral morphology (e.g., high resistivity), while a diffusive (round) transport of nutrients (e.g., high resistivity) implies a branched (channeling) coral morphology (e.g., low resistivity). Similar results are also achieved for plant roots (Miguel 2006).

5.4. Constructal Patterns Formation in Pedestrian Motion

5.4.1. Pedestrian Dynamics: Observation and Models

The interest in the movement of crowds is not a recent one: it has existed ever since large events involving high number of people were organized. Roman amphitheatres, e.g., were built in such a way as to ease the entrance or exit of the venue for spectators. The Coliseum in Rome, for instance, has 80 strategic walkways designed to facilitate access and exit.

Pedestrian dynamics has an important impact on a wide range of applications including transportation (Timmermans et al. 1992; Hankin and Wright 1958), architecture and urban planning (Thompson and Marchant 1995), event organization (Smith and Dickie 1993), emergency exit planning (Donald and Canter 1990, Langston et al. 2006), and crowd control (Hunter et al. 2005).

One is usually convinced that human behavior is unpredictable and the way that people move is chaotic or, at least, very irregular. However, during the last decades, the systematic observation of pedestrians conducted by different researchers (e.g., see Hankin and Wright 1958; Older 1968, Navin and Wheeler 1969; Fruin 1971; Henderson 1971; Helbing et al. 2001) revealed that, in standard situations, individuals employ an optimized behavioral strategy. These observations can be summarized as follows:

(i) As long as it is not essential to move faster in order to get to their destination in time (like, for instance, running to catch a departing bus), pedestrians prefer to progress at the least-energy consuming most comfortable walking velocity. This desired velocity is of about $1.34\,\mathrm{ms}^{-1}$ and a standard deviation of $0.26\,\mathrm{ms}^{-1}$.
(ii) Pedestrians can move freely only at low pedestrian densities. As the pedestrian density increases (e.g., interpersonal distances lessen), walking velocity decreases.
(iii) A group of pedestrians (families, friends, colleagues) behave similarly to single pedestrians (group sizes are Poisson distributed).
(iv) Pedestrians try to keep a certain distance between themselves or to borders (walls, objects, columns, etc.). This behavior helps to avoid contact in case of sudden velocity change and maintain a private area around (territorial effect).
(v) Pedestrians feel uncomfortable when they have to move in a direction opposite to the destination.

FIGURE 5.6. Pedestrians have a preferred side to walk that disappears at high pedestrian densities

(vi) Pedestrians have a preferred side to walk that disappears at high pedestrian densities (Fig. 5.6).
(vii) In general, pedestrians act more or less automatically, they do not reconsider their behavioral strategy when facing new situations.

It is interesting to note that the movement of pedestrians displays many of the attributes of fluid and granular* flows (Helbing et al. 2001, Hughes 2003):

(i) Footsteps in the sand and snow look similar to fluid streamlines (Fig. 5.7).
(ii) Pedestrians organize themselves in the shape of river-like streams (channeling) when a stationary crowd needs to be crossed (Fig. 5.8).
(iii) When moving in crowded places, individuals organize themselves spontaneously in lanes of uniform walking direction (channeling) (Fig. 5.9).
(iv) In dense crowds which push forward, one can observe a kind of shock-wave propagation.

(a) (b) (c)

FIGURE 5.7. Footsteps in the sand (a) and snow (b) look similar to fluid streamlines (c)

* Two-phase flow consisting of particulates and an interstitial fluid that when sheared the particulates may either flow in a manner similar to a fluid or resist the shearing like a solid.

FIGURE 5.8. Pedestrians organize themselves in the shape of river-like streams (channeling) when a stationary crowd needs to be crossed

FIGURE 5.9. In crowded spaces, individuals self-organize in lanes of uniform walking direction (channeling)

(v) At bottlenecks (e.g., doors, corridors) the pedestrian's passing direction oscillates with a frequency that increases with width and declines with the length of the bottleneck. This is similar to granular "ticking hour glasses," in which grains alternate between flowing and not flowing at a constant rate.

A number of empirical models based on observational data are available in published studies (e.g., see Predtechenskii and Milinski 1969; Sandahl and Percivall 1972; TRB 1985; Nelson and MacLennan 1995; Graat et al. 1999). Some models considering the analogies between physical systems and pedestrian motion have also been proposed, a number of which represent crowds as an aggregate of individuals having a set of motivations and basic rules (Table 5.1). The so-called social force model (Helbing 1992; Helbing and Molnár 1995) has its origins in gas-kinetic models and was developed to describe the dynamics of pedestrian crowds. It consists of self-driven people (particles) that interact through social rules—the "social forces." These forces produce changes in the velocities and reflect in turn, a change in motivation rather than in the physical forces acting on the person. The models presented by Hoogendoorn and

TABLE 5.1. Models for pedestrian motion

Authors	Characteristics	Validity
Blue and Adler (1999)	Cellular automata discrete model	Density less than five people per square meter
Fukui and Ishibashi (1999)	Cellular automata discrete model	–
Helbing (1992)	Social force	High and low densities
Helbing and Molnár (1995)		
Helbing et al. (1997)	Active walker	low density
Hoogendoorn and Bovy (2000)	Extremal principle (generalization of the social force)	Density less than five people per square meter
Hoogendoorn et al. (2002)		
Hughes (2002)	Thinking fluid	High and low densities
Langston et al. (2006)	Discrete element discrete model	High and low densities
Muramatsu et al. (1999)	Random walk	–

Bovy (2000) and Hoogendoorn et al. (2002) provide a generalization of the social force model. Early on, Reynolds (1987) presented a model to approach animal motion, like that of bird flocks and fish schools, that bears relevance to crowd dynamics. This flocking model consisted of three simple steering behaviors which described how each individual flocking element (called *boid*) maneuvered based on the positions and velocities of its nearby flock mates.

The "thinking fluid" model (Hughes 2002, 2003) results from the combination of fluid dynamics with three hypotheses which are supposed to govern the motion of pedestrians. These hypotheses, together with those governing the motion of the boids, and the "social forces" are listed in Table 5.2.

5.4.2. Diffusion and Channeling in Pedestrian Motion

The constructal law states that if a system is free to morph in time (evolve), the best flow architecture is the one that maximizes the global flow access (e.g., minimizes the global flow resistances). Thus, the shape and flow architecture of the system do not develop by chance, but result from the permanent struggle between slow and fast for better performance and must thus evolve in time (Bejan 2000).

How and by which mechanism do then pedestrians evolve in space and time? To answer this question, let us consider pedestrian groups that proceed from one point to every point of a finite-size area (territory). According to the constructal law, the best architecture will be the one that promotes the easiest flow of pedestrians. As described earlier, there are two mechanisms for achieving this purpose: diffusion (slow or high resistivity) and channeling (fast or low resistivity). The access time for a diffusive process through a territory of length L is $\sim L^2/D$, Eq. (5.3), and the access time for a channeling flow is $\sim L/u$, Eq. (5.4). To compare both times we need to know the pedestrian's diffusion coefficient, D, and the pedestrian's walking velocity, u.

TABLE 5.2. Hypotheses that support models describing the movement of individuals

Flocking model (Reynolds 1987)	Social forces model (Helbing and Molnár 1995)	"Thinking fluid" model (Hughes 2003)
• Boids try to fly toward the center of mass of neighboring boids • Boids try to keep a small distance away from other objects (including other boids) • Boids try to match velocity with near boids	• Pedestrians move as efficiently as possible to a destination • Pedestrians try to maintain a comfortable distance from other pedestrians and from obstacles like walls • Pedestrians may be attracted to other pedestrians (e.g., family, friends) or objects (e.g., posters, shop windows)	• Walking velocity is determined only by the density of surrounding pedestrians, the behavioral characteristics of the pedestrians, and the ground characteristics • Pedestrians at different locations but with same sense of the task (called potential) would see no advantage to exchanging places • Pedestrians minimize their estimated travel time but also try to avoid extreme densities

In accordance with field surveys, it has been established that pedestrians prefer to move with a walking velocity around $1.34\,\mathrm{ms}^{-1}$, which corresponds to the least energy-consuming velocity (Fruin 1971; Henderson 1971; Helbing et al. 2001). This walking velocity is only reachable if there are no other pedestrians and obstacles in the surroundings. Furthermore, it was noticed that pedestrians in a shopping mall or busy city street (random and nondirectional crowd) exhibit a velocity reduction that is related with the free area available around each individual (Fig. 5.10). Consequently, and as in physics, the pedestrian's diffusion coefficient may be defined as the product of walking (random) velocity, u_{crd}, and the mean available interpersonal distance, λ.

A pedestrian's diffusion coefficient—mean interpersonal distance relationship can be established with the help of the coefficient's definition and the data plotted in Fig. 5.10. This relationship is illustrated in Fig. 5.11. The form of the curve-fitted equation, justified by the correlation coefficient, is

$$D = a_1\lambda - a_2 \tag{5.8}$$

Here a_1 and a_2 are the correlation coefficients, which have been listed in Table 5.3. Consider, e.g., the curve-fitted equation obtained based on Fruin data (Fruin 1971), that when combined with Eqs. (5.3) and (5.4) leads to

$$t_{dif} \sim \frac{L^2}{1.61\lambda - 0.67} \qquad (0.48 \leq \lambda \leq 3.16) \tag{5.9}$$

$$t_{cv} \sim 0.75L \tag{5.10}$$

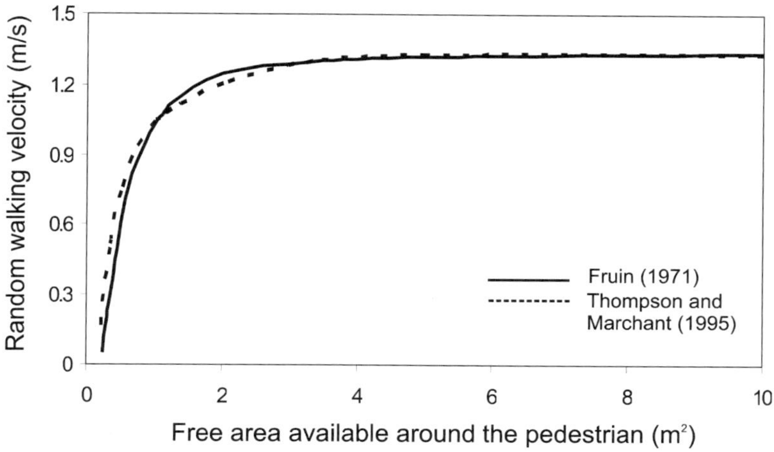

FIGURE 5.10. Effect of the free area available around the pedestrian on the random walking velocity

Based on this, it is straightforward to conclude that channeling enables a better performance if L/λ is larger than $1.2 - 0.5\,\lambda^{-1}$ (or in terms of the pedestrians' density, ρ, larger than $1.2 - 0.5\,\rho^{-1/2}$). Otherwise, diffusion becomes clearly the most competitive transport mechanism.

To summarize, there are two optimal modes of locomotion for pedestrians: channeling which is suitable to distribute pedestrian through a territory (area) and diffusion which becomes more appropriate when accessing space locally (e.g., access to train platforms and bus stops, buildings entrance, etc.). This finding has an impact on not only in the design of new pedestrian facilities but

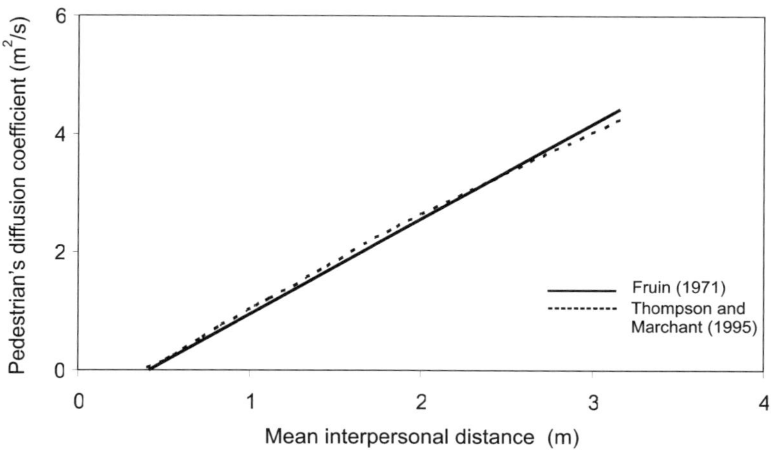

FIGURE 5.11. Pedestrian's diffusion coefficient versus the mean interpersonal distance

TABLE 5.3. Correlation coefficients for pedestrian's diffusion coefficient—mean interpersonal distance relationship

	a_1	a_2	r^2	Validity
Random crowd in a city street Fruin (1971)	1.61	0.67	0.992	$0.48 \leq \lambda \leq 3.16$
Random crowd in area of a building Thompson and Marchant (1995)	1.56	0.56	0.997	$0.40 \leq \lambda \leq 3.16$

also in the improvement of existing facilities. If the target is to promote the easiest access to pedestrians over a large territory where, for some reason, the velocity is lower than the desired walking velocity ($\sim 1.34\,\mathrm{ms}^{-1}$), the placement of gates/lanes/columns/trees along walkways/corridors helps to stabilize the pedestrian flow and make it more fluid (Fig. 5.12). This channeling also allows pedestrians to keep a certain distance from other pedestrians which is highly appreciated (very small interpersonal distances induces contact "collisions" among pedestrians which is seen as "uncomfortable" (TRB, 1985)). The placement of gates/lanes/columns/trees (channeling) is especially important in case of pedestrians walking in opposite directions (Fig. 5.13).

Pedestrians have a preferred side to walk because they profit from it (e.g., moving against the stream is more difficult because it increases interaction and consequently the resistance). Therefore, a baffler along walkways/corridors helps to optimize the flow of pedestrians, as well as to save space that can be used for other purposes.

5.4.3. *Crowd Density and Pedestrian Flow*

Empirical observations have also highlighted that pedestrians can move freely only at very small pedestrian densities. Fig. 5.14 illustrates velocities of pedestrians in the crowds. Pedestrians in low-density crowds ($\rho < 1$ person m^{-2}) are able to walk with an individual desired velocity that corresponds to the comfortable walking velocity ($\sim 1.34\,\mathrm{ms}^{-1}$). In higher crowd densities, interpersonal distances are lessened and the walking velocity is reduced, in order to

FIGURE 5.12. Gates and lanes along walkways help to stabilize the pedestrian flow and make it more fluid

FIGURE 5.13. Pedestrians walking in opposite directions in crowded spaces

keep a certain distance from other pedestrians and to avoid contact. Therefore, pedestrian motion can be summarized as follows:

$$u = u_0 \quad \text{for} \quad 0 \leq \rho \leq \rho_{ct} \tag{5.11}$$

$$u = u(\rho) \quad \text{for} \quad \rho_{ct} < \rho \leq \rho_{max} \tag{5.12}$$

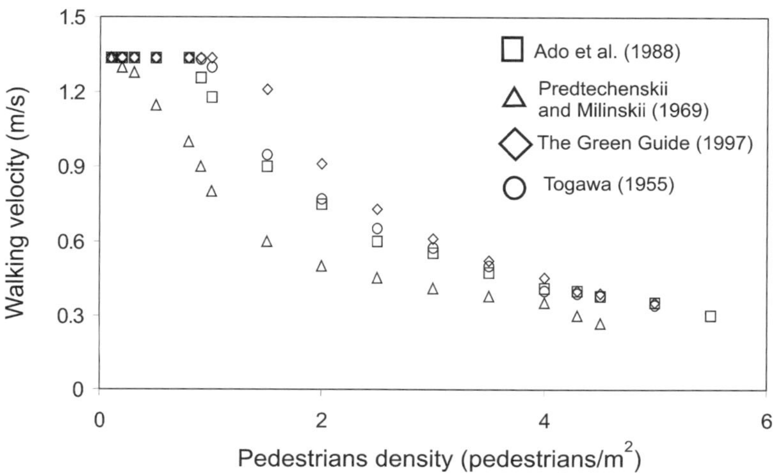

FIGURE 5.14. Walking velocities versus pedestrians' density

where u_0 is the desired velocity (e.g., $\sim 1.34\,\mathrm{ms}^{-1}$), ρ_{ct} is the critical density of pedestrians (e.g., ~ 1 person m^{-2}) and ρ_{max} is the maximum density of pedestrians.

Let us suppose that, above the critical pedestrian density, the temporal change of the pedestrian velocity is the result of "repulsive" forces due to a decrease of interpersonal distances (especially the relative distance to the front person). "Repulsive" forces decrease with the interpersonal distance, and as the area available for the next step is accounted, they are also dependent of the walking velocity (Helbing et al. 2001). According to the Newton's second law of motion

$$\frac{d^2 r}{dt^2} = F_r \tag{5.13}$$

Here r is the position of the pedestrian and F_r is the "repulsive" force per mass affecting the behavior of the pedestrian which is given by (Fang et al. 2003)

$$F_r = \frac{\gamma}{(r - r_o)} \left(\frac{dr}{dt} - \frac{dr_o}{dt} \right) \tag{5.14}$$

where γ is a constant, and $r - r_o (= \lambda)$ and $dr/dt - dr_o/dt$ are the mean interpersonal distance and the mean relative velocity to the pedestrians situated around, respectively. Replacing of F_r by Eq. (5.14) in Eq. (5.13) and integrating the resulting equation, we find that

$$u = \frac{dr}{dt} = \gamma \ln(\lambda) + c \tag{5.15}$$

where c is an integration constant. By definition, the interpersonal distance is minimum (e.g., crowd density maximum) when the walking velocity is zero. Thus, the constant c can be determined and Eq. (5.15) assumes the form

$$u = \gamma \ln \left(\frac{\lambda}{\lambda_{min}} \right) \tag{5.16}$$

and in terms of pedestrian density

$$u = \frac{1}{2} \gamma \ln \left(\frac{\rho_{max}}{\rho} \right) \tag{5.17}$$

In summary, Eqs. (5.16) and (5.17) should hold when the pedestrian's density is between the critical and the maximum densities. Note that experimental walking velocities in this density range show a good agreement with these equations (Fig. 5.15).

The flow of pedestrians may be defined as the product of walking velocity and pedestrian density,

$$\phi = \rho u \tag{5.18}$$

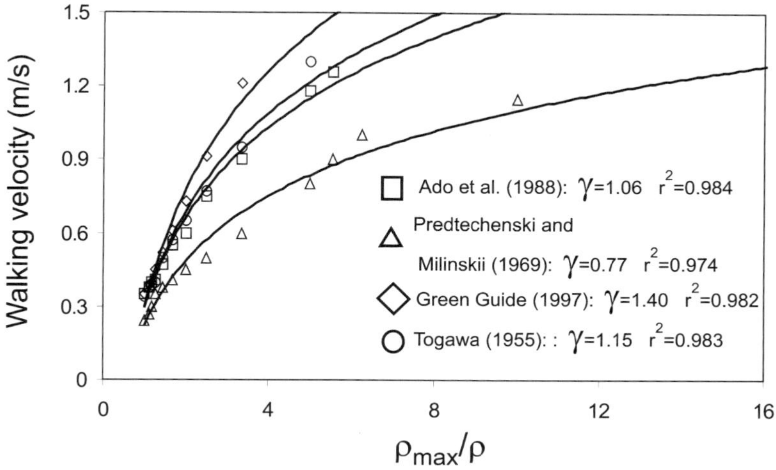

FIGURE 5.15. Experimental walking velocities and the fit with Eq. (5.17)

where ϕ is the pedestrian flow. Therefore, Eqs. (5.11), (5.12), (5.17), and (5.18) combine to form

$$\phi = \rho u_0 \quad \text{for} \quad 0 \leq \rho \leq \rho_{ct} \tag{5.19}$$

$$\phi = \frac{1}{2}\gamma \ln\left[\left(\frac{\rho_{max}}{\rho}\right)^{\rho}\right] \quad \text{for} \quad \rho_{ct} < \rho \leq \rho_{max} \tag{5.20}$$

Based on these equations a flow–density diagram can be drawn (Fig. 5.16). When $\rho \leq \rho_{ct}$, the individual desired velocity is a constant and the maximum flow of pedestrians corresponds to $\rho_{ct} u_0$. For densities above the critical density, there is also an optimal crowd density such that the flow of pedestrians is maximized. If we take Eq. (5.20), the flow has a maximum value of $0.5\gamma\rho_{max}/\exp(1)$ at $\rho = \rho_{max}/\exp(1)$. It is useful to obtain also the relationship between flow and velocity. For $\rho_{ct} < \rho \leq \rho_{max}$ it follows from Eqs. (5.17) and (5.18) that

$$\phi = \rho_{max} \frac{u}{\exp(2u/\gamma)} \tag{5.21}$$

The flow–velocity diagram is depicted in Fig. 5.17. This diagram shows that there is also an optimal velocity such that the crowd flow is maximized. According to Eq. (5.21), the flow has a maximum value of $0.5\gamma\rho_{max}/\exp(1)$ at $u = \gamma/2$. As expected, this maximum flow has the same value as that obtained with Eq. (5.20).

Comparing the flow–density and flow–velocity diagrams (Figs. 5.16 and 5.17, respectively), we note that each velocity/density corresponds to a single flow but, with the exception of maximum flow, the same flow may correspond to two different velocities/densities. We can see how this behaviour can arise. As the

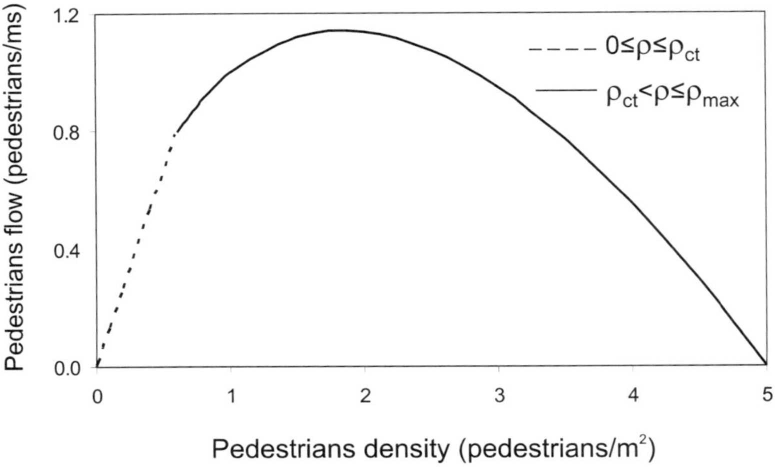

FIGURE 5.16. Flow–density diagram drawn with Eqs. (5.19) and (5.20) ($\rho_{max} = 5$ pedestrians per square meter)

flow progresses to the maximum, the increase of pedestrian velocity (or decrease of density) is offset by a decrease of pedestrian density (or increase of velocity). After reaching the maximum flow, there is a decrease of the pedestrian flow because the growth in velocity (or decrease of density) does not compensate the reduction of pedestrian density (or increase of velocity). Consequently, the same flow may correspond to two different velocities/densities.

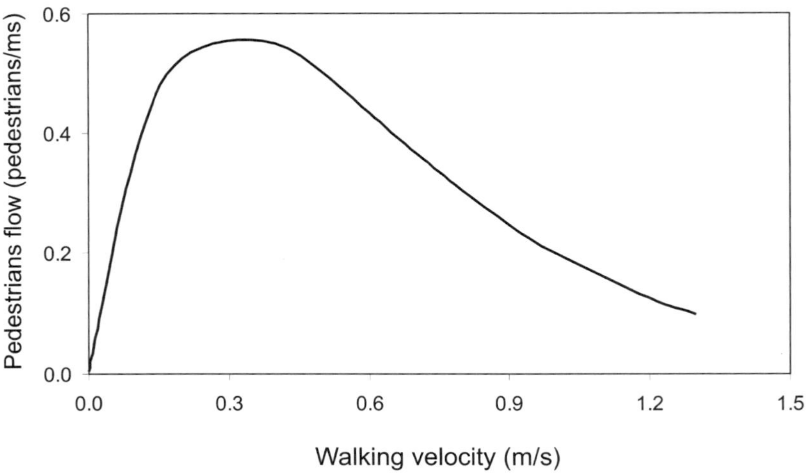

FIGURE 5.17. Flow–velocity diagram drawn with Eq. (5.21) ($\rho_{max} = 5$ pedestrians per square meter)

5.5. Optimizing Pedestrian Facilities by Minimizing Residence Time

5.5.1. The Optimal Gates Geometry

At high pedestrian densities, the walking velocity is drastically reduced and impatient pedestrians trying to use any gap to move on may lead in turn to a complete obstruction of walking paths. Thus, at a high density there is risk of overcrowding and personal injury that should be avoided.

Gates are used at sport facilities, theaters, and so on, to reduce the interpersonal interactions and stabilize the pedestrian flow, by facilitating the access over the territory (Fig. 5.12). Thus, pedestrians must be channeled optimally through gates since channeled flow is characterized by a much lower travel time (Section 5.4.2). The question put forward by the constructal principle is, then, how can the gates be designed in order to ensure that pedestrians flow over the entire space in the shortest time possible?

Consider a crowd approaching gates of the same size and uniformly distributed in space. According to mass conservation equation,

$$\rho u W_g = n\rho_i u_i w_g \tag{5.22}$$

where n is the number of gates, W_g is the total width allocated to the gates, w_g is the width of each individual gate, u is the velocity of the crowd approaching the gates, u_i is the pedestrians velocity within the gates, ρ is the density of the crowd approaching the gates, and ρ_i is the density of the pedestrians within the gates. The goal is the optimal spacing of the gates, w_g, in a fixed territory, W_g, to minimize the travel time of pedestrians defined as

$$t_{tr} = \frac{1}{l_g \phi} \tag{5.23}$$

Let us consider that the length of the gates, l_g, is fixed. The time of travel is minimal when the flow of pedestrians is maximal. It is not that difficult to show that the maximum flow of pedestrians corresponds to $0.5\gamma\rho_{max}/\exp(1)$ (Section 5.4.3) and that the minimum travel time is given by

$$t_{tr} = \frac{2\exp(1)}{l_g \gamma \rho_{max}} \tag{5.24}$$

Therefore, according to Eq. (5.22), nw_g/W_g is

$$\frac{nw_g}{W_g} = \frac{2\rho u \exp(1)}{\gamma \rho_{max}} \tag{5.25}$$

Recalling that flow ρu is given by Eq. (5.20), we can rewrite Eq. (5.25) in terms of the crowd density obtaining

$$\frac{nw_g}{W_g} = \frac{\rho}{\rho_{max}} \ln\left(\frac{\rho_{max}}{\rho}\right) \exp(1) \tag{5.26}$$

Given that ρ_{max} is a constant, the variation of nw_g/W_g is only dependent on the density of the crowd approaching the gates. Fig. 5.18 shows how nw_g/W_g responds to changes of crowd density. It reveals that there is a maximum for nw_g/W_g that occurs when $\rho/\rho_{max} \sim 0.37$. Thus, the optimal number of gates n is W_g/w_g taking $w_g \sim 0.75\,m$ which corresponds to the square root of the free area available around each individual when the flow of pedestrians is maximum (Fig. 5.16).

5.5.2. Optimal Architecture for Different Locomotion Velocities

Constructal theory also predicts the architecture of flow paths that connect a finite territory to a single point when different locomotion velocities are available (Bejan and Ledezma 1998; Bejan 2000). How should architects and urban planners design high performance path systems in these circumstances?

Let us assume that the territory is covered in sequential steps of increasingly larger elements ($A_1 < A_2 \ldots$). According to the constructal theory, the global system (territory) will perform best when all its elements (constructs, portions of territory) perform in an optimal way. Optimal geometry means elements (constructs) that minimize the travel time. Thus, both the shape of the area and the angle between each path and its branches are optimized at each stage. The flow path is constructed starting with the smallest element (construct, portions of territory) and continuing with the larger areas (assemblies, constructs).

Consider pedestrians walking in a rectangular domain with a fixed area $H_1L_1(=A_1)$ at two different velocities: at velocity u_1 in all directions (diffusion) and at velocity u_2 along a centred longitudinal path ($u_1 < u_2$). The goal is

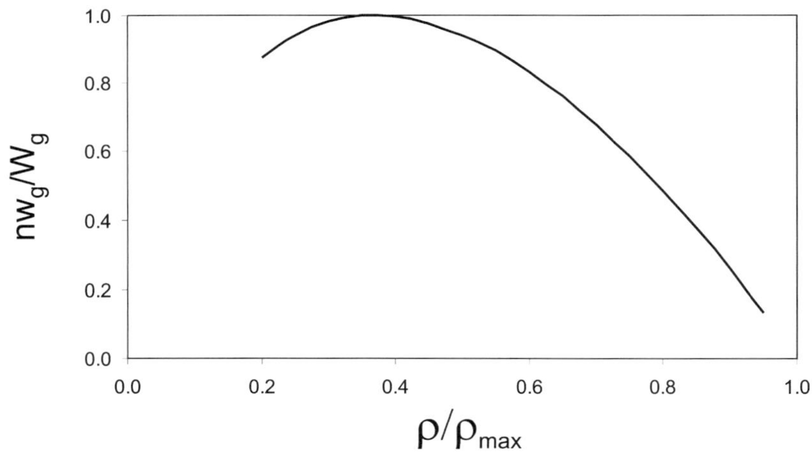

FIGURE 5.18. The variation of nw_g/W_g with respect to the density of the crowd approaching the gates

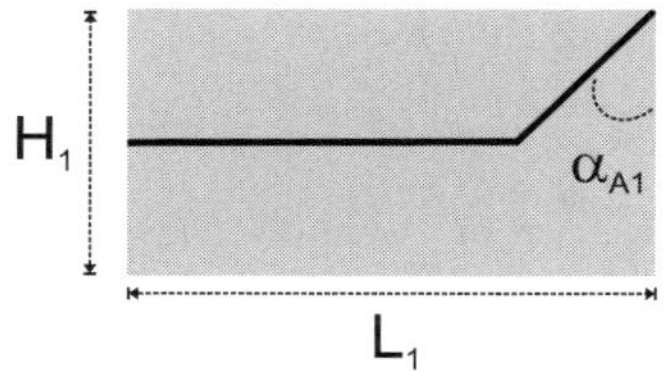

FIGURE 5.19. Rectangular domain with a fixed area H_1L_1 (adapted from Bejan 2000)

to minimize the travel time from anywhere within A_1 to an exit point at its periphery (Fig. 5.19). There are two degrees of freedom in this design: the shape H_1/L_1 and the deviation from the normal for the angle between the slower and the faster paths α_{A1}. Optimizing in view of the stated purpose delivers an optimal geometry and a characteristic travel time t_1 from the most distant point in A_1 until its exit point (Bejan and Ledezma 1998):

$$\left(\frac{H_1}{L_1}\right)_{\text{optimal}} = \frac{2}{\xi_1}\left(\frac{u_1}{u_2}\right) \tag{5.27}$$

$$(\alpha_{A1})_{\text{optimal}} = \cos^{-1}\xi_1 \tag{5.28}$$

$$t_1 = \left(\frac{2\xi_1 A_1}{u_1 u_2}\right)^{1/2} \tag{5.29}$$

with

$$\xi_1 = \left[1 - \left(\frac{u_1}{u_2}\right)^2\right]^{1/2} \tag{5.30}$$

This configuration also optimizes the access from the whole area to the same exit point (Bejan 2000). Consider now a larger fixed area $A_2(= H_2L_2)$ and a faster regime at velocity u_3 (e.g., the desired walking velocity) along a centred longitudinal path ($u_3 > u_2$). Once again one seeks to optimize the access from an arbitrary point of A_2 to a common exit point on its periphery. This problem can be addressed by filling A_2 with optimized A_1 areas, in a manner similar to the foregoing reasoning. The number of elements A_1 assembled into A_2 is given by

$$n_{A2A1} = 2\xi_1\xi_2\left(\frac{u_3}{u_1}\right)\left(1 - \frac{u_2^2}{4u_3^2}\right) \tag{5.31}$$

and the optimal geometry is represented by

$$\left(\frac{H_2}{L_2}\right)_{\text{optimal}} = \frac{1}{\xi_2}\left(\frac{u_2}{u_3}\right) \tag{5.32}$$

$$(\alpha_{A2})_{\text{optimal}} = \cos^{-1}\xi_2 \tag{5.33}$$

with

$$\xi_2 = \left[1 - \left(\frac{u_2}{2u_3}\right)^2\right]^{1/2} \quad (5.34)$$

These results are very similar to those obtained by the optimization of A_1 apart of a constant factor. This optimization can be repeated over larger areas, assemblies being the optimized configuration ratio given by relations similar to the ones obtained for A_2. The final paths' geometry forms a tree network.

Figures 5.20 and 5.21 show how the shape H/L and the optimal angle α_A respond to changes of velocity ratios. When the two modes of locomotion have similar velocities, the optimal angles α_{A1} and α_{A2} are close to 90° and 30°, respectively, and the optimal lengths H/L are maximal. On the other hand, if the faster walking velocity is much larger than the slower walking velocity then paths are perpendicular and the centred longitudinal path L is the much larger than H. We also note that H = L when $u_2/u_1 \sim 0.45$ ($\alpha_{A1} \sim 26.7°$) and $u_3/u_2 \sim 0.9(\alpha_{A2} \sim 26.7°)$.

5.5.3. The Optimal Queuing Flow

Another interesting collective effect of pedestrian dynamics is the formation of queues. Queuing is a common practice is our lives. One queues at the supermarket cashiers, at bus tops, at ticket offices, at stadium entrances, and so on. Once a queue is formed, it acquires dynamic of its own, attracting incomers in a forward motion.

Let q_a be the rate at which pedestrians with a velocity u_a arrive at a certain spot and q_s the rate at which they get served at that spot. When $q_s \geq q_a$, all

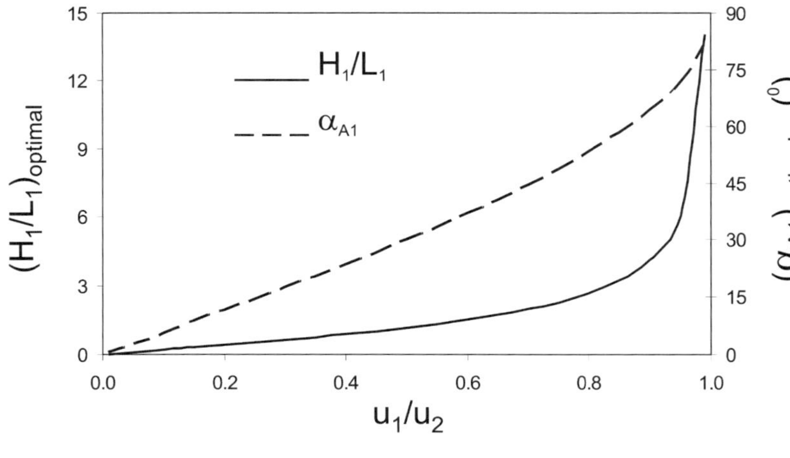

FIGURE 5.20. Optimal values of H_1/L_1 and α_{A1} versus u_1/u_2

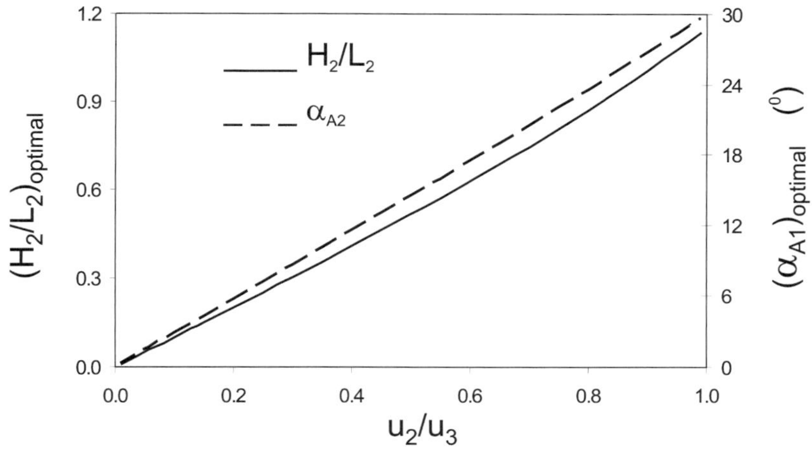

FIGURE 5.21. Optimal values of H_2/L_2 and α_{A2} versus u_2/u_3

individuals get served before queuing. Otherwise, there are individuals who need to wait in order to become served, which leads to the formation of a queue.

Empirical observations reveal that the higher the flow of individuals, the higher the queuing time. The velocity within the queue, v, is (Heidemann 1996; Vandaele et al. 2000)

$$v = 2u_a \frac{\rho_{max} - \rho}{2\rho_{max} + \rho(\beta^2 - 1)} \tag{5.35}$$

Here β is a coefficient that accounts for deviations from the expected service time (e.g., $\beta^2 = 1$ means that pedestrians are served within the expected time). The relation between velocity and flow in queues can be obtained by combining Eqs. (5.18) and (5.35) into

$$\phi = 2\rho_{max} v \frac{u_a - v}{2u_a + v(\beta^2 - 1)} \tag{5.36}$$

Based on this equation a flow–velocity diagram can be drawn (Fig. 5.22). This diagram reveals one flow can correspond to two different velocities (with the exception of the maximum flow) in spite of each pedestrian velocity matching a single pedestrian flow. This reasoning is analogous to the one put forward in Section 5.4.3.

Figure 5.22 also reveals that there is an optimal velocity such that the flow in the queue is maximized (or time of travel is minimized). When, for instance, $\beta^2 = 1$ the maximum flow is $u_a\rho_{max}/4$ and occurs when $v = u_a/2$. We can also observe that an increase of β^2 reduces the maximum value (the peak) of the flow and that this peak occurs at lower velocities. When $\beta^2 < 1$, the maximum flow occurs at higher velocities. After that the pedestrian flow drops very fast until it ceases totally. The reverse occurs if $\beta^2 > 1$. A straightforward explanation can

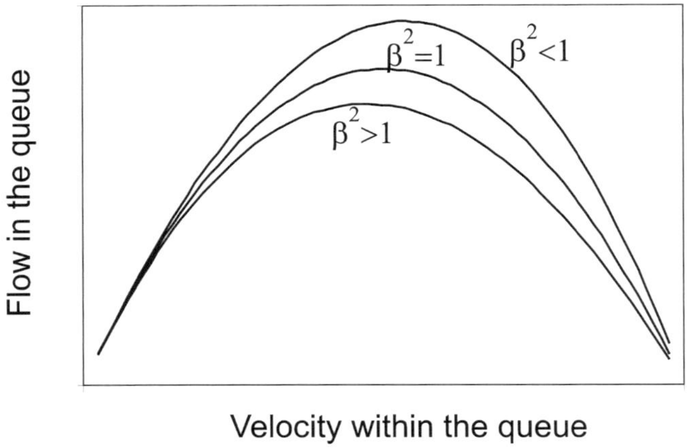

FIGURE 5.22. Flow-velocity diagram drawn with Eq. (5.36)

be provided for this finding. If β^2 is lower than 1, this means that pedestrians get served before the expected time. Consequently, more pedestrians are allowed to enter in the queue, which can be by increasing the arrival velocity of pedestrians to the queue.

5.6. Constructal View of Self-organized Pedestrian Movement

One of the more striking occurrences in pedestrian dynamics is the evidence of spontaneous, self-organized motion. Pedestrians in very crowded open spaces tend to organize in lanes of uniform walking velocity (Fig. 5.9). Similarly, when facing a stationary crowd, pedestrians spontaneously self-organize in river-like streams (rivers of people) in order to cross it (Fig. 5.8). Why do pedestrians spontaneously organize themselves in this type of movement? How and by which mechanism do they evolve in space and time?

The constructal theory's answer to these questions is simple and direct. In line with the access-optimization principle, the optimal flow architecture will be the one that promotes the easiest flow of pedestrian. As described before, there are two locomotion modes for achieving motion: diffusion (slow or high resistivity) and channeling (fast or low resistivity). Diffusion is the preferred locomotion mode as long as it provides the faster pedestrian transport across the territory (e.g., for $L/\lambda < 1.2 - 0.5\lambda^{-1}$, i.e., for large inter-personal distances or low pedestrian densities). Otherwise, channeling becomes the most competitive transport mechanism. Channeling is faster and allows pedestrians to walk with very weak interpersonal interactions. This explains why moving pedestrians form lanes of uniform velocity when the density is high enough. The question is why do lanes of different velocity form? The answer to this is that a group of pedestrians is often composed of many pedestrians with different walking velocities (e.g., young people walk faster than older people). Besides these, pedestrians follow others who are already moving. These factors are then responsible for differences of velocity between lanes.

The movement of pedestrians across stationary crowds (rivers of people) can be deduced from the constructal principle in a way similar to river basin structures (Bejan 2000, 2002). The crowd can be seen as the river basin and the space vacated by the crossing pedestrians as the eroded river bed (path of low resistance). Imagine one of the crossing pedestrians moving toward the stationary crowd. Its successor will see the "open space" vacated near by and will proceed to occupy it in order to be carried away. This means that pedestrians follow others who are already moving, giving rise to channeling networks of pedestrian to appear through the crowd. The lines formed by coalescence of many such paths are the river branches.

In crowds that panic, the streams of people of uniform walking velocity are destroyed because individuals do not know which the right way to escape is. They strive to go forward, thereby reducing interpersonal distances, inducing interpersonal contact (collision) or even loss of balance of other individuals.

5.7. Population Motion and Spread of Epidemics

Large-scale epidemic outbreaks have occurred through the centuries causing major surges in mortality (Anderson and May 1991; Cohn 2002; Suwandono et al. 2006), the worst being the bubonic plague (or Black Death) in the fourteenth century. Historians estimated that the Black Death wiped out a third of the European population in less than four years, no such large-scale outbreak has been reported since (Langer 1964). The last large-scale outbreak was the so-called Spanish Influenza epidemic of 1918–1919 (Oxford et al. 2005). It is estimated that about half of the people living worldwide became infected and 20 million died. An estimated 60,000 died in Portugal, 200,000 in the UK, more than 400,000 in France, and about 600,000 in the USA (Fig. 5.23).

Epidemic diseases still occur in many areas of our planet at a local scale (Barreto et al. 1994; Koopmans et al. 2004; Hsu et al. 2006). There can be little doubt that the improvement of healthcare, a greater vaccine manufacturing capacity, a development of hygiene habits, and an expanded surveillance reduced greatly the impact of an epidemic disease since the second-half of the twentieth century. Nevertheless, the widespread avian influenza outbreaks occurring nowadays throughout Asia show us that a global epidemic is still a possibility (Hsu et al. 2006).

In order to prevent epidemics and minimize disease transmission, it is essential that we are able to evaluate the epidemic spread mechanisms and capability. Models for spread of epidemics have existed since the early twentieth century. Perhaps the most well known are the SIR model (Kermack and McKendrick 1927) and the Noble's plague model (Noble 1974). In microparasite

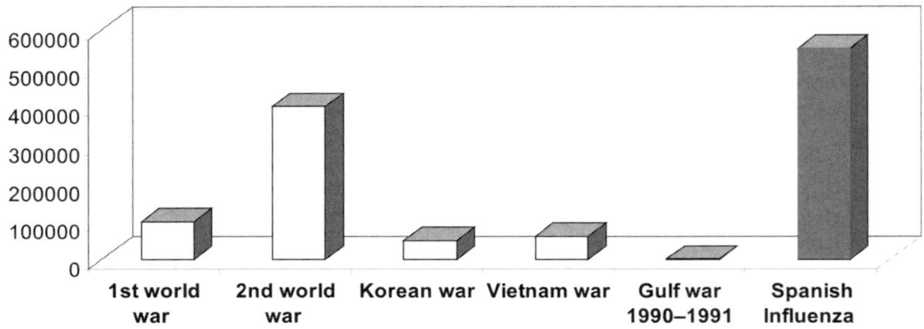

FIGURE 5.23. US military casualties in the wars of the twentieth century and America's deaths from Spanish influenza

infections (mainly viruses and bacteria), individuals are classified as susceptible (S), infected (I), or recovered (R). Susceptible individuals can catch the infection from contact with infected individuals, and the fraction of these individuals recovered is assumed to be immune to the disease. This leads to a set of balance equation that constitutes the SIR model. Extended versions of SIR model have been presented by Anderson and May (1992), Diekmann and Heesterbeek (2000), and Brauer and Castillo-Chavez (2001). On the other hand, the Noble's model considers that the spatial dispersal of individuals can be well approximated by a diffusion process, with the spread of epidemics being modeled as a diffusive–reactive phenomenon. This model was applied to the dynamics of bubonic plague as described throughout Europe in the fourteenth century. Since then reaction–diffusion models have been used to describe, among others, the spatial dynamics of rabies in fox (Kallen et al. 1985) and the Lyme disease transmission (Caraco et al. 2002). In this section we explore two main issues: the mechanics of epidemics propagation and the effect of population motion on the spread of epidemics.

5.7.1. Modeling the Spreading of an Epidemic

Consider two interacting populations of individuals—susceptible and infective. Transmission is the driving force due to which susceptible population becomes infective. The mass conservation of individuals S and I in a territory are governed by:

$$\frac{\partial S}{\partial t} = -\left(\frac{\partial}{\partial x} + \frac{\partial}{\partial y}\right) q_S \pm \psi_S \tag{5.37}$$

$$\frac{\partial I}{\partial t} = -\left(\frac{\partial}{\partial x} + \frac{\partial}{\partial y}\right) q_I \pm \psi_I \tag{5.38}$$

Here S and I are the density of susceptible and infective populations, q_S and q_I are the flux of susceptible and infected, and ψ_S and ψ_I are the sources/sinks of susceptible and infected, respectively. To solve Eqs. (5.37) and (5.38) a representation of the fluxes q_S and q_I and sources/sinks ψ_S and ψ_I are required.

In ancient times, transporting commodities over any significant distances were an expensive and risky enterprise. Thus, travel and commerce was restricted mainly to local markets situated not far from home. Today, different modes of transportation (cars, trains, airplanes) deliver people in business and holidays to the most distant parts of the world in only a few hours. In the preceding section we have shown that diffusion and convection (channeling) are the regimes to distribute individuals through a territory (area). Therefore, we can consider that the flux of individuals is a convective–diffusive process, described generically by

$$q_S = -\left(\frac{\partial}{\partial x} + \frac{\partial}{\partial y}\right) D_S S + u_S S \tag{5.39}$$

$$q_I = -\left(\frac{\partial}{\partial x} + \frac{\partial}{\partial y}\right) D_I I + u_I I \tag{5.40}$$

where D_S and D_I are the diffusion coefficients of the susceptible and infective populations, respectively, and u_S and u_I are the velocities of the susceptible and infective populations, respectively. Combining relations (5.37) to (5.40) yields

$$\frac{\partial S}{\partial t} + \left(\frac{\partial}{\partial x} + \frac{\partial}{\partial y}\right) u_S S = \left(\frac{\partial^2}{\partial x^2} + \frac{\partial^2}{\partial y^2}\right) D_S S \pm \psi_S \tag{5.41}$$

$$\frac{\partial I}{\partial t} + \left(\frac{\partial}{\partial x} + \frac{\partial}{\partial y}\right) u_I I = \left(\frac{\partial^2}{\partial x^2} + \frac{\partial^2}{\partial y^2}\right) D_I I \pm \psi_I \tag{5.42}$$

The sources/sinks of susceptible and infective populations can be modeled based on the following assumptions: (i) natural births and deaths are proportional to the size of the susceptible population (Mena-Lorca and Hethcote 1992); (ii) the infective population has a disease-induced death that is proportional to its size (Nobel 1974); and (iii) the transmission from susceptible to infective is proportional to the size of the susceptible and infective populations (Nobel 1974). This then leads to:

$$\frac{\partial S}{\partial t} + \left(\frac{\partial}{\partial x} + \frac{\partial}{\partial y}\right) u_S S = \left(\frac{\partial^2}{\partial x^2} + \frac{\partial^2}{\partial y^2}\right) D_S S + \mu_{bdS} S - \sigma SI \tag{5.43}$$

$$\frac{\partial I}{\partial t} + \left(\frac{\partial}{\partial x} + \frac{\partial}{\partial y}\right) u_I I = \left(\frac{\partial^2}{\partial x^2} + \frac{\partial^2}{\partial y^2}\right) D_I I - \mu_{dI} I + \sigma SI \tag{5.44}$$

where μ_{bdS} is the net growth rate (e.g., rate of births minus the rate of natural deaths), μ_{dI} is the mortality rate induced by the disease, and σ is the disease transmission coefficient. Several time-scales can be obtained by applying scale analysis (Bejan 2000)

$$t_{dfi} = \frac{L_S^2}{D_S};\; t_{dfi} = \frac{L_I^2}{D_I} \tag{5.45}$$

$$t_{cv} = \frac{L_S}{u_S};\ t_{cv} = \frac{L_I}{u_I} \tag{5.46}$$

$$t_{\mu S} = \frac{1}{\mu_{dbS}};\ t_{\mu I} = \frac{1}{\mu_{dI}} \tag{5.47}$$

$$t_{\sigma S} = \frac{S}{I} t_{\sigma I};\ t_{\sigma I} = \frac{1}{\sigma S} \tag{5.48}$$

where t_{dif} and t_{cv} are the time scales associated with the diffusive and the convective populations motion (Section 5.3.1), $t_{\mu S}$ is the time scale associated with the births and life expectancy of the susceptible population, $t_{\mu I}$ is the life expectancy of an inflective, and $t_{\sigma I}$ is the contagious time of the disease.

Note that the diffusion and convective mechanisms only influence the spread of epidemic after it occurs but they do not play any role on whether the epidemic will occur. The time scale corresponding to the diffusive mode of travel is only smaller than the time scale corresponding to the convective mode if the territory length to access is smaller than D/u. Therefore, diffusion is the optimal travel regime to provide access to all locations while convection is optimal for providing access at large distances.

The growth rate of the susceptible population is strongly associated to $t_{\mu S}$ and $t_{\sigma I}$. There is a positive contribution to the growth of susceptibles since $t_{\mu S}$ is positive and the absolute value is smaller than $t_{\sigma I}$ (e.g., the ratio between μ_{bdS} and the density of infectives has to exceed the disease transmission coefficient). Furthermore, the development of the epidemic is strongly associated to $t_{\mu I}$ and $t_{\sigma I}$. When the life expectancy of an infective is much smaller than the contagious time, the disease tends to disappear. This topic will be detailed further in the next section.

5.7.2. Geotemporal Dynamics of Epidemics

The pattern of propagation of the Black Death and other plagues suggest a wave-like mechanism of propagation (Noble 1974). What are the conditions for the existence of waves? What kind of role do diffusive and convective regimes play on that? How do μ_{dI} and σ affect the propagation of the wave?

For the sake of simplicity, consider a one-dimensional wave traveling in a positive direction λ with a velocity c. Setting $S(x,t) = S(\Lambda)$ and $I(x,t) = I(\Lambda)$ with $\Lambda = \lambda - ct$, and substituting them into Eqs. (5.43) and (5.44) the following coupled differential equations are obtained:

$$\frac{\partial^2 S}{\partial \Lambda^2} + \frac{c - u_S}{D_S}\frac{\partial S}{\partial \Lambda} + \frac{\mu_{bdS} - \sigma I}{D_S} S = 0 \tag{5.49}$$

$$\frac{\partial^2 I}{\partial \Lambda^2} + \frac{c - u_I}{D_I}\frac{\partial I}{\partial \Lambda} + \frac{\sigma S - \mu_{dI}}{D_I} I = 0 \tag{5.50}$$

Since S and I cannot be negative, these equations may represent damped unforced harmonic oscillators under the following conditions: (i) $(c - u_S)/D_S \geq 0$,

$(c - u_I)/D_I \geq 0$ (damping factor) and (ii) $(\mu_{bdS} - \sigma I)/D_S \geq 0$, $(\sigma S - \mu_{dI})/D_I \geq 0$ (oscillatory factor). We are interested in the propagation of the infective population. According to Eq. (5.50), the natural frequency, ω_I, and the damped frequency, ω_{Id}, are

$$\omega_I = \left(\frac{\sigma S - \mu_{dI}}{D_I}\right)^{1/2} \tag{5.51}$$

$$\omega_{Id} = \left[\omega_I^2 - \frac{(c - u_I)^2}{4D_I^2}\right]^{1/2} \tag{5.52}$$

From these equations we note that the convective velocity is exclusively related to the damping factor while the diffusive process is related both with the damping and oscillatory factors. When the damping factor equals zero, $(c - u_I)/D_I = 0$, the system composed by the infective populations reduces to the case of a simple harmonic oscillator: continuous oscillation at the natural frequency ω_I. On the other hand, when $(c - u_I)/D_I > 0$, the system may or may not oscillate, depending on the relation between the damping factor and the natural frequency: If $\omega_I > (c - u_I)/2D_I$, the system is under damped conditions and exhibits transient behavior, oscillating at ω_{Id} with an amplitude that decays exponentially. When $(c - u_I)/2D_I \geq \omega_I$, there is no oscillatory behavior and the system returns smoothly to its equilibrium position. Thus, a wave-like solution can only exist when

$$c \geq u_I + 2\left[D_I(\sigma S - \mu_{dI})\right]^{1/2} \geq 0 \quad \text{and} \quad \sigma S \geq \mu_{dI} \tag{5.53}$$

This result has important implications that deserve to be analyzed. In the case of a wave-like solution, an increase of the diffusion coefficient D_I and velocity u_I implies waves with higher velocities. Besides these, the life expectancy of an inflective $(1/\mu_{dI})$ must be larger than the contagious time of the disease $(1/\sigma S)$. The last result has two consequences: (i) a wave-like solution implies a minimum critical value of the susceptible population density of $\sim \mu_{dI}/\sigma$, and (ii) there is a critical transmission coefficient from the infective to the susceptible population above which an epidemic wave occurs $(\sim \mu_{dI}/S)$. Thus, low-populated territories and rapidly fatal diseases prevent the spread of infection. This explains why very deathly diseases, such as the Ebola virus, do not spread around the world: up to now these diseases show up in poorly populated and remote areas (e.g., σS is very small), and the life expectancy of the inflected population is very short.

The above results have important practical implications. If we reduce travel and commerce (e.g., diffusion and convection mechanisms), we also prevent the spread of the epidemic (an increase of diffusive and convective fluxes of infective population induce traveling waves with higher velocities). In territories having a susceptible population below the minimum critical density, a sudden influx of susceptible individuals may initiate an epidemic. On the other hand, when $S > \mu_{di}/\sigma$, a sudden outflow of susceptible (health) population or immunization of a part of individuals by medical intervention (vaccination, culling) reduces the density of susceptible population and may prevent the spread of disease. Finally, by isolation of the infective population we are able to reduce the transmission coefficient σ and, if the critical transmission coefficient is not exceeded, there is no epidemic outbreak.

References

Anderson, R. M. and May, R. M. (1992) *Infectious Diseases of Humans*. Oxford University Press, London.

Ando, T., Ota, H. and Oki, T. (1988) Forecasting the flow of people. *Railw. Res. Rev.* **45**, 8–14.

Anthony, K. R. N. (1999) Coral suspension feeding on fine particulate matter. *J. Exp. Mar. Biol. Ecol.* **232**, 85–106.

Barreto, A., Aragon, M. and Epstein, R. (1994) Bubonic plague in Mozambique. *Lancet* **345**, 983–984.

Bégué, P. and Lorente, S. (2006) Migration vs. diffusion through porous media: time dependent scale analysis. *J. Porous Media* **7**, 637–650.

Bejan, A. (1997) *Advanced Engineering Thermodynamics*, 2nd edn, Wiley, New York.

Bejan, A. (1999) How nature takes shape: extensions of constructal theory to ducts, rivers, turbulence, cracks, dendritic crystals and spatial economics. *Int. J. Therm. Sci.* **38**, 653–663.

Bejan, A. (2000) *Shape and Structure, from Engineering to Nature*. Cambridge University Press, Cambridge, UK.

Bejan, A. (2002) Fundamentals of exergy analysis, entropy generation minimiza-tion, and the generation of flow architecture. *Int. J. Energy Res.* **26**, 545–565.

Bejan, A. (2005) The constructal law of organization in nature: tree-shaped flows and body size. *J. Exp. Biol.* **208**, 1677–1686.

Bejan, A. and Ledezma, G. A. (1998) Streets tree networks and urban growth: optimal geometry for quickest access between a finite-size volume and one point. *Physica A* **255**, 211–217.

Bejan, A. and Lorente, S. (2005) *La Loi Constructale*. L'Harmattan, Paris.

Bejan, A., Dincer, I., Lorente, S., Miguel, A. F. and Reis, A. H. (2004) *Porous and Complex Flow Structures in Modern Technologies*. Springer, New York.

Bejan, A., Lorente, S., Miguel, A. F. and Reis, A. H. (2006) *Along with Constructal Theory*, University of Lausanne, Faculty of Geosciences.

Ben-Jacob, E., Cohen, I., Shochet, O., Aronson, I., Levine, H. and Tsimering, L. (1995) Complex bacterial patterns. *Nature* **373**, 566–567.

Blue, V. and Adler, J. (1999) Bi-directional emergent fundamental pedestrian flows from cellular automata microsimulation. In A. Ceder (ed.), *Proc. Int. Symp. Traffic and Transportation Theory (ISTTT'99)*. Pergamon, Amsterdam, pp. 235–254.

Brauer, F. and Castillo-Chavez, C. (2001) *Mathematical Models in Population Biology and Epidemiology*. Springer, New York.

Caraco, T., Glavanakov, S., Chen, G., Flaherty, J. E., Ohsumi, T. K. and Szymanski, B. K. (2002) Stage-structured infection transmission and a spatial epidemic: A model for Lyme disease. *Am. Nat.* **160**, 348–359.

Cohn, S. K. (2002) The Black Death: end of a paradigm. *Am. Hist. Rev.* **107**, 703–738.

Diekmann, O. and Heesterbeek, J. A. P. (2000) *Mathematical Epidemiology of Infectious Diseases: Model Building, Analysis and Interpretation*. Wiley, New York.

Donald, I. and Canter, D. (1990) Behavioural aspects of the King's Cross disaster. In D. Canter (ed.), *Fires and Human Behaviour*. David Fulton, London, pp. 15–30.

Fang, Z., Lo, S. M. and Lu, J. A. (2003) On the relationship between crowd density and movement velocity. *Fire Saf. J.* **38**, 271–283.

Fruin, J. (1971) Pedestrian and planning design. Metropolitan Association of Urban Designers and Environmental Planners. Library of Congress catalogue number 70–159312.

Fukui, M. and Ishibashi, Y. (1999) Self-organized phase transitions in CA-models for pedestrians. *J. Phys. Soc. Japan* **8**, 2861–2863.

Gettling, A. V. (1998) *Rayleigh-Benard Convection: Structures and Dynamics*. World Scientific, Singapore.

Graat, E., Midden, C. and Bockholts, P. (1999) Complex evacuation: effects of motivation level and slope of stairs on emergency egress time in a sports stadium. *Saf. Sci.* **31**, 127–141.

Hankin, B. D. and Wright, R. A. (1958) Passenger flow in subways. *Oper. Res.* **9**, 81–88.

Heidemann, D. (1996) A queueing theory approach to speed–flow–density relationships, transportation and traffic theory. *Proc. 13th Int. Symp. Transport. Traffic Theory*, Lyon, pp. 14–26.

Helbing, D. (1992) A fluid-dynamic model for the movement of pedestrians. *Complex Syst.* **6**, 391–415.

Helbing, D. and Molnár, P. (1995) Social force model for pedestrian dynamics. *Phys. Rev. E* **51**, 4282–4286.

Helbing, D., Keltsch, J. and Molnár, P. (1997) Modeling the evolution of human trail systems. *Nature* **388**, 47–50.

Helbing, D., Schweitzer, F., Keltsch, J. and Molnár, P. (1997) Active walker model for the formation of human and animal trail systems. *Phys. Rev. E* **56**, 2527–2539.

Helbing, D. Molnár, P. Farkas, I. J. and Bolay, K. (2001) Self-organizing pedestrian movement. *Environ. Plan. Plan. Des.* **28**, 361–383.

Henderson, L. F. (1971) The statistics of crowd fluids. *Nature* **229**, 381–383.

Hodge, A., Robinson, D., Griffiths, B. S. and Fitter, A. H.(1999) Why plants bother: root proliferation results in increased nitrogen capture from an organic patch when two grasses compete. *Plant Cell Environ.* **22**, 811–820.

Hoogendoorn, S. and Bovy, P. H. L. (2000) Gas-kinetic modeling and simulation of pedestrian flows. *Transp. Res. Rec.* **1710**, 28–36.

Hoogendoorn, S., Bovy, P. and Daamen, W. (2002). Microscopic pedestrian way finding and dynamics modelling. In M. Schreckenberg and S. Sharma (eds.), *Pedestrian and Evacuation Dynamics*. Springer, New York, pp. 123–155.

Hsu, C. C., Chen, T., Chang, M. and Chang, Y. K. (2006) Confidence in controlling a SARS outbreak: Experiences of public health nurses in managing home quarantine measures in Taiwan. *Am. J. Infect. Control* **34**, 176–181.

Hughes, R. L. (2002) A continuum theory for the flow of pedestrians. *Transp. Res. B* **36**, 507–535.

Hughes, R. L. (2003) The flow of human crowds. *Annu. Rev. Fluid. Mech.* **35**, 169–182.

Hunter, K., Petty, M. D. and McKenzie, F. D. (2005) Experimental evaluation of the effect of varying levels of crowd behavior fidelity on the outcome of certain military scenarios. In Spring 2005 Simulation Interoperability Workshop, San Diego, CA.

Kaandorp, J. A. and Sloot, P. M. A. (2001) Morphological models of radiate accretive growth and the influence of hydrodynamics. *J. Theor. Biol.* **209**, 257–274.

Kallen, A., Arcuri, P. and Murray, J. D. (1985) A simple model for the spatial spread and control of rabies. *J. Theor. Biol.* **116**, 377–394.

Kermack, W. O. and McKendrick, A. G. (1927) A contribution to the mathematical theory of epidemics.*Proc. R. Soc. London A* **115**, 700–721.

Koopmans, M., Wilbrink, B., Conyn, M., Natrop, G., van der Nat. H., Vennema, H., Meijer, A., van Steenbergen, J., Fouchier, R., Osterhaus, A. and Bosman, A. (2004) Transmission of H7N7 avian influenza A virus to human beings during a large outbreak in commercial poultry farms in the Netherlands. *Lancet* **363**, 587–593.

Langer, W. L. (1964) The black death. *Scient. Am.* **210**, 114–121.

Langston, P. A., Masling, R. and Asmar, B. N. (2006) Crowd dynamics discrete element multi-circle model, *Saf. Sci.* **44**, 395–417

Mena-Lorca, J. and Hethcote, H. (1992) Dynamic models of infection diseases as regulator of population sizes. *J. Math. Biol.* **30**, 693–716.

Merks, R., Hoekstra, A., Kaandorp, J. and Sloot, P. (2003) Models of coral growth: spontaneous branching, compactification and the Laplacian growth assumption.*J. Theor. Biol.* **224**, 153–166.

Miguel, A. F. (2004) Dendritic growth: classical models and constructal analysis. In R. Rosa, A.H. Reis, A.F. Miguel (eds.) *Bejan's Constructal Theory of Shape and Structure*. CGE-UE, Evora, pp. 75–93

Miguel, A. F. (2006) Constructal pattern formation in stony corals, bacterial colonies and plant roots under different hydrodynamics conditions. *J. Theor. Biol.* **242**, 954–961

Muramatsu, M., Irie, T. and Nagatani, T. (1999) Jamming transition in pedestrian counter flow. *Physica A* **267**, 487–498.

Navin, P. D. and Wheeler, R. J. (1969) Pedestrian flow characteristics. *Traffic Eng.* **39**, 31–36.

Nelson, H. E. and Maclennan, H. A. (1995) Emergency movement. *The SFPE Handbook of Fire Protection Engineering*, 2nd edn, NFPA, Quincy, MA.

Noble, J. V. (1974) Geographic and temporal development of plagues. *Nature* **250**, 276–279.

Older, S. J. (1968) Movement of pedestrians on footways in shopping streets. *Traffic Eng. Control* **10**, 160–163.

Oxford, J. S., Lambkin. R., Sefton, A., Daniels, R., Elliot, A., Brown, R. and Gill, D. (2005) A hypothesis: the conjunction of soldiers, gas, pigs, ducks, geese and horses in Northern France during the Great War provided the conditions for the emergence of the "Spanish" influenza pandemic of 1918–1919. *Vaccine* **23**, 940–945.

Predtechenskii, V.M. and Milinski, A. I. (1969) *Planning for Foot Traffic Flow in Buildings*. Stroiizdat Publishers, Moscow.

Reis, A. H., Miguel. A. F. and Aydin, M. (2004) Constructal theory of flow architecture of the lungs. *Med. Phys.* **31**, 1135–1140.

Reis, A. H. (2006a) Constructal view of scaling laws of river basins. *Geomorphology* **78**, 201–206.

Reis, A. H. (2006b) Constructal theory: from engineering to physics, and how flow systems develop shape and structure. *Appl. Mech. Rev.* **59**, 269–282.

Reis, A. H. and Miguel, A. F. (2006) Constructal theory and flow architectures in living systems. *J. Thermal Sci.* **10**, 57–64.

Reynolds, C. W. (1987) Flocks, herds, and schools: a distributed behavioral model. *Comput. Graphics* **21**, 25–34.

Robinson, D. (1994) The responses of plants to non-uniform supplies of nutrients. *New Phytol.* **127**, 635–674.

Rosa, R., Reis, A. H. and Miguel, A. F. (2004) *Bejan's Constructal Theory of Shape and Structure*. Geophysics Center of Evora, University of Evora, Portugal.

Sandahl, J. and Percivall, M. (1972) A pedestrian traffic model for town centers. *Traffic Q.* **26**, 359–372.

Schweitzer, F. (1997) *Self-organization of Complex Structures: From Individual to Collective Dynamics*. Gordon and Breach, London.

Sebens, K. P., Witting, J. and Helmuth, B. (1997) Effects of water flow and branch spacing on particle capture by the reef coral Madracis mirabilis (Duchassaing and Michelotti). *J. Exp. Mar. Biol. Ecol.* **211**, 1–28.

Smith, R. A. and Dickie, J. F. (1993) *Engineering for Crowd Safety*. Elsevier, Amsterdam.

Suwandono, A., Kosasih, H., Nurhayati, H., Kusriastuti, R., Harun, S., Maroef, C., Wuryadi, S., Herianto, B., Yuwono, D., Porter, K. R., Beckett, C. G. and Blair, P. J. (2006) Four dengue virus serotypes found circulating during an outbreak of dengue fever and dengue haemorrhagic fever in Jakarta, Indonesia, during 2004. *Trans. R. Soc. Trop. Med. Hyg.* **100**, 855–862.

Thaler, P. and Pages L. (1998) Modeling the influence of assimilate availability on root growth and architecture. *Plant Soil* **201**, 307–320.

Thar, R. and Kühl, M. (2005) Complex pattern formation of marine gradient bacteria explained by a simple computer model. *FEMS Microbiol. Lett.* **246**, 75–79.

The Green Guide (1997) *Guide to Safety at Sports Grounds*, 4th edn, HMSO, London.

Thompson, P. A. and Marchant, E. W. (1995) A computer model for the evacuation of large building populations. *Fire Saf. J.* **24**, 131–148.

Timmermans, H., van der Hagen, X. and Borgers, A. (1992) Transportation systems, retail environments and pedestrian trip chaining behaviour: modelling issues and applications. *Transport. Res. B: Methodological* **26**, 45–59.

Togawa, K. (1955) Study on fire escape based on the observations of multitude currents. Report No. 4. Building Research Institute, Ministry of Construction, Japan.

TRB (1985) Pedestrians. In *Highway Capacity Manual*, special report 209, Transportation Research Board, Washington, DC, Chapter 13.

Vandaele, N., Woensel, T. V. and Verbruggen, A. (2000) A queueing based traffic flow model. *Transport. Res. D* **5**, 121–135.

Chapter 6
The Constructal Nature of the Air Traffic System

Stephen Périn

6.1. Introduction

The Air Traffic System (ATS) industry has a great impact on society and many human activities, for examples, through immigration flows, the effect on economy or the health risk due to disease propagation, and the various environmental impacts of air traffic. This world leading industry is facing serious and contradictory challenges for its future development: a threefold increase of traffic in term of passengers and aircraft and an always greater pressure of the overall society concerning the reduction of the environmental impacts of its activity (noise and gas emissions, etc.). In this specific context of a search for new concepts and for a global better efficiency of the Air Traffic System, we propose to study these specific flows as flows optimally distributed both in space and in time on the basis of the constructal law. This law is the basis for the general framework of the constructal theory of global optimization of flows under local constraints. The constructal law of generation of configurations for maximum flow access in freely morphing structures offers an extensible and complementary view to classical approaches and can be used to generate and refine many Air Traffic System flow distribution patterns. In this chapter, we show that complex Air Traffic System flows such as airport and air traffic flows are bundles of refracted paths, which owe their global shape to the maximization of flow access.

The air traffic system has an enormous impact on human activities, especially on economics directly, e.g., through fret activities, catalytically—e.g., a $\sim 4\%$ increase of European gross domestic product during the last ten years has been estimated to having being generated by catalytic effect due to air traffic usage (Cooper and Smith 2005)—and also indirectly, e.g., through emigration and environmental or health issues such as disease propagation (e.g., West Nile Virus or more recently SRA). This impact is of course correlated to the two underlying and interrelated networks of air routes and airports.

From the environmental point of view, we can notice, for instance, that the contrails generated by the aircraft flowing following the North Atlantic routes are responsible of a 7% increase every 10 years of the cirrus occurrences over this area (Penner et al. 1999), and that during the 3 days of no traffic over USA following the 11 September 2001 events, a 1° C increase of temperature was recorded in this country and is supposed to have been caused by the absence of contrails in the North American sky.

Both the Strategic Research Agenda of the Advisory Council for Aeronautics Research in Europe (ACARE 2002) and the NGATS Integrated Plan (Joint Planning & Development Office 2004) emphasize the need for new concepts in the domains of Aeronautics and Air Traffic in order to face the serious and also often antithetical challenges already identified for the future development of the whole Air Traffic System, such as managing a threefold increase in volume of passengers and flights forecasted by the year 2020 (European Commission 2001), the need for reduction of the environmental impacts of the ATS (Penner et al. 1999)—or in other words the reduction of its societal cost—a global reduction of the costs of exploitation and this all along with an enhanced security and safety level of the operations. The cost of inefficiency of Air Traffic Management (ATM) in Europe is estimated to be €4.4 billions, i.e., circa 63% of a total of €7 billions for the total annual cost of the management of air traffic for Europe. The average cost is €0.76 per kilometer, around twice as much as the cost of ATM in the USA (SESAR Consortium 2006).

In this context of a search for a global optimization of the entire ATS under such competing and opposite or clashing constraints and objectives (e.g. traffic increase vs. reduction of costs and impacts), we propose an alternative and complementary approach to classical ways of studying and addressing these flow management problems, such as airport landside aircraft movements and passengers flow modeling using multi-agent software simulations, which can be finely tuned with many parameters including passenger cultural behavior, age, etc. (Airport Research Center 2004), or air traffic network optimization based on graph theory and on a preexisting network (Rivière 2004).

In order to offer new concepts and to meet certain challenges of the Vision 2020 (European Commission 2001), the alternative that we propose is to look at air and airport traffic and processes as flows that are optimally distributed (in space and time) on the basis of the constructal law of generation of configurations for maximum flow access in freely morphing structures (Bejan 1997, 2000a).

6.2. The Constructal Law of Maximum Flow Access

6.2.1. Foundations of Constructal Theory

Constructal theory emerged recently from the field of thermodynamic optimization of engineered systems, and unveiled a deterministic principle of geometry (shape) generation. The universal phenomenon is the generation of flow architecture, such as trees, rivers, and lungs found in natural systems, and the principle is the constructal law: *for a flow system to persist in time (to survive) its configuration must change such that it provides easier and easier access to its imposed (global) currents* (Bejan 1997, 2000a). This law accounts for spatiotemporal flow self-organization and self-optimization—phenomena also referred as emergence in literature—in both natural and man-made flow systems and was initially stated in 1996 by Adrian Bejan (Bejan 1997).

The evolution between 1851 and 2003 of the geometry of the sail of the America's cup racing yacht illustrates perfectly the constructal idea of flow system morphing through time in order to achieve a better efficiency. From the aerodynamic point of view, a rounded tip is effectively better designed than the older pointed tip—this first and more recent design avoiding both local flow separation and stalling (Chatterjee and Templin 2004).

Another and maybe more direct and experimental approach which can be put into practice by everyone in order to understand and visualize on "real time" the concept of constructal law is the self-organization occurring in a boiling plate of rigatoni (see also Bejan 2000a, Sections 7.7 and 7.9). After some minutes of cooking, the initial random distribution of tube-shaped pasta morph toward a flow-oriented geometry, i.e., the tubes align themselves with flow structures, as visible in Fig. 6.1. The incomplete number of vertical tubes here can be attributed to the short duration of the experimentation (some minutes) and to the non-uniform heating from below. The same kind of flow organization is found, for instance, in the sideways drift of freely deriving icebergs in open sea (see Bejan 2000a, Section 7.2, p. 150).

A first innovative and unifying aspect of this theory is that it applies to a wide range of flow systems, regardless of their nature or scale. It can be—and in fact has been already—applied to optimization studies of natural or engineered systems ranging from the size of packet of electronics to the one of macroscopic flows such as meteorological flows, i.e., a range from 10^{-6} to 10^{6} meters (Bejan 2000a).

In a flow system, easier access means less thermodynamic imperfection (friction, flow resistances, drops, shocks) for what flows through the system (river basin, animal, airport). The optimal distribution of these numerous and highly diverse imperfections is the flow architecture itself (river basins, blood

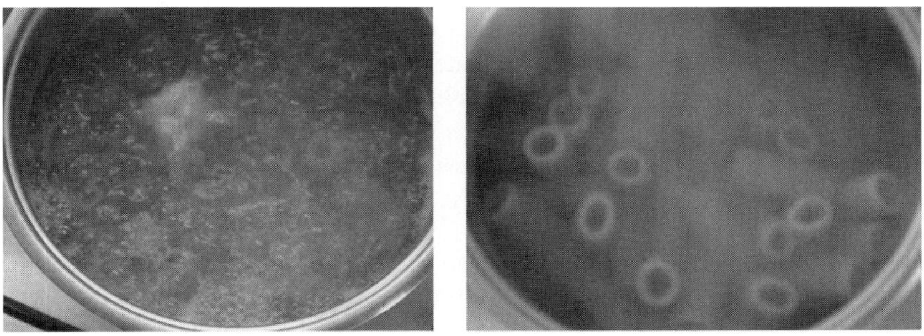

FIGURE 6.1. The self-organization (emergency phenomena) in a plate of boiling rigatoni, auto-aligning with themselves with the flow structure, is an easily reproducible way of illustrating the main concept of constructal theory

vascularization, atmospheric circulation, etc.) If the flow system exhibits at least two dissimilar regimes, one highly resistive (diffusion, walking, climbing, and descending or regional air traffic) and one less resistive (stream, streets, en-route air traffic) (Périn and Bejan 2004), the region with dissimilar regimes can be balanced in such way that at the global level the system flows with minimum but finite imperfection.

In this least imperfect configuration, the flow system destroys minimum useful energy (fuel, food, exergy). This optimized movement is a bouquet of refracted paths in the sense of Fermat: the law of refraction governs the movement of goods in economics, where it is known as the law of parsimony (Lösch 1954; Haggett 1965; Haggett and Chorley 1969). The history of the development of trade routes reveals the same tendency. We often hear that a city grew because "it found itself" at the crossroads—at the intersection of trade routes. We believe that it was the other way around: the optimally refracted routes defined their intersection, the city, the port, the loading and unloading site, etc.

Previous work based on constructal theory showed that tree-shaped structures are typical patterns arising from this optimal distribution of imperfections in a flow system. Some domains of application with the corresponding flow regimes, tree-shaped channels, and interstitial spaces are respectively the following (Bejan 2000a):

- urban traffic, people, low-resistance car traffic, and street walking
- circulatory system, blood, low resistance blood vessels (capillaries, arteries, etc.), and diffusion in capillaries tissue
- river basins, water, low-resistance rivulet and rivers, and Darcy flow through porous media.

6.2.2. *The Volume-to-Point Flow Problem*

More complicated flows are assemblies of paths, optimally refracted such that the global flow access is maximized. A river basin under the falling rain is like an area inhabited by people: every point of the area must have maximum access to a common point on the perimeter. Let us consider that there are two media, one with low resistivity (channel flow) and the other with high resistivity (Darcy seepage through wet river banks). The shape of the basin comes from the global maximization of flow access. This generic problem of optimization of a flow between a volume (or surface) and a point was theorized on the basis of the constructal theory (Bejan 2000a). The resulting optimized path was deduced in a pure deterministic way and following a sequence of steps consisting of geometric optimizations starting at the smallest scale of the system (elemental volume) and of progressively larger assemblies of building blocks at greater scales—and from this constructive process also came the name of the theory: constructal, from the Latin verb *constrŭĕre*, to build.

Figure 1.5 illustrates this process: it shows how to derive optimal shape from the minimization of travel time between a surface and a point by aggregating the small and optimized building blocks shown in Fig. 1.2 in high-order constructs such as the one displayed in Figs. 1.4 and 1.5.

The more basic problem illustrating one of the main concepts of constructal theory, i.e., the optimal balance in a flow system with at least two different regimes, is displayed in Fig. 6.2. There are two speeds V_0 and V_1 available during a travel between a point M situated in the middle of a square of side length L and a point p at one of the corners of the opposite side. The optimal path minimizing the travel time will pass through the point R, where the change of speed will occur (Bejan 2000a). A general solution of the optimal x abscise of point R is given by

$$x = L/2[(V_1/V_0)^2 - 1]^{1/2} \tag{6.1}$$

It is interesting to notice that dogs naturally conform to this algebraic formula to minimize the time to retrieve a ball sent, e.g., from the shore to the water of a lake (Pennings 2003). In this case, dogs have two means of locomotion due to the two media: running on the ground (the fastest) and swimming (the slowest). In most cases the animal chooses a path in good agreement with the optimal one. So, despite the fact that dogs don't do any calculation, its "behavior is an example of the uncanny way in which nature (or Nature) often

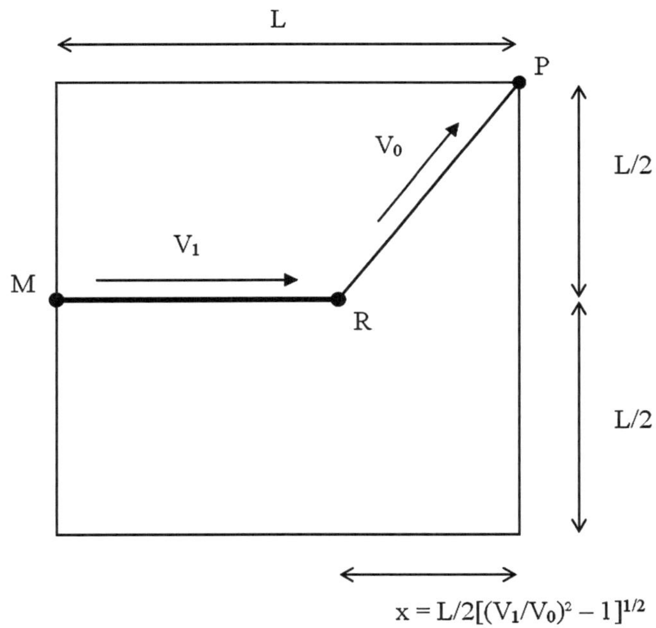

FIGURE 6.2. The more basic problem of Constructal theory is the optimal balance between two speeds V_0 and V_1 in a travel through a point M in the middle of a square to one of the corners of the opposite side P. The optimal path minimizing the travel time will pass through the point R, where the change of speed will occur (Bejan 2000a)

FIGURE 6.3. "Aibopolis"–French Aibo's owner contest organized by www.aibo-fr.com at Paris, February 20, 2005, and sponsored by Sony and Robopolis. The Sony Parisian Lab recently developed a self-learning algorithm enabling an AIBO (dog-like) robot to discover by itself how to coordinate its movements for locomotion (Oudeyer et al. 2005)

finds optimal solutions" (Pennings 2003). The apparition of this behavior in a biological system is of course itself understandable through Natural selection. In an afferent domain, constructal theory was used to demonstrate that all kinds of animal locomotion (running, swimming, flying) can be understood and unified from a single principle: the optimal balance between the horizontal and vertical energy loss. Recently, the Sony Computer Science Lab (SCSL) at Paris developed a mechanism of intelligent adaptive curiosity. This program was implemented in an AIBO robot (cf. Fig. 6.3) which demonstrated its autonomous mental development and self-learning capacities (Oudeyer et al. 2005). One of the most impressive achievements is the experiment, where, through the discovery of sensorimotor affordance, the robot learns by itself how to move its members, then how to coordinate them to get up, and finally to move. At this point, it is interesting to note that what accounts for natural system (animal) account also for engineered system like AIBO or other robots: using constructal theory, an optimal means and speed of locomotion could be predicted for exploratory robot, for instance moving into remote terrestrial or extra-terrestrial environments (Bejan and Marden 2006b).

6.3. Relevant Results for Aeronautics

6.3.1. Aircraft Design

Constructal theory is not only concerned by tree-shaped flow, even if it shows that this kind of network geometries are typically generated in natural and engineered systems by a competition between at least two dissimilar regimes of a same flow (e.g. laminar vs turbulent). As an example of a different way of using constructal theory, we propose to look first at its results in the field of locomotion systems. Scaling laws are typical of many biological systems—where they are also known as allometric laws, relating breathing rhythm with body mass, heart beating frequency with body mass, or flight cruising speed with body mass. These experimental laws were neither clearly understood nor theorized since their discoveries, but constructal theory brings a new light on this domain. For instance, the proportionality between the optimal flight speed V_{opt} and the mass M_b of a flying body was demonstrated using the theory on the basis of a very simple scaling law (Bejan 2000a):

$$V_{opt} \sim M_b^{1/6} \tag{6.2}$$

Where V_{opt} (in ms^{-1}) is the optimal flight speed, i.e., the speed minimizing the useful energy spent to suspend the body (exergy, food), and M_b is the body mass (in kg) of the considered flying body. From this point of view, we could talk of a "sustainable" speed. Furthermore, the same principle was invocated recently to unify all animal movements in the same framework, unveiling that animal locomotion (running, swimming, flying) is an optimal balance between the vertical and horizontal energy loss during the movement occurring in a gravity field (Bejan and Marden 2006a,b). Why is this allometric law of optimal flight speed V_{opt} particularly relevant for aeronautics—as long as we discussed until then only biological systems? Because not only insects and birds conform to it but also man-made machines such as aircraft. Fig. 1.17 shows the cruising speed of insects, birds, and airplanes compared with the prediction based on constructal theory (approximation).

The studies performed so far on the subject are just approximation and in-depth studies should perform complete thermodynamic optimization taking into account the geometric characters of the flying body.

A completely different but interesting application of this work on optimal flight could be paleontology. Among others, the estimation of pterosaur's flight speed is subject to many uncertainties, because of the multiple missing data. For instance, the flight speed estimation depends on the body mass of the pterosaur, and this body itself is unknown and must be estimated, e.g., on the basis of allometric relations, determined by regression using data on living species. For example, Fig. 6.4 shows the good agreement between the theoretical optimal speeds (Eq. 6.2) of seabirds (measured) and the estimated pterosaurs' cruising flight speeds calculated from Chatterjee and Templin (2004). Constructal theory offers, then, now an alternative method in the domain of locomotion in paleontology, which could be used in conjunction with classical methods.

It should be noticed that on the basis of the same simple model of interaction of a flying body discussed above and used for the optimal flight speed, it is also possible to demonstrate that the sizes or weights of on board heat and fluid flow systems of aircraft or other vehicles can be derived from the maximization of global system performance (Ordonez and Bejan 2003), and to predict the minimum power requirement for environmental control systems of man-made machines such as aircraft (Bejan 2000b).

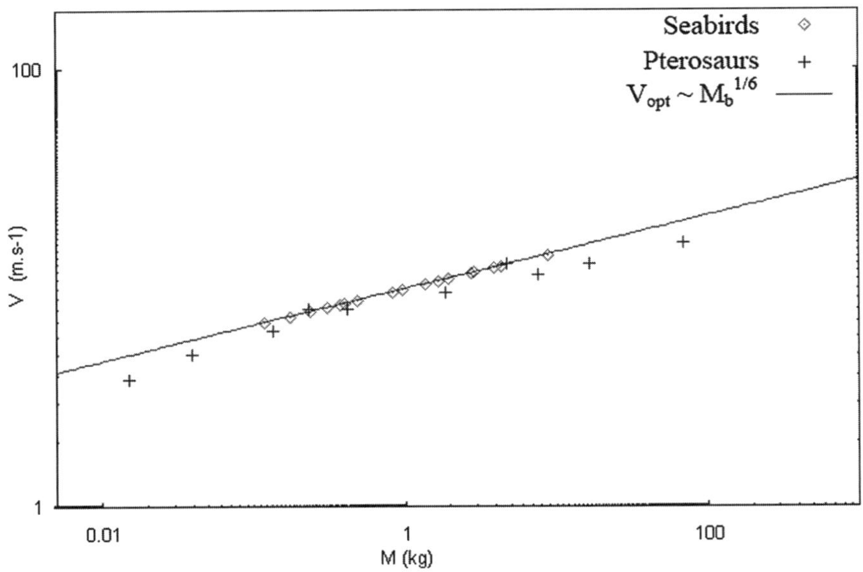

FIGURE 6.4. Measured seabirds' and estimated pterosaurs' flight speed (Chatterjee and Templin 2004), compared to the Constructal optimal and theoretical speed V_{opt} (Bejan 2000a)

6.3.2. Meteorological Models

Meteorology has an important impact on the activities of the ATS, both on airside and on landside. Meteorology and meteorological modeling are considered—with reason—as hard science, governed by the well known but complicated equations of fluid dynamics (Navier-Stokes) requiring long calculation time.

Here again, and surprisingly, constructal theory delivers really accurate predictions concerning the modeling of Earth's climate, despite the simplicity of the model used. By considering successively the Earth as an optimal collector and radiator—respectively through its illuminated and dark hemispheres—this constructal model of earth climate delivered results in good agreement with observational data, e.g., concerning the Earth's surface partitions (Hadley cells, Ferrel cells, and Polar cells), the Eckman number, and several other optimized variables (average temperatures, wind amplitude) (Reis 2004; Reis and Bejan 2006). From the constructal point of view, because the flowing Earth is a constructal heat engine, its flow configuration has evolved in such a way that it is the least imperfect that it can be. It produces maximum power, which it then dissipates at maximum rate. The constructal law is so offering a brand new class of meteorological models that we could expect it to be developed in the following years. It could be interesting for ATM to introduce the air traffic contrails in the constructal model in order to study its impact, which is different on the illuminated (collector) face and on the dark (radiator) face.

Concerning weather and climate, constructal theory could also be applied to optimize deicing/icing (and heating/cooling) processes, another domain of application of high interest for the airport landside activities (deicing of aircraft, building, taxiways, etc.) and about which several works have already been done (ASHRAE 1987, 1990).

6.4. Application to the Air Traffic System

The previous paragraphs show us the relevance of constructal theory to study a wide range of disciplines, including domains like aeronautics and meteorology, linked to our purpose, the only requirement being to be applied to systems with dissimilar regimes, regardless of their nature and scale. As long as the ATS is also a complex flow system, it is logical that this section will now look at the possible application to two sub-domains of it, namely Air Traffic Flow Management and airport systems.

6.4.1. Air traffic flow

For these reasons in particular, it is of great interest to understand the generating mechanism of these air traffic networks. Figure 6.5 displays a partial view of the current network of air traffic routes above Western Europe. By looking at this map, it could not be intuitive to understand the possible relationships with the typical constructal tree-shaped network previously presented. If the traffic demand graph display is filtered in order to display only the city-pairs (direct links) with at least 10 flights per day, as Fig. 6.6 shows us, a more precise organization of the flow demand appears. But neither the graph of demand (direct links) nor the air route network is the flow: this last is much more like a "flow fossil," an instantaneous "picture" of the whole one-to-many interacting flows, the static counterpart of the dynamic streams. If we then look at the real incoming or outgoing flow to a particular airport (i.e., trajectories, not direct links), this relation will become clearer. For example, Fig. 6.7 shows both incoming and outgoing flight trajectory arriving at and departing from the Nice airport, on the French Riviera (LFMN ICAO code), for one day of traffic.

The incoming and outgoing flows of this airport display the characteristic tree-shaped pattern and branching mechanism specific to many constructal tree networks. At the moment, we can just make hypothesis on the mechanisms generating this typical geometry. On the basis of constructal theory and existing results in the field of flow optimization, we can suspect that this typical geometry comes from a balance between competing mechanisms, e.g., the minimization of length of each specific city-pair (or the minimization of the number of hub connections for the city-pairs) and the minimization at the same time of the cumulated length of the edges of the whole network—both mechanisms driven by economics and under geopolitical constraints (Bejan and Lorente 2004; Guimerà and Amaral 2004; Guimerà et al. 2003, 2005). The whole European air route

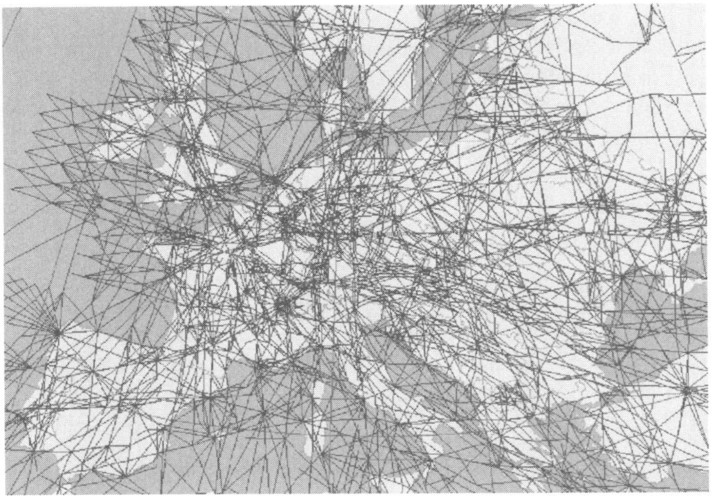

FIGURE 6.5. Partial view of the current air route network above Western Europe

network itself can be viewed as composed of all superimposed segments of each of these constructal trees, sourcing from every European airport. The same kind of radial tree structures as displayed in Fig. 6.8 was in fact obtained following a deterministic constructal optimization process for the design of a flow structure connecting a central point sourcing a stream sunk into several equidistant points disposed on the periphery of a disk (Bejan and Lorente 2004).

FIGURE 6.6. European Traffic Demand (city-pairs) for one day of traffic, filtered in order to display only the city-pairs with at least 10 flights per day. The color and size of the segments are both proportional to the city-pair traffic

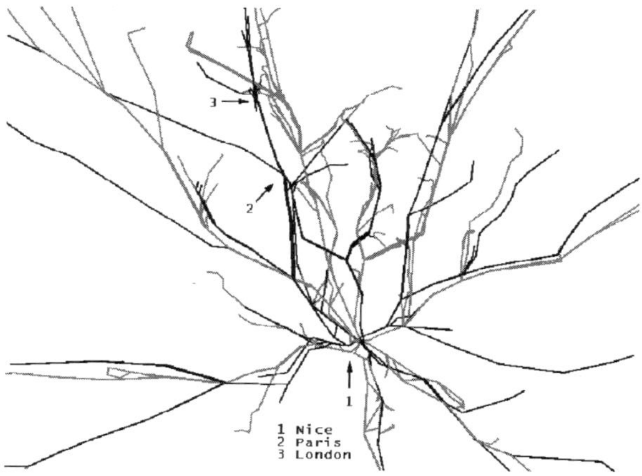

FIGURE 6.7. Outgoing and incoming flows of the French Nice airport (LFMN ICAO code) for one day of traffic, displaying the typical branching of constructal tree-shaped flow structures

Furthermore, the worldwide airport network was recently studied in order to investigate its overall efficiency and growth mechanism and it was found not only that this airport network is a "small-world" network—i.e., a hub-structured network, following power-law distribution for the number of non-stop connections from a given city and for the number of shortest paths going through

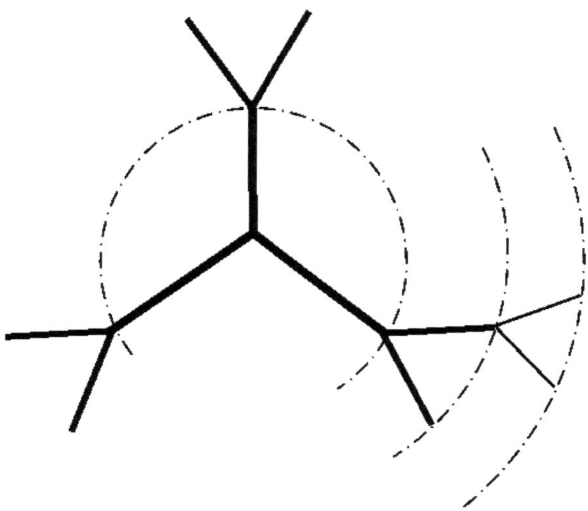

FIGURE 6.8. Such kind of radial tree structure between a central point and equidistant points were obtained through a process of optimization based on constructal theory (Bejan and Lorente 2004)

a given city—but, more interesting for our purpose, that "the formation of this multi-hub structure can be explained in terms of an attempt to maintain the overall efficiency of the network" (Guimerà et al. 2003).

In other words, the current airport network has a hub and spoke topology driven in particular by the minimization of the number of waiting flights for a given flight frequency. This observation has a direct consequence on the temporal schedule of the flight, as discussed below.

On the basis of these observations, we can conclude that the air route network and the related airport network both follow the constructal law of global flow optimization under local constraints, and their geometries are the result of the collision between several objectives such as minimum travel time, minimum number of legs, minimum cost, and minimum network length.

The airport network also displays an interesting community structure: using recent network complexity algorithms, cluster of closely interrelated nodes can be individualized inside a given network, such as the worldwide airport network. These community blocs are clearly subject to geopolitical constraints (Guimerà et al. 2005). This mechanism differs from the one implied, for instance at a different scale, in ethnic residential segregation, but it could be viewed as an extension of the autocorrelation found in Blau Space, from human social networks to the domain of the worldwide airport networks.

The hubs and spokes structure of the airport network is in fact similar to the topology of the Internet network, both of them being "small world" network, especially if compared to planar graph like road network (Gastner and Newman 2004). The parallel between the Internet network backbones and long-haul flights could be pushed further, e.g., by regrouping upper airspace Air Traffic Control (ATC) centers in "backbone"-like components, and could result in Air Traffic Flow Management reorganization allowing "a completely new kind of air traffic control where the airspace design depends on the flow rather than on the geographical shape of the ATC centers" (Benderli 2005). The similarities of this approach with the Paradigm SHIFT concept of Dual airspace (Guichard et al. 2005), highlighted in Benderli (2005), are in good agreement with the constructal theory, the latter focusing on the optimal distribution of the air traffic flow regimes like short-haul and long-haul flights (Périn and Bejan 2004). The Dual airspace concept proposes a regional vs en-route traffic discrimination based on the vertical evolution trend of the air traffic flow, then organized in districts (regional traffic) and highway (en-route traffic) structures.

6.4.2. The Constructal Law and the Generation of Benford Distribution in ATFM

City-rank in terms of population was one of the original set of data studied by Benford, and certainly one of the set best fitting his distribution known as Law of Anomalous Numbers (Benford 1938). This phenomenological law—also called first digit law or Benford's law—states that in data set commonly used in physics, engineering, economy, etc., the probability of occurrence of a number

TABLE 6.1. Benford distribution

Digit	1	2	3	4	5	6	7	8	9
Benford law	30.1%	17.6%	12.5%	9.69%	7.92%	6.69%	5.80%	5.12%	4.58%

belonging to {1, .., 9}, in base 10, as first digit is given by the following P_D probability, displayed in Table 6.1:

$$P_D = \log_{10}(1 + 1/D) \tag{6.3}$$

The "reality" of this law and its origin are questioned still today, and a X^2-based test was proposed to check if a data set is eligible as a Benford distribution (Scott and Fasli 2001). A more in-depth study based on this X^2 criteria shows that in fact half of the data set considered by Benford doesn't fit its distribution and that a great number of data sets conform to an exaggerated Benford-like distribution, i.e., the first digit distribution follow a (quickly) monotonically decreasing function (Scott and Fasli 2001).

Usual explanations for or arguments over Benford law found in the literature are commonly based on Central limit theorem, and consider the multiplication of random factors as the generator of this distribution. Scale invariance, power-law distributions, and fractal geometries were also proposed as explanations (Pietronero et al. 1998, Scott and Fasli 2001).

By considering that constructal theory predicts the existence of many scaling laws and power-law distribution in natural and engineered systems such as Hack's Law, optimal flight speed, Kleiber's Law, etc., that it provides scale-covariant flow system architectures and replaces the assumption that nature is fractal, we propose to consider the possibility of constructal law as the origin of scale invariance and of the generation of Benford or Benford-like distributions. This consideration could bring a new light on this intriguing phenomenon: "The understanding of the origin of scale invariance has been one of the fundamental tasks of modern statistical physics. How a system with many interacting degrees of freedom can spontaneously organize into critical or scale invariant states is a subject that is of up-surging interest to many researchers" (Pietronero et al. 1998).

The trend of distribution of cities' size is in fact well predicted by constructal theory (Bejan et al. 2006) (see also Section 1.3 and Figs. 1.7 and 1.8)—and we could even suppose in first approach that airport size is also correlated to city size (excepted maybe for "artificial" hubs). Do aeronautical and airport-network related data conform to a Benford distribution? Figures 6.9–6.12 display the result of the computation of first-digit frequency of extensive aeronautical data (circa 11,000 flights and 3,700 routes) such as the number of flights per city-pair route, the geodesic route lengths, the flown trajectory lengths, and the number of route segments. All data show good agreement and a future study will precisely determine their adherence to a Benford law distribution. On the four sets of

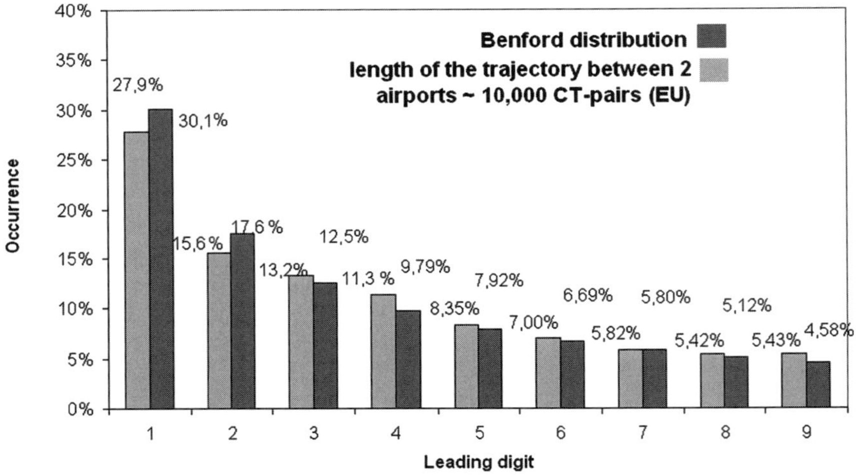

FIGURE 6.9. Distribution of the leading digit of the length of the real trajectory between two airports for circa 10,000 city-pairs in Europe, compared with the Benford distribution

data considered, only the number of flights per city-pair route seems to conform to an exaggerated Benford-like distribution, very quickly decreasing, where the cumulated first-digit probability of the {1, 2, 3} sub-set approaches 85%. This last kind of behavior is in fact well known and easily understandable, e.g., due to the inherent upper and lower bounds of many phenomena.

A potential application of Benford-like distributions in Air Traffic Flow Management could concern the generation of more realistic traffic data, especially for exploratory studies based on, e.g. 2020 or 2030, forecasted data sets and previsions of traffic.

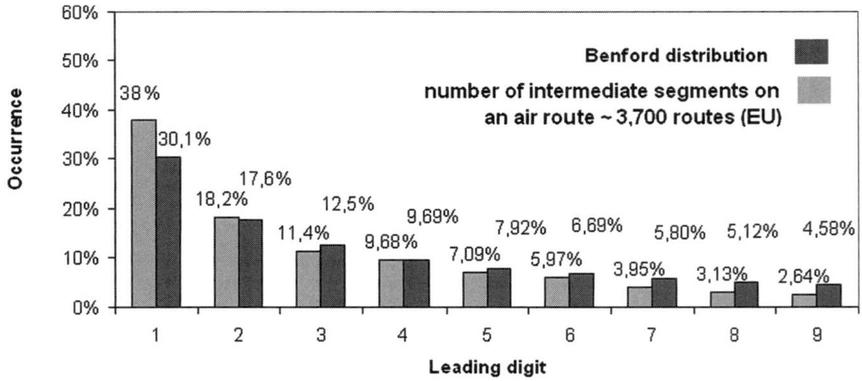

FIGURE 6.10. Distribution of the leading digit of the number of intermediate segments on a European air route, compared with the Benford distribution

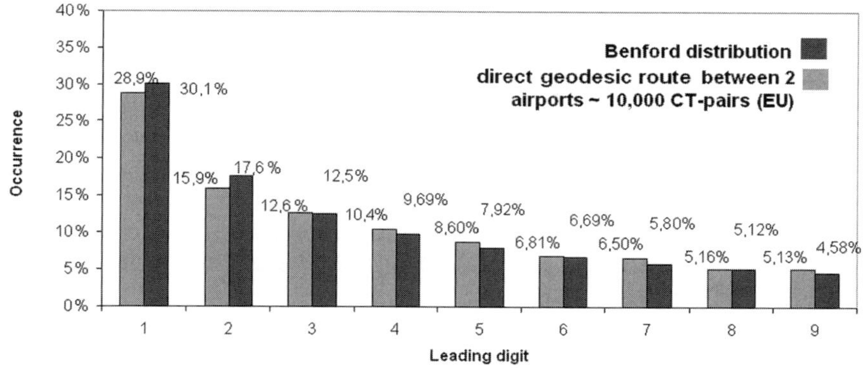

FIGURE 6.11. Distribution of the leading digit of the direct geodesic route (orthodrome) between two airports for circa 10,000 city-pairs in Europe, compared with the Benford distribution

6.4.3. *Spatial Patterns of Airport Flows*

Airport systems house many different kinds of flows (aircraft, passenger, luggage, fuel, etc.) that could be studied and optimized using constructal theory, but the following paragraphs will focus mainly on the passenger flow point of view. Since the 1960s extremes in airport design, well illustrated by the futuristic and artistic design of the TWA terminal at NY airport (1956–1962) by Eero

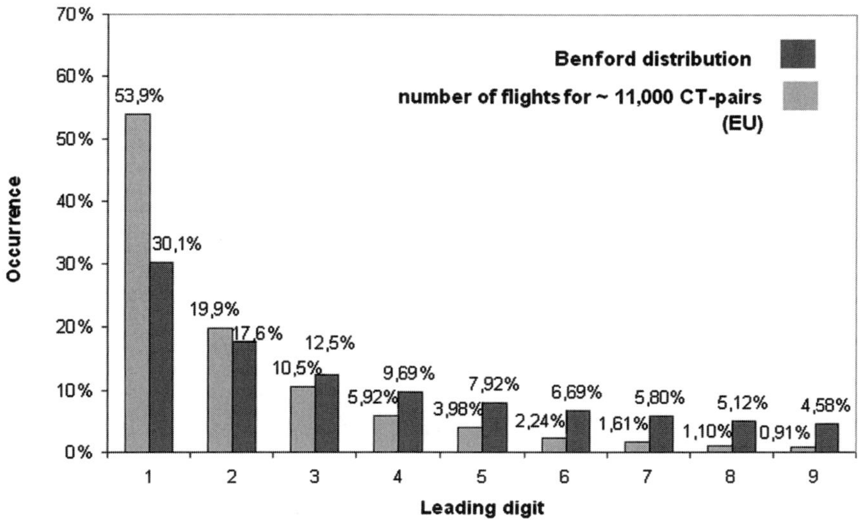

FIGURE 6.12. Distribution of the leading digit of the number of flights per city-pair for circa 10,000 city-pairs in Europe, compared with the Benford distribution

Saarinen and looking like a bird lifting off, this domain came down to Earth due to the operational and financial constraints (Gallagher 2002).

After the deregulation of the US aviation market occurred, at the end of the 1970s, the highly functional airport design well known as "midfield" building started to appear. This architecture is directly linked to the synchronous reorganization of the airlines companies' strategy based on hub system allowed by the market deregulation. This airport building architecture is also frequently criticized, especially by some architects who found that it leaves too few possibilities from their artistic designer point of view (Gallagher 2002). The following paragraphs will especially take a look on this question from the constructal theory point of view.

Consider the rectangular territory of length L and height H on the top of Fig. 1.11 where the objective is to have access to point M. If every inhabitant Q has two modes of transportation, walking with speed V_0 and riding on a faster vehicle with speed V_1, then the average travel time between all the points of the area and M is minimum when the area shape is $H/L = 2V_0/V_1$. This optimally shaped rectangle is a bundle of an infinite number of optimally refracted paths of type QRM. Constructal theory has shown that tree-shaped flows such as river basins, lungs, and vascularized cooling of electronics can be deduced by starting with the optimally shaped elemental area, e.g., the wet river bank between the smallest rivulets, or the alveolus of the lung.

Why is the constructal rectangle of Fig. 1.11, top, relevant to the discovery that the shape of the airport can be optimized? Because optimally shaped rectangular elements are everywhere in city living, even though the builders of these elemental structures did not rely on the constructal law to design them. They balanced two dissimilar efforts (V_0 vs V_1) in order to minimize the global effort.

When the shape H/L is optimal, the travel time from P to N is the same as the travel time from N to M, namely $H/2V_0 = L/V_1$. In constructal theory this is known as the equipartition of time (resistance), or the optimal distribution of imperfection (Bejan 1997, 2000a). This is illustrated by modern edifices such as the Atlanta airport drawn in Fig. 1.11, bottom.

Several objectives were pursued in the development of this tree-shaped airport flow structure: the minimization of travel time for pedestrians and the minimization of time and transportation cost for the goods flowing between the terminal and each gate, etc. (De Neufville 1995).

The black line in Fig. 1.11, bottom, is the high-conductivity stem serviced by a two-way train. The perpendicular bars are the five concourses of Atlanta airport, along which travel is much slower (walking, carts). Because the train speed (V_1) is higher than walking (V_0), the shape of the rectangular territory $H \times L$ can be selected such that the occupants of the area have maximum access to one point (M). When the shape is optimal, the time of walking is the same as the time of riding on the train. Because of this, the rectangular shape of the Atlanta airport is optimal. In agreement with constructal theory, the time to walk on a concourse is the same (~5 minutes) as the time to ride on the train (Bejan and Lorente 2001)

But a "hybrid" midfield airport like the Atlanta airport does not optimize only the transfer time for the passengers, in fact the whole design was conceived to ease at the same time the passenger flow, the access of aircraft to the embarking gates from the runways (because the airport stands in the middle of the taxiway and runway systems), and the maneuvers of aircraft to accost and depart from each gate, facilitated by the impossibility for any aircraft to be blocked by any other maneuvering aircraft due to no dead-end apron as can be found in other airport designs (like "X" shaped terminal). Again and again, the same causes produce the same effects: the midfield airport is the consequence of the collision of clashing objectives—the simultaneous minimization of both passenger and aircraft travel (De Neufville 1995). There is more: "hybrid" midfield airports like Atlanta perform better than "pure" midfield designs such as Denver because "they give a higher level of service to different passenger types. Their more balanced load distribution reduces overall walk distances and travel times" (De Neufville 1996).

From the passenger's point of view, the layout of the Atlanta airport could be refined such that any peripheral point (here, the extremities of the concourses) could be reached at a constant time. A simple way to achieve this goal is to increase the complexity by adding one more degree of freedom, e.g., allowing a continuous diminution of the concourse length as displayed in Fig. 6.13. See Section 11.5 and chapter 12 of (Bejan 2000a) for a more complete discussion of these optimization aspects and the resulting transport cost reduction. But this solution does not take into account the two other constraints of the maximization of access for both arriving and departing aircrafts. The competition between passengers and aircraft objectives, with the constraint of the specific runway configuration chosen—i.e., the aircraft source and sink points—for the Atlanta airport, certainly finally resulted in the optimized rectangular geometry. Runway

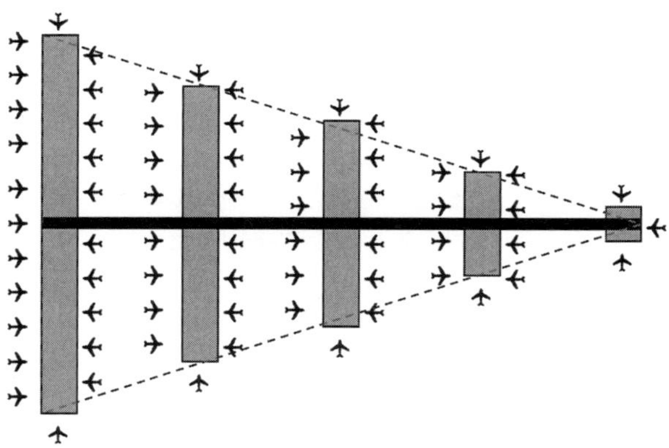

FIGURE 6.13. From the passenger's point of view, adding a new degree of freedom to the external shape of the hybrid midfield Atlanta airport, in this case a continuous variation of the concourse length, allows to reach the extremities of each concourse at a constant time

orientation is in fact a critical parameter, depending mainly on the dominant wind conditions and topographic configuration. The runway's layout constrains the other facilities implantation, and among them especially the airport buildings.

A future contribution of constructal theory could concern the optimal dimensioning of (midfield) airports, propositions of alternative layouts based on this architecture, or even sustainable (i.e., optimized) design of airport system—taking into account new thermoeconomical metrics. For instance, the mile-long concourse of the Northwest Airlines' midfield terminal at Detroit Metro, which "embodies the current state of airport design in ways both good and less good" (Gallagher 2002), is questionable on the basis of the raw analysis developed in the previous paragraphs. A more in-depth study is of course required to develop more precisely the arguments, but it could be thought that the great length of the concourse could be advantageously replaced by a more scalable architecture based on several similar concourses of smaller length, like in Atlanta's midfield design.

The objective here is not to search for the "designer" of this optimized rectangular shape found in midfield airports: large numbers worked on it, and they used time, freedom, and memory (culture) to construct it. The point is that all these observations account for the fact that complex flow structures of airport systems have an optimal shape, and this optimal design can be anticipated by using the constructal theory.

The usefulness of constructal theory is also to be investigated in the domain of directional information presentation to airport passengers. The good presentation of this kind of information, such as the one visible in Fig. 6.14, is of great importance in order to avoid the creation of passenger concentration and the resulting traffic jam. Passengers should be not only canalized but directed. A more speculative application could concern the optimization of the complexity of the information display by itself, but the level of complexity concerned certainly does not reach the same degree of complexity as the one reached in ideographic languages (such as Chinese).

FIGURE 6.14. Information panel at Brussels National airport. The good placement and presentation of directional information is a key point to prevent passenger's traffic-jam and avoiding passenger's hot-spot concentrations inside an airport infrastructure

6.4.4. *Temporal Patterns of Airport Flows*

As stated above, the puzzling allometric laws of proportionality between the heart beating frequency or the breathing rhythm and the animal body mass was rationalized thanks to constructal theory (Bejan 2000a). Optimized pulsating flows are common in both natural and engineered systems, and these temporal structures are all supporting the constructal principle of flow access maximization.

The same kind of temporal organization can be found in temporal patterns of flight schedules of airlines operating to a specific airport. It is known as wave-system. A wave-system structure consists of three parameters: (1) the specific number of waves, (2) the timing of the waves, and (3) the structure of the individual waves. The principle of this scheduling is to connect any incoming flight to an outgoing flight in order to maximize the possibilities of connection and minimize the waiting time of passengers, and occur specifically in the hub-and-spoke network of airports, due to the constraints on the flow generated by this spatial organization of the network (spatial concentration).

This kind of temporal organization exists in among all European airlines and had the opportunity and the freedom to develop, after the deregulation of the European air transport market in 1988. In the USA, this temporal concentration of flight was initiated 10 years before, in 1978, when the same deregulation occurred in the American market. Why does the flight scheduling of airlines evolved toward such a wave-system? In this configuration, this money-driven transition in scheduling occurred because "airline hubs with wave-system structures perform generally better than airline hubs without a wave-system structure in terms of indirect connectivity given a certain number of direct connections" (Guimerà et al. 2005).

The wave-system of schedule is in fact truly acting as a "pump" of flights. We can draw an analogy of this with the blood pump, the heart, and the associated tree-shaped network of incoming flows (veins) and outgoing flows (arteries). Synchronization between distinct "pumps" can also occur in the airport network system, in order to optimize the connections and synergies, for instance between the three waves of Lufthansa at Frankfurt airport and Munich airport (Gastner and Newman 2004).

The evolution of this system is clearly visible in Fig. 6.4 of Burghouwt and De Wit (2003, p. 15): The flight schedule of Air France (AFR) at Paris CDG airport in 1990 morphed from an unstructured flow to a clear wave-system in 1999. The two pictures in Fig. 6.4 of Burghouwt and De Wit (2003) represent the artificially reconstructed "waves" whereas Fig. 6.15 shows us the real temporal schedule of the flights (in 2005). This last picture unveils the hidden mechanism of the wave-system, illustrating the temporal dephasing of the arriving and departing flights.

From the constructal theory point of view, there is an optimal way of balancing the arrival and departure flow. This class of problem is also known as Monge-Kantorovitch problem (MKP) of optimal mass transport between two "densities" (Benamou et al. 2000), in our case in space-time: the arrival flights' (i.e., passengers) density and the departure flights' density.

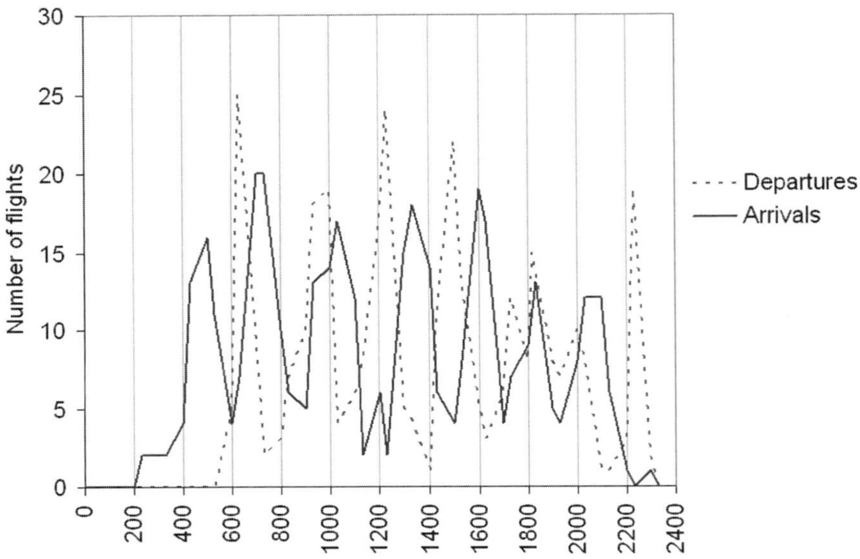

FIGURE 6.15. Wave-system temporal pattern of the Air France flight schedule of Paris CDG airport. This figure displays the number of flights' arrival and departure cumulated for every 15 seconds. It unveils the temporal dephasing of the arriving and departing flights in order to maximize the connections

Due to the resistivity of the system (landing, moving on runways, debarking, going to the connection gate, embarking, etc.), we can logically conclude that if the two wave peaks are synchronous, the first departing passengers cannot correspond to the first arrival flights, and the last arrival passengers will need to wait for the next wave to find the correct connection. The mapping of M(x) in Fig. 6.16 is a solution to the problem of connecting every passenger of an incoming flight of the first wave to a flight belonging to the successive departing wave.

The two "densities" in Fig. 6.15 seem furthermore to be effectively different for this specific day: departure peaks being more dense (more high, less wide) than arrival, maybe because departure time was globally a parameter more under control this day for this airport than the arrival time (uncertainty due to weather, airspace congestion, etc.), so departures should have been scheduled a little bit more precisely—but a consideration based on one day of traffic, in this highly context-sensitive domain, cannot allow us to confirm or extend such trend at a greater time-scale at this point.

Passenger's hub connection optimization can then be viewed as a problem of optimization of access in a geometrical space, and this geometrical space is also economical product space (Carone et al. 2003), the products proposed by carriers being flight schedules to several destinations from given airports. Dephasing peaks allow them to offer a greater product surface to their clients,

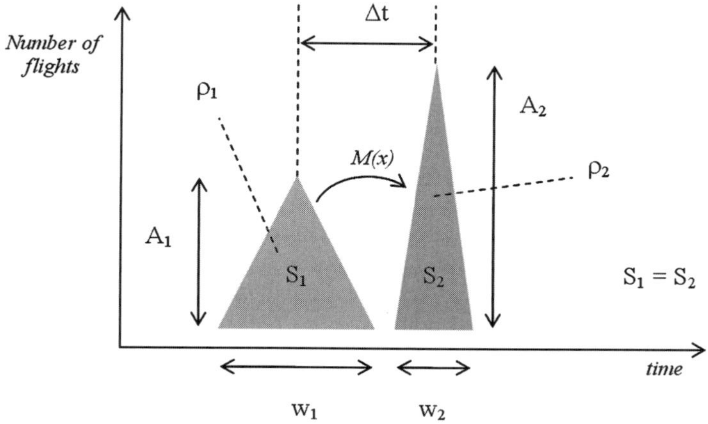

FIGURE 6.16. The wave system of flight's schedule can be viewed as a MKP problem of optimal passenger transport mass, i.e., mapping M(x) between the arrival (W1, blue) and departure (W2, pink) flow densities

without adding new flights, but just by shifting the departure and arrival waves' peaks by at least the minimal connection time for a passenger. In addition to the dephasing time, the geometrical characteristics of the waves (amplitude, wavelength) could be themselves the subject of an optimization, depending on the dimensional constraints of a specific airport.

There is much to interrelated process: the on-land optimization, i.e., the building infrastructure (hybrid midfield airport like Atlanta) facilitating passenger and aircraft movements, and the temporal optimization (optimal dephasing), in order that every connecting passenger coming with an arrival wave can leave with a flight of a departing wave.

6.4.5. *Aircraft Fleets*

The trend during evolution toward larger body size for biological system was first formalized by Cope in 1887 for several lineages (Chatterjee and Templin 2004). The reasons and mechanisms of this evolutionary tendency are always subject to speculations. Of course, we propose here to consider the constructal law has a prevalent cause in this biological trend. Bigger thermodynamic systems perform in fact better for many reasons, such as heat retention, longer duration of life, greater chance of reproduction, and increased predation efficiency (Chatterjee and Templin 2004). This effect is obviously itself also counter-balanced by other aspects, e.g., the growing need in terms of food (resources) of the system. For flying animals, flight in V-formation is also an answer to the problem of creation of a bigger system, performing better in its whole than each of its part taken independently.

In the economical niche of air traffic market, two species (two market plans) are defended by their own supporters: fleet of small and large aircraft (mega-jumbo). The Darwinian economical selection will decide which of them is the fittest for the future aviation market (Hansson et al. 2003; Smith 2002). A true and better comparison between these two options should consider not only their economical aspects but should integrate all relevant extra-financial aspects such as environmental and social impacts.

In brief, larger aircrafts are considered as more adapted to a hub-and-spoke network system, whereas smaller aircrafts are more fitted to a point-to-point network strategy. But we could suspect also that these two different kinds of network and their exploitation would also have a different environmental impact, e.g., regarding the contrails' impact, due to the spatial concentration of hub-and-spoke network opposed to the very more widespread geometry of the point-to-point network. Anyway, the emergence of fleet of mega-jumbo jets such as the Airbus A380 is a clear and current tendency in aeronautics.

The A380 is in fact for aircraft the equivalent of the *Quetzalcoatlus northropi* for pterosaurs (Fig. 6.17). In the case of pterosaurs, despite the inherent fuzziness of the data, a global trend toward bigger systems, i.e., bigger pterosaur, emerges clearly, through evolution. The last and biggest species of pterosaur, *Quetzalcoatlus northropi* (*Qu*), 70 kg (estimation), is separated by 160 million years from the oldest and smallest, *Eudimorphodon ranzii* (*Eu*), 0.015 kg (estimation), and is effectively 4,700 times heavier. Following the law of optimal flight speed, *Qu* should have been also four times ($\sim 20\,\text{ms}^{-1}$) faster than *Eu* ($\sim 5\,\text{ms}^{-1}$), where $4\,\text{ms}^{-1} \sim 4,700^{1/6}\,\text{ms}^{-1}$ (cf. Eq. 6.2). As specified above, bigger systems perform better, and the several allometric laws (Kleiber, etc.) justified on the basis of constructal theory provide several arguments for this evolutionary trend.

Bigger body means smaller global metabolic rate. And larger aircraft, like larger biological system, have many advantages: better economical ratio (e.g. cost per seat), etc. To understand this trend toward bigger system, we should consider that the theoretical framework of constructal theory offers now a new time arrow: the constructal law, or "how" everything flows: "configurations morph toward easier flowing architectures, toward animal designs that are more

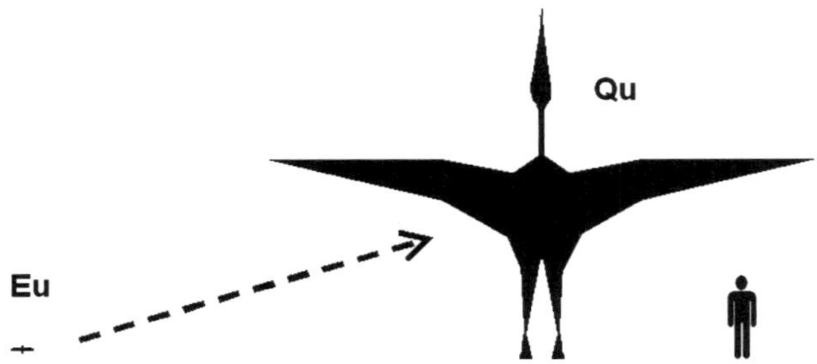

FIGURE 6.17. Global trend in body size evolution for pterosaurs between Upper Triassic and Upper Creaceous (160 million years): *Quetzalcoatlus northropi* (*Qu*) is 4,700 bigger than *Eudimorphodon ranzii* (*Eu*). Drawings of *Eu*, *Qu*, and human species—this last for comparison—are displayed at scale

fit, [. . .] and toward man + machine species that are more efficient" (Bejan 2005). The evolution toward larger living systems is so implied by this new time arrow. Constructal theory just "put Darwin into Physics" (Bejan 2005, Bejan and Marden 2006a,b). And what accounts here for pterosaurs also accounts for aircraft.

Which of the two business plan, smaller vs larger aircraft's fleet, will be the "winning" option? Maybe the two: like biological systems, each of them is more specifically adapted to certain part of the aviation market, i.e., of the passenger's flow system demand, and the answer may consist of the optimal balance between the two choices. But the point that we would like to address here is that the trend toward bigger thermodynamic systems can be identified both in biological and in engineered systems and, again, this tendency could be better understood through the theoretical framework offered by constructal theory.

Hub-structured networks, wave-systems of flight scheduled, midfield airport, and fleets of mega-jumbo jets should not be considered as independent systems but rather as interrelated components and trends of a same self-optimizing flow system (De Neufville 1995, Burghouwt and De Wit 2003). The aviation market deregulation offers carriers the freedom to reconfigure their flight schedule in a more optimal spatial and temporal structure, as briefly described in Fig. 6.18, and it can be viewed as a manifestation of the constructal law of maximization of flow access and the evolution of self-optimizing system evolving in a "Performance vs Freedom do morph" space.

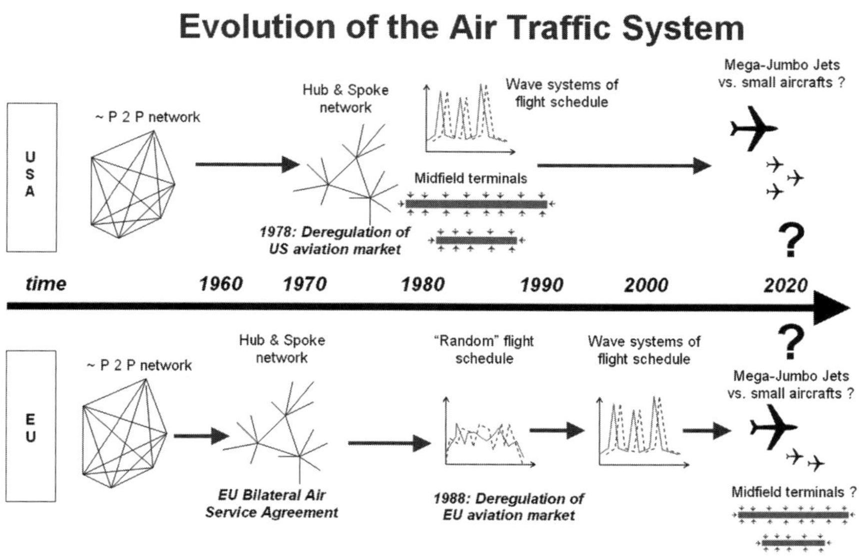

FIGURE 6.18. Evolution (sketch) of the Air Traffic System in USA and Europe during the second half of the 20th century until now

6.5. Conclusions

The purpose of this chapter was to introduce the constructal ideas, concepts, and philosophy, to provide some already existing results in domains relating to our purpose, the dynamic social system which is the Air Traffic System, and finally to look at the potential application of this recent theoretical framework to this ATS.

The previous paragraphs show us the relevance of constructal theory to study a wide range of disciplines, including domains like aeronautics and meteorology. As we could have expected—on the basis of the fact that the ATS is a complex flow system—a brief look at the current geometries of ATS sub-systems, such as airports network and air traffic flow and landside airport flows, shows us that not only seems the Constructal theory to be relevant to study these existing flow patterns or to design future flow systems but that these various flows already obey the Constructal law of maximum flow configuration and evolved toward geometries offering a facilitated access to the imposed streams that flows through them.

Constructal theory encounters in fact an always-greater interest and is extending its domain of application far beyond than its original domain: mechanical engineering. But, like any paradigm, constructal theory faces also some inherent resistivities to "flow" to the scientific community. The diffusion of new scientific theory is in fact by itself a social process. We could also remember the 19th century where a great development of the Gas Kinetic theory took place, especially through the impulsion of James Clerk Maxwell and Boltzmann, based on statistical description of physical phenomena. It is interesting to note that at this time Maxwell encouraged physicians to take example on the social sciences and to use their statistical description methods. But social physics, on the basis of analogies such as the comparison between, e.g., social revolutions and fluid transition, does not allow lot of progress in its own domain (Fuller 2005).

This workshop and book Constructal Theory of Social Dynamics offers a great opportunity for social science to renew its old alliance with physics, on the basis of a new theory of flow optimization, and offered new keys to understand the nature and development of social flows, and also contributed to the "diffusion" of the concepts of constructal theory into the "interstitial spaces" of the scientific community.

In the domain of Air Traffic system and Aeronautic, the application of constructal principle is expected to generate new concepts, algorithms, and generally applicable methods for understanding, modeling, and designing complex structures found in these domains. We envisage particularly to push further the studies of airport flow and air route network optimization, among many other kinds of ATS flows that could be envisaged.

The contribution of constructal theory in this domain is to know that each of the existing ATS flows has an optimal shape, and that this optimal shape is anticipated by the constructal law.

References

ACARE (2002) Strategic Research Agenda, Vol. 2, Chapter 5, The Challenge of Air Transport System Efficiency.

Airport Research Center (2004) CAST Presentation, Comprehensive Airport Simulation Tool, Simulation of passenger flows and processes within airport terminals.

ASHRAE (1987) *ASHRAE Handbook, HVAC Systems and Applications*, ASHRAE Press, Atlanta, GA, Chapters 7, 46 and 55.

ASHRAE (1990) *ASHRAE Handbook, HVAC Systems and Applications*, ASHRAE Press, Atlanta, GA, Chapters 25 and35.

Bejan, A. (1997) *Advanced Engineering Thermodynamics*, 2nd edn, Wiley, New York.

Bejan, A. (2000a) *Shape and Structure, from Engineering to Nature*, Cambridge University Press, Cambridge, UK.

Bejan, A. (2000b) Thermodynamic optimization of geometry in environmental control systems for aircraft, *Int. J. Heat Technol.* **18**, 3–10.

Bejan, A. (2005) The constructal law of organization in nature: tree-shaped flows and body size, *J. Exp. Biol.* **208**, 1677–1686.

Bejan, A. and Lorente, S. (2001) Thermodynamic optimization of flow geometry in mechanical and civil engineering, *J. Non-Equilib. Thermodyn.* **26**, 305–354.

Bejan, A. and Lorente, S. (2004) The constructal law and the thermodynamics of flow systems with configuration, *Int. J. Heat Mass Transfer* **47**, 3203–3214

Bejan, A. and Marden, J.H. (2006a) Unifying constructal theory for scale effects in running, swimming and flying, *J. Exp. Bio.* **209**, 238–248.

Bejan, A. and Marden, J.H. (2006b) Constructing animal locomotion from new thermodynamics theory, *Am. Scientist* **94**, July–August, 342–349.

Bejan, A., Lorente S., Miguel, A.F. and Reis, A.H. (2006) Constructal theory of distribution of city sizes. Section 13.4 in Bejan, A., *Adv. Eng. Thermodyn.*, 3rd edn, Wiley, Hoboken, New Jersey.

Benamou, J.-D., Brenier, Y. and Guittet, K. (2000) The Monge-Kantorovitch mass transfer and its computational fluid mechanics formulation, *Int. J. Numer. Meth. Fluids* **40**, 21–30.

Benderli, G. (2005) ATM Market Evolution due to Deregulation, SEE Note No. EEC/SEE/2005/008, EUROCONTROL Experimental Centre, France.

Benford, F. (1938) The law of anomalous numbers, *Proc. Amer. Philos. Soc.* 78, pp. 551–572.

Burghouwt, G. and De Wit, J. (2003) *The Temporal Configuration of European Airline Networks*, Agora Jules Dupuit, Publication AJD-74, Université de Montréal, July.

Carone, M.J., Williams, C.B., Allen J.K. and Mistree, F. (2003) An application of constructal theory in the multi-objective design of products platforms, ASME Paper DETC2003/DTM-48667, *Proceedings of DETC'03 Design Engineering Technical Conferences and Computer and Information in Engineering Conference*, Chicago, 2–6 September.

Chatterjee, S. and Templin, R.J. (2004) Posture, locomotion and paleoecology of pterosaurs, *Geological Society of America*, special paper 376.

Cooper, A. and Smith, P. (2005) *The Economic Catalytic Effects of Air Transport in Europe*, SEE Report N° EEC/SEE/2005/003, EUROCONTROL Experimental Centre, France.

De Neufville, R. (1995) Designing Airport Passenger Buildings for the 21st Century, *Transport Journal*, UK Institution of Civil Engineers, Vol. **111**, May, 83–96.

De Neufville, R. (1996) Optimal configuration of complexes of large airport passenger buildings and their transport systems, Proceedings, Airports of the Future, International Symposium Nov. 1995, Académie Nationale de l'Air et de l'Espace, Cépaduès Editions, Paris, pp. 342–352.

European Commission (2001) *European Aeronautics: A Vision for 2020, Meeting society's needs and Winning Global Leadership*, Report of the group of personalities, January, p. 13.

Fuller, S. (2005) Faces in the crowd, *NewScientist* **186**, No. 2502, 4 June, p. 21.

Gallagher, J. (2002) *Midfield Terminal Architecture: A design That's Down-to-Earth—Understated Style is Reminiscent of Other modern Airports*, Feb. 1, http://www.freep.com/money/business/airprt1_20020201.htm

Gastner, M.T. and Newman, M.E.J. (2004) *The Spatial Structure of Networks*, arXiv:cond-mat/0407680, Vol. 1, 26 July.

Guichard, L., Guibert, S., Hering, H., Dohy, D., Grau, J.Y., Nobel, J. and Belahcene, K. (2005) *Paradigm SHIFT: Operational Concept Document*, EEC Note No.01/05, EUROCONTROL Experimental Centre, France.

Guimerà, R. and Amaral, L.A.N. (2004) Modeling the world-wide airport network, *The European Physical Journal B* **38**, No. 2, March, pp. 381–385.

Guimerà, R., Mossa, S., Turtschi, A. and Amaral, L.A.N. (2003) *Structure and Efficiency of the World-Wide Airport Network*, arXiv:cond-mat/0312535, Vol. 1, 19 Dec.

Guimerà, R., Mossa, S., Turtschi, A. and Amaral, L.A.N. (2005) The worldwide air transportation network: Anomalous centrality, community structure, and cities' global roles, *Procedings of the National Academy of Sciences*, n° 22, Vol. 102, May 31, pp. 7794–7799. http://www.pnas.org/cgi/doi/10.1073/pnas.0407994102

Haggett, P. (1965) *Locational Analysis in Human Geography*, Edward Arnold, London.

Haggett, P. and Chorley, R.J. (1969) *Network Analysis in Geography*, St. Martin's, New York.

Hansson, T., Ringbeck, J. and Franke, M. (2003) Flight for survival: A new business model for the airline industry, *Strategy + Business*, issue 31.

Joint Planning & Development Office (2004) Next Generation Air Transport System, Integrated Plan. http://www.jpdo.aero

Lösch, A. (1954) *The Economics of Location*, Yale University Press, New Haven, CT, 1954.

Ordonez, J.C. and Bejan, A. (2003) Minimum power requirement for environmental control of aircraft, *Energy* **28**, 1183–1202.

Oudeyer, P.Y., Kaplan, F., Hafner, V.V. and Whyte, A. (2005) The Playground Experiment: Task-Independent Development of a Curious Robot, *Proceedings of the AAAI*, Spring Symposium on Developmental Robotics, San Francisco.

Penner, J., Lister, D., Griggs, D. and Dokken, D. (1999) *Aviation and the Global Atmosphere*, A Special Report of the Intergovernmental Panel on Climate Change (IPCC), Cambridge University Press, June.

Pennings, T.J. (2003) Do dogs know calculus?, *The College Mathematics Journal* **34**(3), May, 178–182.

Périn, S. and Bejan, A. (2004) Constructal ATS, First application of the Constructal Theory to Air Traffic System, Technical Note n° CATS-TN-012804-AM , Ars Magna Universalis, January 28.

Pietronero, L., Tosatti, E., Tosatti, V. and Vespignani, A. (1998) Explaining the uneven distribution of numbers in nature, arXiv:cond-mat/9808305, Vol. 2, 12 Nov.

Reis, A.H. (2004) Constructal view of global circulation and climate, Proc. Symposium *Bejan's Constructal Theory of Shape and Structure*, Evora Geophysic Center, University of Evora, Portugal, pp. 171–193.

Reis, A.H. and Bejan, A. (2006) Constructal theory of global circulation and climate, *Int. J. Heat Mass Transfer* **49**, 1857–1875.

Rivière, T. (2004) Redesign of the European route network for Sectorless, *Proc. 1st Int. Conf. Res. Air Transport.*, Zilina, Slovakia.

Scott, P.D. and Fasli, M. (2001) Benford's law: an empirical investigation and a novel explanation, CSM Technical Report 349.

SESAR Consortium (2006) The Current Situation, Air Transport Framework, Deliverable 1, SESAR Definition Phase, DLM-0602-001-03-00, July, p. 9.

Smith, C.J. (2002) The road to profit improvement: larger aircraft or smaller aircraft? An independent case study of two competing airlines, IATA, Madrid, September. http://www.sh-e.com/presentations/IATA-Madrid_092002.pdf

Chapter 7
Sociological Theory, Constructal Theory, and Globalization

Edward A. Tiryakian

7.1. Introduction

Constructal theory, as formulated by Bejan and his associates (2000, 2005), is a bold endeavor "to develop a predictive theory of social design" in the form presented as the *constructal law*: "for a flow system to persist in time (to survive) its configuration must change in time such that it provides easier access to its currents." The invitation of an engineer/physicist for dialogue with the social sciences, particularly my own discipline of sociology, is highly welcome, since such invitations and settings conducive for interaction are all too rare.[1] The flows and structures of complex social systems are very much sociological concerns, since social change has been a primary focus of sociological theory from its 19th century foundation to the present. A continuing feature of modernity is social change, and sociologists have utilized a variety of models and modes of explanation to cope with social change, even, to predict some of its outcomes. On the contemporary scene it has become increasingly recognized in the past 15–20 years that the problematic for *macrotheory* is to provide the conceptual framework for dealing with the emergent phenomenon of *globalization* as the unit of analysis, and with it a set of attending questions: the relation of globalization to nation-states (that have served as the units of analysis at the macro level) and the bearing of globalization on the "new international order" promised at the beginning of the 1990s with the implosion of the Soviet Empire and the triumph of neo-liberalism in the economic and political realm that followed the successful close of the first Gulf War. Tacitly, globalization as a master process seemed to entail the global distribution of democracy at the political level and open free markets at the economic level. The vast transformation of the rest of the world from the implicit model of modernity of the West (and, in particular, the United States) seems ready made for relating the mechanisms of this transformation to

[1] E.A. Wilson in *Consilience, The Unity of Knowledge* (1998) professed the desire for the social sciences to share with his sociobiology a common general theory of human knowledge; however, in that work he makes known that only economics at best has anything rigorous to offer to his synthesis.

the development of a general theory of society "as a conglomerate of mating flows that morph in time in order to flow more easily: people, goods, money, information, etc."

I will later give a more specific delineation of "globalization", but for the present I take globalization not as a state of the social world having permanent forms but rather as a set of flows which alters the social landscape and its environments, and in turn is altered by the changing landscape. The complex and dynamic aspects of globalization certainly make a rapprochement of sociology and physics/engineering feasible, with overlapping interests, both in seeking new fields of application for each, and in seeking the refinement of the respective theories. Thus, a basic point of this chapter is that Adrian Bejan's constructal theory, if it seeks to widen areas of applicability from the flows in the "natural" world to the global flows in the "social" world, has an important domain to explore in globalization. Conversely, given the amorphous aspects of globalization and its multiple themes rich in empirical descriptions but lacking theoretical rigor, constructal theory might provide heuristic models for more sophisticated analyses than additional conceptual differentiation and articulation.

I approach this encounter of sociology with engineering/physics not only with keen anticipation of the fruits of dialogue but also with a cautionary stance based on my reading of earlier encounters. A model from engineering regarding societal flows may easily offer a heuristic metaphor, perhaps even for selected areas, a partial causal explanation, but on methodological grounds, I am agnostic as to whether a general causal explanation of social change can be provided.

7.1.1. Physics and Engineering in Previous Sociology

Although interaction between sociology and physics/engineering has, until Adrian Bejan's overture, been quite limited in recent years (and this may reflect more my ignorance than actual conditions), a quick historical glimpse offers a better vista, starting right with the founder of sociology as a scientific enterprise, Auguste Comte.

Comte was trained as an engineer at the elite Polytechnic School, and the imprint of his early training shows in his aim to make the new discipline one that would not only generate empirical "positive" knowledge about society but also bring about social engineering for the reconstruction of a social order. The latter had, in the wake of the French Revolution and the skeptical "negativism" of the Enlightenment, become in what today's terminology might be called "chaotic"—in Comte's lifetime, the turbulence of the Revolution of 1830 and the two Revolutions of 1848 sandwiched between the despotic rules of Napoleon I and Napoleon III. The history of sociology acknowledges readily Comte's adaptation of his early training in the formulation of his "law of three stages" and his taking the social engineer as in charge of the reconstruction of society. But we seldom get to see how this might reflect the tremendous achievements of physics and engineering in the first half of the 19th century: the reconstruction of the infrastructure of modern society (canals, bridges, railroads, highways) and the formulations of modern physics and mathematics in parsimonious principles and laws. I would argue that engineers, even at an elite French school, are trained to solve empirical problems and Comte did view sociology as a problem-solving science, the problem of social order, as not only being of theoretical interest but eminently as being of an urgent, practical kind.

Comte initially (in the 1820s) labeled his new science "social physics" (to distance himself from his former employer Saint-Simon's invention of a science studying society as a complex phenomenon of social change, or "social physiology"). But this involved him in a controversy regarding priority of the name with a Belgian, Adolphe Quételet, who published in 1835 *Social Physics* (revised in two volumes in 1869 and reissued in 1997). For Quételet, social physics, based on mathematics and probability, was a new field of inquiry in which to apply statistical tools for understanding "moral and intellectual man." Quételet in the 1869 edition provides an important historical overview of the evolution of statistics and probability theory, and their application to social phenomena, initially to insurance tables and in the early 19th century with Poisson, Fourier, and Gauss, among others, to applying statistics to governmental affairs. The revised edition he prepared was presented in 1869 at the joint meeting of the British mathematical statistical association and the International Association for Practical Statistics.

The other 19th person who was trained in engineering before making his mark in the social sciences is Vilfredo Pareto, recognized to this day by economists for his theory of optimization and mathematical application to equilibrium theory. Sociologists in the 1930s, especially at Harvard which had an interdisciplinary faculty seminar on Pareto, made much use of his later sociological theorizing regarding the nature of social systems.[2] For a variety of reasons that need not detain us, he has dropped out of the sociological canon (though still highly esteemed in economics with his theory of optimality) but given his rejection of linear theories of progress, he might be ripe for a comeback in the postmodern era.

Physics has had in the past—especially the distant past—an interesting interplay with sociology and the social sciences. Still useful as a reference work, P.A. Sorokin's *Contemporary Sociological Theories* (1928) devotes an entire chapter to the "Mechanistic School", which includes an extensive treatment of Pareto. Sorokin argues cogently that the 17th century saw "an extraordinary effort to interpret social phenomena in the same way that mechanics had so successfully interpreted physical phenomena" (1928: 5). Shedding the anthropomorphism, moralism, and other metaphysical apparatus, Hobbes, Leibnitz, Descartes, Spinoza, and others set themselves to studying social phenomena as rationally and objectively as they and their peer set to study physical phenomena.

[2] It is not by chance that the core doctrinal theoretical work of post-World War II sociology bore the title *The Social System* (1951). Its author, Talcott Parsons, had been a young faculty member of that famed Pareto seminar conducted by L.J. Henderson, and Parsons subsequently readily acknowledged the importance of Pareto. In turn, his own general analytical approach to social systems greatly stimulated empirical inquiry.

Human activity could be taken as instances of mutual gravitation or repulsion, similar to the regularity of physical movement and which could be interpreted by the principles of mechanics (1928: 7).

One need not search wide to see the tremendous impression made on contemporaries of Newton and Leibnitz by their discoveries of the calculus and the principles of astronomy. Whence, as Sorokin notes,

> The social physicists of the seventeenth century... constructed the conception *of a moral or social space* in which social, and moral, and political movements go on ... Physical mechanics explains the motions, also, of physical objects by the principles of inertia and gravitation. Similarly, social mechanics regarded the social processes as a result of the gravitation and inertia of human beings or groups (1928: 8f).

Sorokin traces the "mechanistic school" and its various attempts in the 18th, 19th, and early 20th centuries to "transfer the conceptions and terminology of mechanics into the field of social phenomena" (1928: 17). He is hard on all of them as "pseudo-scientific", disregarding specific characteristics of social phenomena, misusing analogies and unable to make the predictions they promise. What the school has contributed, he grants, is to have influenced the social sciences in the use of quantitative and causal studies of social phenomena.

Sorokin might have included a later instance of a "mechanistic school", one published by his Harvard colleague 20 years later. George Zipf in 1949 proposed a very general "natural" law he termed *The Principle of Least Effort* (PLE), based on a very large number of empirical data, as governing human behavior. PLE, he claimed, is "the primary principle that governs our entire individual and collective behavior of all sorts, including the behavior of our language and preconceptions" (1949: viii). What is that principle?

> Each individual will adopt a course of action that will involve the expenditure of *the probably least average of his work* (by definition, *least effort*) (1949: 543, emphasis author's).

Zipf's conclusion is very relevant to my antecedent discussion:

> A systematic social science will make possible an objective social engineering ... PLE will provide an objective language in terms of which persons can discuss social problems impersonally, even as physics is a language for the discussion of physical problems (ibid.).

For all the impressive data that Zipf gathered, I am not aware that the Principle of Least Effort is recognized today. It may still have some applications in social engineering[3] but one might think of an equal amount of expenditure of income

[3] Indeed, it might even be shown to be a special case of constructal theory regarding optimal flows destroying the least energy. This is clearly recognized by Bejan: "all the domains in which Zipfian distributions are observed (information, news) are homes to tree-shaped flow systems (point-area, point-volume)...In this way information and news are brought under the great tent of constructal theory: geometry, geography, tree architectures, freedom to morph and optimized finite complexity (hierarchy) are constructal properties of the flow of information" (Bejan et al. 2006, 13.4, p. 6).

and energy at the individual and collective levels which seem the opposite of PLE but reflect individual and collective value preferences (e.g., commuting to the exurbs, school busing, etc.).

Probably the greatest impact of models/theories from physics and engineering upon sociology (and other social sciences) in the past half century came from the confluence in the 1950s and 1960s of new perspectives theorizing *systems* and their *control* or regulation. Two key figures working two metro stops apart, one at MIT and the other at Harvard, developed and stimulated key works.

At MIT Norbert Wiener, with a strong background in physics, mathematics, and engineering, attracted wide attention in an early period of automated data to the place of information in organization and communication. Although his work was interdisciplinary and collaborative,[4] he received widespread recognition in 1948 for invoking the term "cybernetics" [coined much earlier in the 19th century by Ampère, the founder of electrodynamics, to denote the science of *regulating* or *governing* human affairs (Cybernetics 2006)]. To reach a wider audience, Wiener followed this a few years later with *The Human Use of Human Beings* (1967) highlighting that "society can only be understood through a study of the messages and communication facilities which belong to it...the theory of control in engineering, whether human or animal or mechanical, is a chapter in the theory of message. It is the purpose of Cybernetics to develop a language and technique that will enable us to attack the problem of control and communication in general" (1967: 25). Wiener and his associates unofficially heralded the "information age," which a generation later at the century's end received a comprehensive sociological treatment by Castells (1996–1998). Wiener's perspective, emphasizing the crucial concept of *feedback* from engineering, is tacitly more optimistic than Castells, as Rosenblith discusses cybernetics in the "Afterword" of *The Human Use of Human Beings*. Feedback may occur at a higher level of organization "when information of a whole policy of conduct or pattern of behavior is fed back, enabling the organism to change its strategic planning of future action" (in Wiener 1967: 276).[5] Feedback (especially negative feedback) is thus instrumental to realizing the ends sought (or the purpose) in the process of action.

At Harvard, Talcott Parsons had articulated a mode of conceptualizing at a high level of abstraction a general approach to any and all social systems, including

[4] For an important reference work tracing the series of interdisciplinary conferences between 1946 and 1953 that led to cybernetics stemming from using models of electronic circuitry in the wartime experience of complex automatic fire-control systems developed by engineers and mathematicians, see Heims (1991).

[5] To illustrate cybernetics from our own society, the Federal Reserve Bank, with the primary mandate of preserving the soundness of the currency, is continuously adjusting interest rates based on economic data it receives, some of which are outputs of its earlier decisions which are fed back as data from which new changes in interest rates will be decided. The Fed seeks to avoid positive feedback which might throw the economy into a runaway inflation, yet not apply monetary brakes so hard as to cause a depression. The setting of interest rates by the Fed may thus be seen as a prime "regulator" of the economy.

societies as a special instance due to their self-sufficiency and complexity. A key feature of his analysis was to view social systems as differentiated in terms of some key functions necessary for their viability in their environment (Parsons had made earlier use of the notion of homeostasis taken from physiology). After several stages of theorizing (Parsons 1970), Parsons proposed that there seemed to be four paramount differentiated functional areas of social (and, in particular, societal) systems, involving the adaptation, goal attainment, integration, and latent pattern-maintenance, with each of these four treated as a sub-system and collectively designated as the AGIL paradigm. These are not sub-systems operating in isolation from one another; they are, Parsons affirmed, interdependent and interrelated via symbolic media of exchange (Treviño 2001).

Social systems are analytical components of what Parsons developed as a theory of social action addressed to the Hobbesian problem of social order. To explicate Parsons's elaboration of the social system and its relation to other analytical components of action—the physiological, cultural, and psychological or personality sub-systems[6]—would take us too far afield. However, it is relevant to note that his theorizing of the social system, structurally and dynamically, was a continuous enterprise and that he was always ready to incorporate into the basic structure of his analysis new models from biology, physics, and engineering. Among these, he found cybernetics providing an important increment.[7]

In particular, the element of *control* seemed crucial in relating information theory to analyzing the evolutionary integration of social systems and their environments. Parsons readily acknowledged this. In pursuing goal-directed behavior, control of environmental conditions is made possible hierarchically with systems high in *information* essentially regulating development with *energy* provided by lower order systems (Parsons 1969: 33; Treviño xlix). In framing this to an evolutionary view of human society, Parsons regarded the cultural system of value-orientations as a sort of pinnacle in social organization, resting on the internalization of values to provide social control and minimize the use of force and coercion in the functioning of modern society. Cultural systems that carry cultural codes (e.g., higher education) are predominantly concerned with processing information, while physiological systems at the lower end of the cybernetic hierarchy are more concerned with energy processing.[8] It might

---

[6] Parsons also indicated an additional "environment" for human action, that of "ultimate reality" or the transcendental, which he was eventually to term the "telic."

[7] As of this writing, I do not know whether Parsons had personal acquaintance with Wiener, however plausible this is. In *The Social System* (1951) there is no index entry for "cybernetics" and his intellectual biographical account (1970) does not indicate when or how he became acquainted with cybernetics or information theory, but a footnote (Parsons 1969: 10) shows his familiarity with Wiener's basic texts.

[8] Some may recognize in Parsons's abstraction of cybernetic hierarchy a model that Plato had graphically discussed in the metaphor of the bodily appetites governed by reason as the rider.

be suggested that the social adaptation of cybernetics led to Parsons and Wiener having a convergent optimistic view of the benign regulation of social systems. Implicitly, self-regulation at the macro level is a model best applicable to democratic, liberal society which allows and makes use of feedback.[9]

Although Parsons did not single-handedly generate a "systems" approach to sociological theorizing, his towering stature in American sociology for nearly a quarter of a century did much to promote and refine sociological use of "system analysis," including invoking cybernetics. Further development and refinement continued well beyond, e.g., in the writings of Etzioni's theory of the "active society" in which organizations are elements of a larger system of social control and mobilization (Rojas 2006: 55), Buckley on feedback loops and negative entropy in action theory (1967; 1968), Luhmann's "autopoietic" differentiation theory (1982), and Bailey's even more ambitious "new systems theory" (1994).[10]

Undeniably, the broad range of writings we have just touched on regarding cybernetics and its derivatives is an impressive array of contacts that sociology has had with physics and engineering. Still, the interaction does not seem to have been sustained in the past two decades, at least not in having provided empirical research for sociology nor in generating a new interdisciplinary corpus, say like biochemistry. Such also seems to be the judgment of Etzkowitz in his insightful review of Heims:

> After a brief rise into the intellectual firmament, [cybernetics] deflated back into electrical engineering... It never really emerged as an independent discipline in the United States, although it did to some extent in Europe ... [It] was a precursor of computer science... The goal of the founders of cybernetics, to develop a synthesis of elements of the natural and social sciences to answer questions that cross-cut human and physical nature, is yet to be achieved (Etzkowitz 1993: 495).

This, of course, is not to say that the "natural history" of cybernetics came to an end because the interaction between engineering and the social sciences was not sustained; cybernetics undoubtedly has evolved with computer science and work in automation and Artificial Intelligence. But I am not aware that mainstream sociology *today* in its research program or conceptual frameworks has integrated cybernetics, and I defer to the engineers and computer scientists whether the social sciences have had an important "feedback" function in advancing latter day cybernetics.

[9] To be sure, there is no guarantee that the hierarchy of controls will not be subject to severe and at times disruptive strain, as was manifestly the case in the second part of the 1960s when the cultural normative system in the United States, and in European countries such as France, Germany, and Italy, was severely challenged by youth movements "high in energy."

[10] I am indebted to Ken Land for bringing Bailey's work to my attention. Although I am skeptical as to the heuristic or theoretical import of Bailey's synthesis of mainstream sociology and systems theory, it is an excellent primer of major contributors to the latter, including Bertalanffy, Ashby, Prigogine, and James Miller.

My point in this section is that one must be very cautious in adopting "laws" and "principles" from one domain of inquiry to another, especially if there may be a claim of complex social systems and social phenomena being explained causally. Yet, I do not intend to advocate the opposite, i.e., that no set of social phenomena can be accounted for by non-social phenomena. It is entirely possible, even desirable, that principles or models from the natural sciences (including physics and engineering) may be heuristic in the codification of empirical social science research. In the past this has been much more the case for biological and genetic models having stimulated sociological theory than physics and engineering, but that may reflect the rarity of social scientists, including sociologists, having proximate contacts with physicists and engineers. And conversely, it is always possible that the feedback of sociologists may offer the engineer and physicist empirical and theoretical materials to sharpen, modify, and test their general theory. What is a basic prerequisite, in my judgment, is some sustained interaction over a series of meetings or conferences, rather than an episodic encounter.

With these considerations, I now turn to the applicability of constructal theory, namely the very complex set of phenomena designated as "globalization." It will be necessary to have a preliminary consideration of "global" as both an intersubjective reality (qua an apprehension of the world as a totality) and a setting in which and toward which human activities take place with increasing frequency and increasing volume. We can then expand this to a more direct discussion of globalization and its possible nexus with constructal theory.

7.2. Theorizing the Global

Various significant events, material and non-material, have generated in the past quarter of a century or so a general *global consciousness* supplementing, but not replacing, the more differentiated consciousness of the *Umwelt* of our experience and its regions. (The latter has been for the most part a *national consciousness* of the world structured by the nation-state and its agencies, such as the mass media and educational institutions). These events include *at the technological level*, continuous global communication (multidirectional flows of information) by the Internet and the satellite system enhanced by photos of the whole earth from outer space; at one *political level*, the implosion (1988–1991) of the Soviet empire which had been like a huge medieval fortress keeping a moat that prevented the development of modernity from flowing across East and West; at *the economic level*, continuous capital flows and trading, along with the rise of East Asia first, now followed by South and Southeast Asia, as economic giants capable in global manufacturing, trade, and financial markets of competing with established advanced economies and accumulating surplus capital to such an extent as to have the American dollar depend on their purchase of US Treasuries.

Finally, one other politico-cultural event has in this decade added an unexpected dimension of globalization: the attacks of September 11, 2001, on American soil that prompted the present American administration to declare, in effect, a global war against terrorism, where the theater of the war has no specific territorial boundaries and de facto where nothing is "off limits." From a retaliatory strike against Afghanistan in South Asia, the conflict has found a central node in the Middle East, with Palestine, Lebanon, and, likely, Iran becoming drawn in and part of the violent turmoil, which has an "outreach" in terrorist attacks not only on European targets such as Madrid and London, but also in exotic vacation places such as Bali. The obverse side of the war on terrorism is the American exportation by means of "hard power" if not "soft power" of "democracy" to unspecified target areas—theoretically most of the world's authoritarian if not totalitarian regimes, but in actual practice greatly limited by an unstated *Realpolitik*. All in all, in the present decade, the bipolar world of the postwar era in which Russia and the United States were, respectively, the *primus inter pares* of their respective alliances has morphed into a single strategic *globalscape* as far as American security interests are defined. In effect, this is the latest and most far-reaching evolution of the Monroe Doctrine, justifying an around-the-world system of surveillance and preventive actions.

7.2.1. Globalization

Since the early 1990s, and with accelerating speed, "globalization" has become a theme of great interdisciplinary interest, for the social sciences and the humanities. What it refers to is rather amorphous, though there is a general recognition it refers to some qualitative and quantitative changes having the whole world as arena, changes which have impact on political, economic, ecological, and cultural aspects of social organization at the local (some have talked about "glocalization" as the production of new forms of adaptation stemming from the interaction of the global and the local level[11]), regional, national, and transnational levels.

If nothing else, globalization denotes and refers to vast *flows* that cover very large areas, and which take place in a much more compressed fashion than previously (taking into account that previous periods, such as the 19th century, also had waves of globalization cut short by unanticipated breakdowns, such as World War I). The flows are flows of capital, goods and their production, information (from new technologies not known 20 years ago such as the fax and the Internet), and, just as much, people (both in the form of voluntary socioeconomic migrations and in the form of equally massive involuntary displacements such as refugees).

---

[11] Roland Robertson, one of the early formulators of globalization analysis in cultural terms, has also made extensive reference to "glocalization," most recently in an analysis of migrants and sports identities (Giulianotti and Robertson, 2006). Urry provides a broad field of applications for glocalization in developing his approach to complexity theory of non-linear systems viewed as "unstable, dissipative structures" (Urry 2003).

Moreover, the flows are multidirectional, rendering the emergent global system exceedingly complex and turbulent (Rosenau 1990; Papastergiadis 2000), subject to discontinuities—hence perhaps a factor in the appeal of "chaos" and perhaps even more "complexity" theory. The recent theorizing of John Urry (2003) merits attention at this point.

In proposing that "complexity" is a more accurate approach to global reality than a causal model of invariant relationships, Urry emphasizes its *emergent* properties at the global level: "For complexity, emergent properties are irreducible, interdependent, mobile, and non-linear" (2003: 77f). There is no single global power but an interdependence of global organizations and institutions and civil society which seek to formulate "rules of the game" for the emergent global order; the latter is evoked in multiple signifiers of what Benedict Anderson proposed as an emergent "imagined community"—a global community instead of a national community, one having such "signifiers" as the Olympic Flag, the blue-green Earth, Nelson Mandela, Earth Day, Mother Teresa, etc. (Urry: 81).

We need, however, to go beyond a simple designation of "complexity." In culling various perspectives regarding an emergent global social system, we present two schemas (Figs. 7.1 and 7.2) derived from Parsons's approach in the well-known AGIL paradigm. Figure 7.1 (globalization I) depicts a set of interchanges or flows that enhance the adaptive capacity of the emergent global system. Overall, the economic, political, and cultural framework of globalization is that this is a master process of development which will increase the happiness and welfare of the greatest number. This is the basic *telos* of modernity, with four autonomous components of development, of which global economic development is one but not the only one; the others are termed "human development" (as set forth in annual surveys of the United Nations), "citizenship development," and lastly the emergent global community development with generalized value commitments such as human rights. However, there is also operative mirror-image processes of the first and the latter (globalization II) utilizing globalization for ends that, in effect, increase misery and subvert the processes and channels that globalization has put in place. The latter set has received scant attention in the literature but needs to be incorporated into a more general theory.

It will be important for constructal theory to apply itself beyond the "intelligence of nature" (Poirier 2003) to globalization by mapping the various flows and interchanges that are suggested in these schemas, since such an application might point to new ways of optimizing the flows of energy that drive globalization. In such an application, it will be important for constructal theory to take into account obstacles or blockages that disrupt the intended flow of "energy." For example, one major aspect of globalization is to increase global economic growth and general economic abundance. In the more recent (post-World War II and even more recently) period, there has been concerted efforts at the regional and global level to reduce trade and investment barriers and make markets more efficient with a global rationalization of production and services (e.g., "outsourcing"). At the same time, there has been channels set up for countries

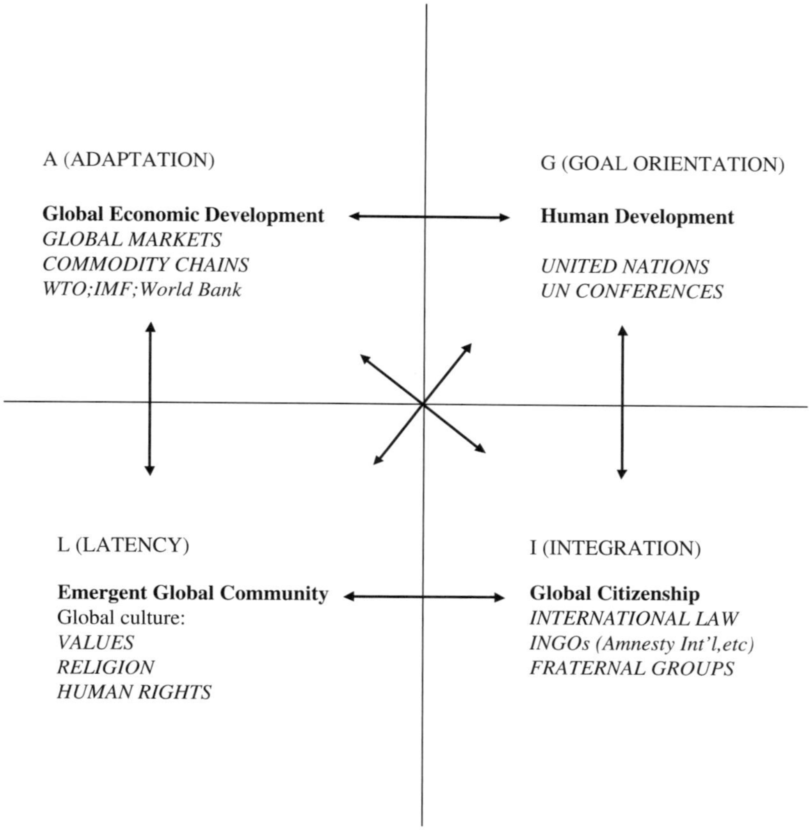

FIGURE 7.1. A Parsonian perspective on globalization I

with a capital surplus to provide either at the level of direct assistance or through collective organizations (such as the World Bank) charged with providing capital and knowledge to countries lacking the infrastructure necessary for development. Organizations have been set up to regulate these flows and establish criteria of eligibility for recipient countries (e.g., the IMF, the WTO). In theory, at least, global economic development is viewed optimistically as a win-win situation for the two major regions formerly designated as "North" and "South." Economic globalization, thus, is viewed as providing the integration of the world in a very large marketplace, and a recent paper by members of the IMF point to the financial success of globalization, even in a turbulent world: the total international financial assets of the advanced economies increasing by 67% in the four years from 2000 to 2004 and by 60% for emerging markets (Kose, et al. 2006).

The real challenge for constructal theory is not so much to map the flows of growth but also to take into account the obstacles and deviations that take away from optimal development. At the economic level, the lack of transparency

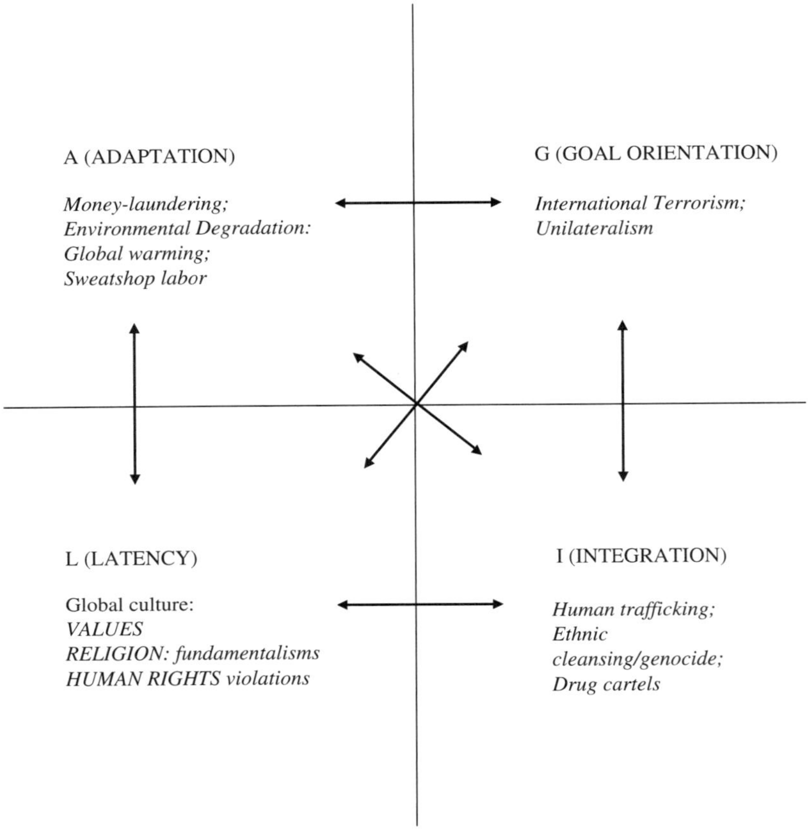

FIGURE 7.2. A Parsonian perspective on globalization II

(Holzner and Holzner 2006) regarding governance and high levels of corruption and bribery, especially in the "South" as well as in many ex-Communist countries, constitutes a severe drain on economic performance, most telling for countries in sub-Sahara Africa which have had negative growth rates, unlike the global aggregate total. But globalization has generated resistance and criticism, from those speaking on behalf of the "South" who view its regulations as making for an unfair game (Sassen 1998), or even from some of its own administrators (Stiglitz 2002).

A more complex set of flows is that of ideas and institutional structures to put in place values of modernity such as human rights, the empowerment of minorities, and democratic participation in the electoral process. Broadly speaking, this is socio-political globalization of the normative framework of the United Nations and the international community it seeks to represent. Again, these flows meet resistance and obstacles, often very violent ones, with new forms of social movements, sometimes seeking identity in new space, sometimes seeking to maintain identity in contested space (McDonald 2006).

In concluding these considerations, I wish to emphasize that the field of globalization is in my judgment highly appropriate for a new encounter between engineering, as represented by Adrian Bejan and his various associates, and sociology. The workshop or seminar setting that stimulated the present book is very appropriate to extend the interaction between sociology and engineering, and not only with globalization as a focus.

References

Albrow, M. (1996)*The Global Age: State and Society Beyond Modernity*, Polity, Cambridge.

Bailey, K. D. (1994) *Sociology and the New Systems Theory*, State University Press of New York, Albany, NY.

Bejan, A. (2000) *Shape and Structure from Engineering to Nature*, Cambridge University Press, Cambridge.

Bejan, A. and Lorente, S. (2005) *La loi constructale*, L'Harmattan, Paris.

Bejan, A., Lorente, S., Miguel, A. F. and Reis, A. H. (2006) Constructal Theory of Distribution of City Sizes, 13.4 in A. Bejan, *Advanced Engineering Thermodynamics*, 3rd edn., Wiley, Hoboken, NY.

Buckley, W. (1967) *Sociology and Modern System Theory*, Prentice-Hall, Englewood Cliffs, NJ.

Buckley, W. (1968) *Modern Systems Research for the Behavioral Scientist: A Sourcebook*, Aldine Publishing, Chicago, IL.

Castells, M. (1996–1998) *The Information Age: Economy, Society and Culture*, Blackwell, Oxford.

Cybernetics (2006) http://en.wikipedia.org/wolo/Cybernetics. Last modified 04:09, 10 September 2006.

Etzkowitz, H. (1993) What Happened to Cybernetics? *Contemporary Sociology* **22**(4), pp. 493–495.

Giulianotti, R. and Robertson, R. (2006) Glocalization, Globalization and Migration: The Case of Scottish Football Supporters in North America, *International Sociology* **21** (March), pp. 171–198.

Heims, S. J. (1991) *The Cybernetic Group*, MIT Press, Cambridge, MA.

Holzner, B. and Holzner, L. (2006) *Transparency in Global Change. The Vanguard of the Open Society*, The University of Pittsburgh Press, Pittsburgh, PA.

Kose, M. A., Prasad, E., Rogoff, K. and Wei, S-J (2006) Financial Globalisation: A Reappraisal, reported in Martin Wolf, "Forces Behind Globalisation with the day," *Financial Times*, September 13, 2006, www.imf.org

Luhmann, N. (1982) *The Differentiation of Society*, Columbia University Press, New York.

McDonald, K. (2006) *Global Movements. Action and Culture*, Blackwell, Malden, MA.

Papastergiadis, N. (2000) *The Turbulence of Migration: Globalization, Deterritor-Realization, and Hybridity*, Polity, Cambridge.

Parsons, T. (1969) *Politics and Social Structure*, Free Press, New York.

Parsons, T. (1970) Some Problems of General Theory in Sociology, in John McKinney and Edward A. Tiryakian, eds., *Theoretical Sociology*, Perspectives and Development, pp. 27–68, Appleton-Century-Crofts, New York.

Poirier, H. (2003) Une théorie explique l'intelligence de la nature, *Science et Vie* **1034**, pp. 44–63.

Quételet, A. (1997) (1869, 1835) *Essai sur le développement des facultés de l'homme*. Reissue edited by Eric Vilquin and J-P Anderson. Académie Royale de Belgique, Brussels.

Rojas, F. (2006) The Cybernetic Institutionalist, in Wilson Carey McWilliams, *The Active Society Revisited*, pp. 53–70, Rowman & Littlefield, Lanham, MD.

Rosenau, J. (1990) *Turbulence in World Politics: A Theory of Change and Continuity*, Princeton University Press, Princeton.

Sassen, S. (1998) *Globalization and its Discontents*, New Press, New York.

Sorokin, P. (1928) *Contemporary Sociological Theories*, Harper and Brothers, New York.

Stiglitz, J. E. (2002) *Globalization and its Discontents*. W.W. Norton, New York.

Treviño, A. J. (ed) (2001) *Talcott Parsons Today. His Theory and Legacy in Contemporary Sociology*, Rowman & Littlefield, Lanham, MD.

Urry, J. (2003) *Global Complexity*, Polity, Cambridge.

Wiener, N. (1965) *Cybernetics: or, Control and Communication in the Animal and the Machine*, MIT Press, Cambridge, MA.

Wiener, N. (1967) *The Human Use of Human Beings: Cybernetics and Society* (with an Afterword by Walter A. Rosenblith), Avon Books, New York.

Zipf, G. K. (1949) *Human Behavior and the Principle of Least Effort: An Introduction to Human Ecology*. Addison-Wesley, Cambridge, MA.

Chapter 8
Is Animal Learning Optimal?

John E. R. Staddon

Optimization theory has always been the first choice for understanding adaptive systems. The behavior of animals, both learned and instinctive, is clearly adaptive. Even before Darwin, biologists noted the close match between the form and behavior of animals and the "conditions of existence," the environment—the *niche* as it would now be called—in which they live. And utility maximization was until quite recently the predominant approach in neo-classical economics. But since Darwin, biologists have known that the form and behavior of organisms is quite often not optimal, in any reasonable sense. The human appendix, the peacock's ungainly tail, and the vestigial legs of snakes are clearly not optimal: humans, peacocks, and snakes would clearly live longer and move more efficiently without these redundant appendages.

But what of *behavior*? Economics is a pretty successful social science and, in the 1970s, psychologists and behavioral ecologists had high hopes that these techniques could shed light on animal learning and behavior (Krebs and Davies 1978; Staddon 1980). But this hope also had to be given up. This article is a brief summary of why, and what the alternatives are.

8.1. Reinforcement Learning

Reinforcement learning is the study of how behavior is guided by its consequences—either *reinforcement*, which increases the probability of behavior it follows, or *punishment*, which decreases the probability. Most learning of "higher" animals—mammals, birds, and humans—falls into this category. It is easy to see how under normal circumstances it will result in organisms getting more of what they like and less of what they dislike, in a sort of hill-climbing optimization process. And indeed, two decades ago, much theoretical research supported the idea that learned behavior maximizes the rate of reinforcement. Apparent failures could usually be attributed to cognitive limitations. Rats and pigeons cannot learn even quite simple sequences, for example. Thus, they cannot master a task that requires double alternation (LLRR) to get food reinforcement. But once *cognitive constraints* of this sort were taken into account, it looked as if all *operant behavior* (as behavior guided by consequences is called) could be accommodated by optimization.

8.1.1. Instinctive Drift: Do Animals "Know" What to Do?

But even in those early days, some striking violations of optimality were known. Keller and Marion Breland were students of B. F. Skinner, coiner of the term "operant" and pioneer in the experimental study of reinforcement learning in animals. They learned to train pigeons to guide missiles on Skinner's "Project Pelican" during World War II, and later went into the business of training animals for commercial purposes such as advertising. They found that animal behavior is not so malleable as Skinner taught. They found that Skinner's memorable phrase "reinforcement shapes behavior as a sculptor shapes a lump of clay" wildly exaggerates the power of reward and punishment to mould behavior. In an article entitled *The Misbehavior of Organisms* (1961), the Brelands described numerous violations of the reinforcement principle. One of the more dramatic involved a raccoon (Fig. 8.1).

> Raccoons condition readily, have good appetites, and this one was quite tame and an eager subject. We anticipated no trouble. Conditioning him to pick up the first coin was simple. We started out by reinforcing him for picking up a single coin. Then the metal container was introduced, with the requirement that he drop the coin into the container. Here we ran into the first bit of difficulty: he seemed to have a great deal of trouble letting go of the coin. He would rub it up against the inside of the container, pull it back out, and clutch it firmly for several seconds. However, he would finally turn it loose and receive his food reinforcement. Then the final contingency: we put him on a ratio of 2, requiring that he pick up both coins and put them in the container.
>
> Now the raccoon really had problems (and so did we). Not only could he not let go of the coins, but he spent seconds, even minutes, rubbing them together (in a most miserly fashion), and dipping them into the container. He carried on this behavior to such an extent that the practical application we had in mind—a display featuring a raccoon putting money in a piggy bank—simply was not feasible. The rubbing behavior became worse and worse as time went on, in spite of nonreinforcement (Breland and Breland 1961).

The raccoon's problem here isn't really a cognitive one. In some sense he "knows" what is needed to get the food since he learned that first. Yet this other "washing" behavior soon takes over, even if it blocks further rewards. The Brelands termed this aberrant behavior and others like it in other species, *instinctive drift*. The fact that animals may learn effective Behavior A successfully, yet revert irreversibly to ineffective Behavior B, suggests that cognition isn't everything. An organism may be cognitively capable of an optimal pattern of behavior, yet fail to persist in it.

8.1.2. Interval Timing: Why Wait?

Here are two other, more technical, examples from the extensive experimental literature on reinforcement learning.

FIGURE 8.1. A raccoon at night, its usual active time

8.1.2.1. Ratio Schedules

Reinforcement schedules are simply rules that relate an organism's behavior to its reinforcing consequences. For example, the behavior may be lever-pressing by a hungry rat, and the rule may be "one lever press gets one food pellet", this is called a *fixed-ratio 1* (FR 1) schedule. Obviously, there is nothing magical about the number one. Animals will respond on ratio values as high as 50 or more. And the ratio need not be fixed; it can vary randomly about a mean from reinforcer to reinforcer: *variable ratio* (VR).

After a little exposure, behavior on most reinforcement schedules appears to be pretty optimal: animals get their reinforcers at close to the maximum possible rate with minimal effort. On ratio schedules, e.g., they respond very rapidly, but expend little energy. But FR schedules do offer a small puzzle. When the ratio value is relatively high, say 50 or so, animals don't start responding at once after each food delivery. Instead, they wait for a while before beginning to respond.

The wait time is approximately proportional to the ratio value: the higher the ratio, the longer the wait. Indeed, if the ratio is high enough, the animal may quit entirely.

The obvious explanation is some kind of fatigue. Perhaps the animal just needs to take a breather after a long ratio? But no, this can't be the explanation because they don't pause on a comparable *variable* ratio. For the right explanation, we need to look at *interval* reinforcement schedules.

8.1.2.2. Interval Schedules

Interval schedules use a more complex rule than ratio schedules—although the principle is still simple enough. An example is "a lever press 60seconds after the last pellet gets another pellet." This is termed a *fixed-interval 60* (FI 60) schedule.

Behavior on FI schedules is also close to optimal. Animals wait before beginning to press the lever and lever-pressing thereafter accelerates up to the time when food is delivered. Figure 8.2 shows a typical "scalloped" record of cumulative responding (in this case, pecking on a disk by a hungry pigeon), rewarded with brief access to grain. The figure also shows the typical *wait time* after food delivery before pecking begins and the accelerated peck rate thereafter. Wait time is proportional to the interfood interval: if the animal waits 15seconds before beginning to respond on an FI 30second schedule, it will wait 30second on an FI 60. Thus, it wastes relatively few responses and gets the food as soon as it is available: altogether a pretty optimal pattern.

Contrast that behavior with what they do on a very similar procedure called a response-initiated delay (RID) schedule (Fig. 8.3). A RID schedule is almost the same as an FI schedule. The only difference is that the time, instead of being measured from the last reinforcer, is measured from the first response after a reinforcer: the organism starts the clock itself, rather than the clock restarting

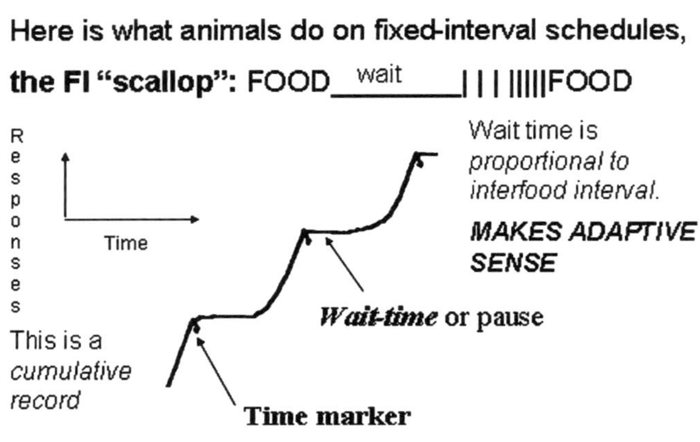

FIGURE 8.2. Adaptive behavior—responding at an accelerating rate, adjusted to the usual time of food delivery—on fixed-interval reinforcement schedules

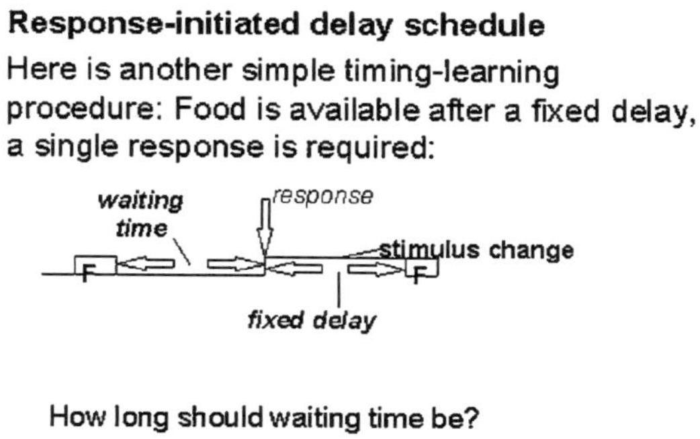

FIGURE 8.3. Another time-based reinforcement schedule: response-initiated delay. Food is delivered at a fixed time after the first post-food response. The subject should respond as soon as possible—but animals (and humans, sometimes) do not

after each reinforcement. The optimal behavior here is similar to that on FI: wait after the first response, a time proportional to the delay time. But the first response should not be delayed, since any delay adds to the total time between reinforcers. Thus, the optimal behavior is simple: respond at once after each reinforcer, then quit until food is delivered.[1]

This is not what most animals do. Instead, they ignore the clock-starting response requirement and treat the schedule as if it were fixed interval, waiting for a time proportional to the actual interfood interval before the first response.

The process can be analyzed formally in a very simple way as follows. On fixed interval, the wait time, WT, is approximately proportional to the interfood interval, I (this is termed *linear waiting*). Hence,

$$\mathrm{WT} = \mathrm{kI} \tag{8.1}$$

where constant k is around 0.5 and I is the interval value. On an RID schedule, the interfood interval is necessarily the sum of the wait time plus the imposed delay. So if the same process operates, $\mathrm{WT} = \mathrm{k}(\mathrm{WT} + \mathrm{I})$, which yields

$$\mathrm{WT} = \frac{\mathrm{kI}}{(1 - \mathrm{k})} \tag{8.2}$$

Because the quantity $1 - \mathrm{k}$ is less than one, this yields a wait time that is much too long. Equation (8.2) in fact describes what pigeons do on RID schedules (Wynne and Staddon 1988).

This is the answer to the fixed-ratio schedule puzzle. Evidently, pigeons have a built-in automatic timing mechanism that suppresses responding during

[1] On some RID schedules, a second response is required after the delay to produce the reinforcer. The optimal behavior in this case is to pause after the first response for a time proportional to the delay before responding again.

a predictable delay in between reinforcer deliveries—even when pause is maladaptive, as it is on ratio and RID schedules. It also explains the lack of a post-reinforcement pause on *variable*-ratio schedules, because here the interreinforcement interval is *not* predictable. It is as likely to be short as long.

So what can we conclude from these examples:

1. That animals behave optimally only under restricted conditions.
2. That behavior often follows quite simple, mechanical rules—such as linear waiting.
3. That these rules have evolved to produce optimal behavior only under *natural* conditions, i.e., conditions encountered by the species during its evolutionary history. Fixed-interval schedules are to be found whenever food occurs in a regular temporal pattern, but RID schedules, in which the timing of the food is initiated by the animal's own behavior, never.

8.2. What are the Alternatives to Optimality?

Historically, there are three ways to understand adaptive behavior:

1. *Normative*: Behavior fits the environment: it minimizes energy, maximizes food intake, reproduction, flow access (constructal theory), or, in economics, utility.
2. *Mechanistic*: Behavior is explained by a cause–effect process, which may be either physiological, as in Sherrington's account of the reflex, the Hodgkin-Huxley equations of nerve action, or conceptual/computational, as in Lorenz's hydraulic model of instinct or neural-network models of behavior.
3. *Darwinian*: This is also a causal approach, but more open-ended than option 2. The basic idea is simply that adaptation is best understood as the outcome of two complementary processes, *variation*, which generates behavioral options, and *selection*, which selects from among the resulting repertoire according to some criterion.

Option 1 is still the dominant approach in economics, but recent developments there—experimental economics, game and prospect theory, bounded rationality—increasingly call it into question as a general theory. Constructal theory, the topic of this book, is an optimization approach that has proven very successful in describing many physical and some biological systems. Given the serious problems encountered by other versions of optimality theory in describing learned behavior, it may be less successful there. On the other hand, nerve growth in the brain—a "morphing system" of the sort where constructal theory applies—maybe an area where these ideas will prove useful.

Option 2, a fully mechanistic account, is obviously the most desirable, since it provides a causal account of behavior that in principle may be related directly to brain function. Such accounts are not yet available for most interesting behavior.

Option 3 is a useful compromise. The Darwinian, selection–variation approach can be applied to any adaptive behavior even if details of the process are uncertain. All operant behavior can be usefully regarded as selection from a repertoire of behavioral variants (or, more precisely, a repertoire of generative *programs* or subroutines of which the observed behavior is the output: Staddon 1981). Selection—reinforcement—is then limited by the variants offered up by the processes of behavioral variation. In the raccoon and instinctive drift, e.g., once the animal had learned to manipulate the token, once it became predictive of food, the processes of variation generated the instinctive "washing" behavior, which in fact delayed reinforcement.

The same process has been studied in pigeons in a Pavlovian conditioning procedure called autoshaping. In autoshaping, a disk is briefly illuminated just before food is delivered. The hungry pigeon soon comes to peck the lit disk. The pecking seems to be elicited by the light much as Pavlov's dogs salivated to the bell. Moreover, the pigeon continues to peck, albeit at a reduced rate, even if the experimenter arranges that pecking turns off the disk and prevents the delivery of food. Apparently, a food-predictive stimulus powerfully elicits species-specific behavior which will occur even if it prevents the food.

This variation-selection process can be formalized as a causal model (e.g., the Staddon–Zhang model of assignment of credit in operant conditioning, which also offers an account for instinctive drift: Staddon 2001, Chapter 10). But it is useful even if not enough is known to permit causal modeling.

References

Breland, K. and Breland, M. (1961) *The Misbehavior of Organisms*, http://psychclassics.yorku.ca/Breland/misbehavior.htm.

Krebs, J. R. and Davies, N. B. (eds.) (1978) *Behavioral ecology*. Sunderland, MA.

Staddon, J. E. R. (ed.) (1980) *Limits to Action: The Allocation of Individual Behavior*. Academic Press, New York.

Staddon, J. E. R. (1981) Cognition in animals: Learning as program assembly. *Cognition* **10**, 287–294.

Staddon, J. E. R. (2001) *Adaptive Dynamics: The Theoretical Analysis of Behavior*. MIT/Bradford, Cambridge, MA: Pp. *xiv*, 1–423.

Wynne, C. D. L. and Staddon, J. E. R. (1988) Typical delay determines waiting time on periodic-food schedules: static and dynamic tests. *J. Exp. Anal. Behav.*, **50**, 197–210.

Chapter 9
Conflict and Conciliation Dynamics

Anthony Oberschall

9.1. The Natural and the Social Sciences

The endeavor to create a science joining the social sciences to the natural sciences into a single framework has a long and distinguished history, but it has also been punctuated by repeated setbacks and loss of interest (Lecuyer and Oberschall, 1968). I am not referring to facile analogies by the likes of Karl Marx who claimed that his "scientific" socialism had discovered the "laws" of historical change, or others who drew parallels between the human and the social body, e.g., they are both born, develop, mature, age, and die, be they individual persons or empires and societies. Examine the concept of "social structure" in all its variants: class structure, organization structure, and most recently social networks. The idea of such structures is based on analogies with physical entities—buildings, bridges, airplanes, and animals. A blueprint for a bridge is subject to well-known laws and principles, and a certain amount of modeling and testing is realizable and in fact realized. Imagine a civic textbook explaining the structure and functioning of the US government: constitutional separation of powers, president as commander in chief, Congress with its oversight responsibilities, treaty and budget powers; the statutory powers of various departments of government (state, defense) and executive agencies (CIA, NSC); the legacy of the federal courts' decisions on these rights and responsibilities. Knowledge of this constitutional and statutory blueprint will not explain how the Bush administration decided on the Iraqi war in 2003 and how it implemented those decisions (Woodward 2004). Why is that?

There is certainly a huge difference between the structure of a giraffe and of a hippopotamus. It is inconceivable for a giraffe to have short stubby legs like a hippo and for a hippo to have long thin legs like a giraffe. For a hippo to float in water and feed on grass nearby, it has to have a huge mass that the legs of the giraffe couldn't support, whereas a giraffe with short stubby legs could not feed on the leaves of tall trees, nor cover long distances between tree clusters in a timely manner. Natural selection and adaptation in the theory of evolution explains the mechanisms for the structure of the giraffe's and hippo's legs. The path to war chosen by the Bush administration was not constrained by the constitutional and statutory structures of the US government. It chose alternative structures—bypassing the UN Security Council, control of information on WMDs provided to Congress, news media and the public; shifting power from the State Department and the CIA to the Department of Defense and from Congress to the White House, etc. Nor can it be said that the Bush administration was making these changes because of some adaptive process due to lessons learned from the first Gulf War and other post-cold war US interventions to change authoritarian regimes. To the contrary, experts on the Middle East and on foreign and military policy warned about the mode of diplomacy, warfare, and post-war reconstruction chosen by the Bush administration. There was no optimization, adaptation, or evolution in these decisions, and no constraints in ways that laws discovered by science apply to the legs of giraffes and hippos. It is rather a case of the "art of muddling through" which is characteristic of political decisions. It is somewhat more successful than the famous random walk of the drunken man or of the blind leading the blind for getting to a destination, but it also often gets you to the wrong destination, as Serbian President Milosevic discovered about his Greater Serbia goal during the break up of Yugoslavia and the Bush administration discovered about urban guerilla warfare and terrorism in post-war Iraq. Ideology, groupthink, false beliefs, the arrogance of power, self-deception, underestimation of one's adversaries, uncertainty from contingent decisions, and other processes capturing the non-rational dimensions of human decision-making play a major role in these war and peace decisions.

Going beyond metaphor and imagery, some mathematicians and scientists in the 18th and 19th centuries came to believe that human behavior in aggregates exhibited causal patterns that could be described and explained by probability theory, among them Condorcet, Laplace, Lavoisier, Fourier, and Quetelet (Oberschall 1968). The principal difference between physical and human nature is the intentionality of human actions. In the aggregate, however, it was thought that dispositions, intentions, and emotions were distributed randomly and cancelled one another in the same way that errors of observation had a Gaussian (bell shape) distribution. The scientists believed that although individual behavior was difficult to predict, in the aggregate human behavior exhibited lawful regularities that could be tested with social science data of the sort that was becoming available in the first half of the 19th century from census data, crime statistics, public health statistics, military records on conscripts, and the like. Yet what started with high expectations on the application of science in human affairs proved elusive.

As an example, consider population dynamics and policy where these efforts were pushed the furthest. We have a pretty good understanding for what makes couples choose the number of children they wish to have as a family (Becker 1981). At the micro-level, as couples are more affluent, they will have fewer children but invest more in the quality of children (schooling, music classes, sports, etc.). At the macro-level, as a society becomes more affluent and various social programs for pensions, health care, and welfare are instituted, these programs become a substitute for what grown children's responsibility to their parents used to be, and one would expect the demand for children to decrease. Thus, at the micro and the macro levels, the population explosion set off by better nutrition and health that lowered mortality (especially infant mortality) will be offset by lower fertility rates as we, individually and collectively, become more prosperous. We also know the long-term consequences of aging on population size, on labor force participation, and on tax contributions for pension and health programs. The consequences are aging populations in most of Europe, an expectation of declining population size and of an increasing dependent population. Technology can be substituted for labor; older people can get incentives for working more years by withholding pension benefits, and young couples can be provided with incentives (free day and health care, child allowances) for more children. More likely, immigrants from societies caught in the poverty trap who want to live and work in Europe can be increased. All such moves are fraught with considerable uncertainty on consequences beyond the size of the population that matter for a country's welfare. But what exactly is an optimum population policy for such countries? Is there an optimum population (or range of population), and what exactly is to be optimized by such a policy? Do we know the ramifications and consequences of trying to assimilate non-European people into European populations and cultures, or of the rapid progress in the health sciences for longevity? These are matters of political debate and choice, but will the choices be wise?

Why there is no tendency for social welfare to be optimized by some combination of individual and political choice has been explained by the "tragedy of the commons," which is an instance of the Prisoner's Dilemma (PD) that rules human affairs: even though it is in everyone's interest (individuals, corporations, states) in the long run to change their behavior, which is personally costly, for avoiding a collective bad, it is in their short-run interest to let others change their behavior and to free ride on them (by not changing). Hence, we have an aging population, global warming, congested highways, arms races that merely reproduce a balance of power at greater cost to both adversaries, speculative crazes in financial markets, and other processes of competition (increased spending on political attack ads by rivals for office) that do not add to individual or social benefit despite increased costs to all.

Olson (1968) has shown that the only escape from a PD is coercion or a political agreement on rationing behavior which is enforced. Is that likely? There is no sovereign world government for enforcing limits on state's behavior. Within states, democratic governments have no incentive for imposing costs on the very people that elect them and keep them in power. Although the aggregate consequences of these behaviors and practices are known to be harmful in the long run, there is no mechanism, domestic or international, for avoiding the "tragedy of the commons." Instead of a law of gravity or entropy that forces humans to adapt, there is no limit to the individual and collective capacity of humans for self-deception, denial, rationalizations, false beliefs, and of course selfishness.

A major intellectual advance in the social sciences occurred with Von Neumann and Morgenstern's (1947) game theory (see also Rapoport 1970) which explains the dynamics of strategic choices for human transactions (individual persons, groups, states, political leaders, military commanders, chess players), including the tragedy of the commons. The simplest games have analytic solutions, if one makes a number of assumptions about rational choice, transparency, implementation and other matters which may or may not be approximated in some real-life applications. More complex games have been analyzed using more realistic assumptions about information, commitment, threat, and deception (Schelling 1963). For games played repeatedly in large populations, computer simulation has yielded insights into viable and winning strategies, also under limiting assumptions (Axelrod 1984).

In what follows, I will apply "game-theory thinking" to the real world of war and peace, of insurgency and peace making, to explain the failed "Oslo" peace process between Israel and the Palestinians. In the conclusion, I will reflect on the extent to which this sort of methodology intersects with constructal theory, and what can be learned about promising links between the natural and the social sciences.

9.2. Conflict and Conciliation Dynamics (CCD)

Conflict and conciliation dynamics explains the inner dynamics and eventual outcome of conflict and conciliation processes typical of peace processes in divided societies, e.g., the Oslo peace process between Israel and the Palestinians (failure at the Camp David summit in summer 2000) and the Northern Ireland "Troubles" from 1968 to 1998 that ended successfully with the Northern Ireland Peace Agreement.

CCD has four components: players, issues, strategies, and states of the process. *Players* are the stakeholders in the conflict. They are governments, insurgent organizations, political parties/leaders, and external actors such as states and international organizations (United Nations, NATO). There is a minimum of two players, called adversaries, but there may be more. A player can be divided into factions: Likud and Labor each competed with one another to be the incumbent Israeli government. The strategy of a player can change depending on its internal components, e.g., the Israeli government can be Likud led, Labor led, or a coalition with both. Similarly, the Palestinians are divided into the Palestinian Liberation Organization (PLO), Hamas, and other militant groups.

The *issues* are what the conflict is about: state formation, power sharing, refugee return, borders, and water rights. During the Oslo process, Labor wanted a two-state solution with the Palestinians. Likud and its allies wanted annexation of much of the West Bank to Israel. The PLO wanted two states with the 1967 borders, along the so-called "Green Line." Hamas and its allies wanted all of Palestine including the entire state of Israel, minus all the Jews living there.

Strategies are decision rules for pursuing the players' goals. The strategies take into account the adversaries' expected choices and responses. The intersection of the two strategies can be described with an outcome matrix in which the rank order of each players outcome preference is entered as a number or plus and minus signs. The simplest way to describe choices is by a dichotomy C and D, where C stands for conciliatory moves and D is a hostile move. Examples of C are making a concession, implementing a reform, starting a ceasefire, releasing detainees, reaching an agreement, and implementing the terms of an agreement. Examples of D are shootings, bombings, ceasefire violations, walking out on negotiations, outlawing a group, targeted assassination, economic sanctions, and the like. A strategy is a rule for choice of moves: e.g., always D, or tit for tat (TFT).

Game theory is a formal method for studying the players' strategic interactions. When both players have an incentive for reciprocated C moves (C,C), but are tempted to achieve a greater gain at the expense of the other player by moving D (D,D), the game is called a Prisoner's Dilemma (PD). An arms race is an example of a PD. PD and the iterated PD (repeated game) has a long and distinguished history. Although some CCD situations approximate an iterated PD, some are not, and most CCD processes consist of several interconnected games. For instance, a government can play against its principal adversary while simultaneously playing also against an internal rival.

The state of a CCD process can be characterized by stages or phases, such as low or high level of insurgency, talk/fight, ceasefire, negotiations for a peace agreement, etc. In the Oslo process, the adversaries were for the most part in a talk/fight stage, mixing some negotiations with some armed fighting, a mix of C and D moves. In the CCD analysis of the Oslo process, the main conflict is between the Israeli government and the Palestinians represented by the PLO. But the Israeli government, which is a coalition of parties led by either Labor or Likud, keeps being challenged by its political rivals who want to become the government, Likud replacing Labor or vice versa. Among the Palestinians, the PLO gets being challenged by Hamas and other militant groups who want to become the dominant political group. Advancing the peace process with conciliatory C by the Israeli government and by the PLO gets derailed when both make hostile D which keeps them in power against their internal rivals. The game between the two principal adversaries becomes an interconnected web of games, within each adversary between rivals, and between the Israeli government and the Hamas militants and between the PLO and the West Bank settlers.

Putting these ideas into a general form applicable to other peace processes, I highlight two dynamics within the larger flow of CCD. The first is the *mobilization dilemma*. It refers to the fact that each adversary's leadership keeps mobilizing its supporters and the public against internal rivals in a contention for power during the peace process. The consequence is that the leaders are constrained from making certain or too many accommodations to their adversary for fear of appearing weak, and therefore vulnerable to a hardline rival. Thus, some Ds are mixed in with Cs, and that sends a contradictory signal to the adversary about one's commitment to a peace process. The adversary is similarly constrained. The result is that reciprocated conciliatory Cs leading to a peace agreement keeps getting sidetracked by D moves from both adversaries. I call the second dynamic the **coercion paradox.** Many D moves in civil strife are collective punishments which affect innocent bystanders, not just the militants who are the primary targets of D: curfew for an entire town, closing of a border crossing, checkpoint at which everyone is impeded, suspension of work permits and labor migration. The paradox is that although raising the cost of opposition to one's adversary does inhibit it to some degree, it also outrages many neutral bystanders who become supporters of the militants and are recruited into the ranks of the militants, which in turn feeds the conflict. On balance, repression (tit-for-tat coercive moves) can increase conflict instead of advancing the peace process.

9.3. CCD Flow Chart Representation of a Conflict and Peace Process

The database for CCD is generated by a content analysis of the mass media, documentary sources, and scholarly literature which monitor the players, issues, strategies, and events. By way of illustration, Fig. 9.1 is a flow chart of the Israeli–Palestinian conflict and peace process from 1990 to the aftermath of the Camp David peace talks in the summer of 2000. I only recorded the most important events, and show here only a simple dichotomy of C and D incidence (not the duration and intensity of events), coded C and D. In the late 1980s the Israeli government was a Likud–Labor coalition that was internally divided over trying to negotiate with the PLO, and consequently there were only D moves by both sides, indicated by the red squares on the flow chart. In October 1990, the first intifada or Palestinian revolt against the Israeli occupation broke out in Gaza and then in the West Bank. There were violent incidents on a daily basis which I don't record in a flowchart for more important conflict and peace events. Daily coercive events are the background noise on which the recorded events are embedded. In September 1991 the PLO suffered a setback from backing Saddam Hussein in the first Gulf War, and the Israelis were apprehensive about the growing popularity of Hamas during the intifada. The Israeli government and the PLO agreed under external stakeholder pressure to participate in the Madrid peace conference—marked by C squares. The Madrid talks deadlocked, were suspended, restarted and deadlocked in the next year and a half. A major change occurred with the June 1992 Israeli elections won by Labor, and Rabin who became prime minister. Hamas violence was answered by Israeli repression, December 1992 to May 1993, but Rabin realized he was caught in a coercion paradox (the bloc of D squares): the more he collectively punished, the stronger the Palestinian militants became. The Rabin government initiated a new conciliatory strategy of dealing directly with the PLO (hitherto labeled a terrorist organization) and the PLO recognized the state of Israel for the first time since the founding of Israel in 1948. That got the Oslo peace process started (blocs of C in May 1993 to October 1993).

From 1994 to end of 2000, visual inspection shows repeatedly a series of Cs interrupted by a series of Ds whose inner dynamic I explain below through a

Year	Month	Day	Israel	Palestine	Event
	1	14	D		Russian immigrant to OT's
	2				
	3				
	4	11	D		Settlers in Gaza
	5	20		D	Rioting in Gaza
	6				
1990	7				
	8				
	9				
	10			D	Attack on beaches
	10	8	D		OT borders closed
	11	24			
	12				
	1				
	2				
	3	31	D	D	Repress intifada; knife attacks in Jerusalem
	4	16	D		Settlement expansion
	5				
1991	6				
	7				
	8				
	9	23	C	C	Agree to participate in peace conf.
	10	30	C	C	Madrid conference opens
	11				
	12	15	D		East Jerusalem settlement expansion
	1				
	2				
	3				
	4				
	5				
1992	6				Madrid talks deadlock
	7				
	8				
	9				
	10				
	11				
	12		D	D	Mass deportation of Hamas supporters; Arabs leave talks
	1		D	D	Civil strife over deportations
	2		D	D	"
	3		D	D	"
	4		D	D	"
	5		C	C	Resume Madrid talks
	6		C	C	"
1993	7				
	8	15, 30	C, C	C	Agree on return of deportees; Secret agreement on peace talks
	9		C	C	Oslo accords
	10		C	C	Talks start
	11				
	12	13	D		Israel delays Gaza, Jericho exit
	1				
	2	25	D	D	Settler kills 30 at Hebron mosque; Arabs suspend negotiations
	3				
	4	24	C	C	Paris Protocol
	5	4; 13	C	C	Gaza Jericho agreement; IDF exit Gaza
	6				
1994	7	25			Israel and Jordan end state of war
	8	29	C	C	Oslo accord on government transfers
	9	30			Gulf states end boycott of Israel
	10	20	D	D	Tel Aviv suicide bomb; Israel closes border
	11		D		Israel closes border
	12				

FIGURE 9.1. Israeli–Palestinian Conflict (C = Cooperation; D = Non-cooperation)

Year	Month	Day	Israel	Palestine	Event
1995	1	22		D	Suicide bomb attack
	2				
	3				
	4		D		New settlements in East Jerusalem proposed 7000 units
	5		D		"
	6				
	7				
	8				
	9	28	C	C	Oslo B. Interim peace agreement
	10	11	C		Prison release
	11	4			Rabin assassinated
	12				
1996	1	5	D		Hamas leader assasinated by Israeli agents
	2			D	Suicide attacks in Israel
	3				
	4				
	5	29			Likud, Netanyahu win elections
	6				
	7				
	8	2; 29	D	D	Settlement expansion (I); general strike (P)
	9				
	10	25; 28	D	D	Tunnel civil strife; battle of Nablus
	11				
	12	10; 29	D		Settlement expansion; settler shoots Hebron market
1997	1	17	C	C	Hebron agreement
	2	26	D		Har-Homa East Jerusalem settlements
	3	21	D	D	Tel Aviv suicide bomb; Israel closes border
	4	29	C		Work permits for migrants
	5		C		Peace process resumed
	6				
	7	22	D	D	Jerusalem suicide bombs (P); border closures and sanctions (I)
	8	18	C		Tax funds released
	9	1; 14	C		Borders opened; prisoner release
	10	1; 8	C	C	Veto of new settlement; Netanyahu–Arafat direct talks
	11				
	12				
1998	1		C, D	C, D	Clinton sponsored talks on Oslo timetable end in acrimony
	2				
	3				
	4				
	5		D	D	Settlement construction; boundary clashes
	6		D		
	7		C	C	Direct peace talks
	8				
	9				
	10	19	C, D	D	Wye River talks and memorial; settlement expansion; Hamas grenade attack
	11	20	D	D, D	Clinton visits, riots; clashes on prisoner release (P)
	12		D, D	D	Clinton summit; failure of Wye implementation; Intensified settlement activity
1999	1				
	2				
	3				
	4				
	5	17			Labor wins elections; Barak PM
	6	3		D	Day of rage against settlement expansion
	7				
	8				
	9	4; 9; 10	C, C, C	C	Wye Two agreement; Prisoner release (I); WB territory transfer
	10				
	11	8	C	C	Final status talks begin
	12				

FIGURE 9.1. (*Continued*)

Year	Month	Day	Israel	Palestine	Event
2000	1				
	2	3	D		Barak blocs on illegal settlement roads
	3	21	C		Israeli redeployment done
	4	30		D	Palestine exit talks over settlement
	5	15-18		D	Days of rage
	6				
	7	11-25	C, D	C, D	Camp David final status talks; talks fail
	8				
	9	25	C	C	Agree to resume peace talks
	10		D	D	Sharon Temple Mount expedition; Civil strife
	11				
	12				

FIGURE 9.1. (*Continued*)

game theory analysis as consisting of four simultaneous iterated, games: (1) the Oslo peace game, (2) the coalition game, (3) the Militant game A, and (4) the Militant game B.

9.3.1. Oslo Agreement Game (1993)

Land for Peace is chosen by PLO and Israeli government (+, +) even though Hamas is opposed and the Israeli settler movement wants settlement expansion in the Occupied Territories (OTs) and East Jerusalem (S = settlement expansion PP = peace process).

The payoffs are marked with + and – signs, the row player's payoff is first, the column player's payoff is second. The payoffs are based on statements made by the adversaries about their goals, preferences, and policies at news conferences, news interviews, in party platforms, manifestos and charters, and other public documents.

Israeli Government (Labor)	PLO: Peace	PLO: No Peace
Land	+, + Oslo Agreement	– –, ++ Hamas
No Land	++, – – Likud	–, – Pre-Oslo status quo

The payoff matrix also indicates that Hamas would like to get land without making peace, and Likud would like to get peace without surrendering land.

9.3.2. Coalition Game

"More settlements, or we oust government": to appease the right-wing coalition partners and the settler movement, the Israeli government decides to permit some settlement expansion.

		Right-wing partner	
		Support Government	Oust Government
Israeli Government	More S	+, ++	– –, – –
	Stop S	++, –	–, +

Outcome: (+, ++) Government stays in power, more settlements built

9.3.3. Militant Game A

Israeli government versus Palestinian militants. The Israeli government threatens to suspend the Oslo process if the militants engage in violence.

		Militants	
		Violence	Noviolence
Israeli Government	More S	–, +	+, – –
	Stop S	– –, –	++, ++

The outcome of this game should be (++, ++), "stop S, no violence" but the Israeli government also plays the coalition game, and that game is more important; thus the Israeli government chooses "more S", and the militants respond with "violence."

9.3.4. *Militant Game B*

Israeli government tells the Palestine Authority (PA) to crackdown on violent militants or else Israel will crackdown on the militants and suspend the peace process. This sets off a PA versus Palestinian Militants game.

	Militants: Continue violence	Militants: Stop violence
PA: Crackdown	– –, +	–, –
PA: No crackdown	+, ++	++, – –

If PA cracks down, it risks a Palestine civil war, and possibly losing to militants. If PA does not crackdown, it jeopardizes the peace process with Israel, but stays in power. The PA prefers staying in power to risking ouster. The militants prefer confrontation with Israel over confrontation with the PA. The outcome is therefore (+, ++), no crackdown and violence continues.

The outcome of Militant Game B feeds into the Oslo Game, where the PLO "no peace" (since the Militants continue violence) will intersect with the Israeli government "no land." Thus, what starts as "peace" for "land" at Oslo keeps breaking down into continuation of the conflict. Insurgency and counterinsurgency dominate what is left of the peace process.

To restart the peace process, strong pressure by external stakeholders (especially the US) on both adversaries is necessary, but the games remain the same and the process keeps reverting to "no land, no peace."

9.4. Empirical Checks and Discussion

The peace process is more likely to continue and to be reciprocal when the effects of the mobilization dilemma and the coercion paradox are dampened. In the Israeli–Palestinian conflict, that is the case when the Labor party and its allies win a comfortable Knesset majority in elections so that they do not need to include a right-wing coalition partner, thus not engage in the "coalition game." Also, due to external stakeholder pressure, the Israeli government commits to a freeze on settlement expansion, usually temporary and conditional, which should reduce the militant violence by suspending the Militant Game A. When militant violence decreases, one would expect the Oslo Game 1 to proceed.

There were two moments of opportunity for the peace process. From mid-1995 to mid-1996, the Oslo B exchanges generated a lot of reciprocated Cs, support for violence hit a low among Palestinians (as measured in opinion polls) and support for the Oslo peace process among Israelis was at an all time high. But Prime Minister Rabin was assassinated by a right-wing fanatic, Hamas and Islamic Jihad engaged in suicide bombings within Israel (so-called "spoiler violence"), Israel responded with collective punishments and suspended Oslo B implementation, and Likud won the Israeli election on the security issue (stop suicide bombings) by promising to react tougher than Labor, and Benjamin Netanyahu became the prime minister on a platform of crackdown and settlement expansion. The second opportunity was from mid-1999 to July 2000 when Prime Minister Barak pushed for final status negotiations but approved settlement expansion in order to hold his shaky government coalition together, which was responded to by militant violence and Israeli coercive retaliation. The peace process leading to the Camp David summit became entangled in coercive moves by both sides which led to accusations of bad faith and doubts on whether any agreement would be adhered to.

CCD is less than a causal theory familiar from the natural sciences yet is more than an ad hoc description of the conflict process. In the social sciences, theory has to deal with the strategic dimension of human action. If one takes an umbrella to work, it is not likely to change the chances of raining. Games against "nature" have that inevitable property, or as the statesman-philosopher Sir Francis Bacon wrote, "to command nature she must be obeyed." In human conflict, if one carries a weapon, it is definitely going to influence the chances of one's adversary carrying a weapon (we call it an arms race). Strategic interaction dynamics are more complicated to grasp than games against nature. Nevertheless, CCD can be used to make some *ex ante* predictions about the increased versus decreased chances of outcomes for the conflict.

Game theory methodology is useful but in its current state can't handle formal solutions for interconnected simultaneous games, which are more convoluted than so-called "supergames." Another difficulty is the characterization of C moves when deception is practiced, e.g., the Israeli government announces a settlement "freeze" but it actually continues construction and expansion which it refers to implementing plans that have already been approved, or the PA cracking down on militants but releasing them from prison after a short incarceration.

Much simulation work about strategies in iterated PD games indicates that TFT under a variety of conditions is a viable strategy. TFT is basically cooperative but it also resists being taken advantage of. It rewards C with C, initiates with a C, and continues with C until adversary moves a D, then it responds with D until the adversary resumes with a C. Looking over conflict data from content analyses of real conflicts, it strikes me that adversaries use "Pavlov" rather than TFT. In Pavlov, one chooses D until the adversary retaliates with a D, and then one moves a C. If C gets reciprocated, one does Cs for a time, but then one sneaks a D to test the adversary and check whether one gets away with non-cooperation. In real-peace processes, external stakeholders put pressure on the adversaries, by using both positive and negative inducements, for discouraging the use of Pavlov by the adversaries and for encouraging the choice of TFT. On both the empirical and the conceptual side, CCD is promising but only the start of a theory of a conflict and peace process. A more detailed exposition of CCD applied to peace processes is in Oberschall (2007).

9.5. Conclusions

As I understand from statements by Adrian Bejan on constructal theory, it predicts how flows that are constrained by natural laws will result in specific structures in accordance with some optimization principle.

As far as the relationship of CCD with constructal theory goes, what flows in CCD are C and D events resulting from human decisions. The goal of the flow is the adversaries' preferred outcome of the war and peace process. The flow configuration, its geometric structure, is the structure of the games that capture the dynamics: the number and types of games and adversaries that confront one another and the interdependencies of the games. The constraints on the adversaries' moves (which generates flow) are the strategies each uses against the other(s), and sanctions imposed by external stakeholders that affect payoffs and thus the desirability of certain outcomes. These are analogies one can make between CCD and constructal theory, but I am uncertain about a deeper connection.

Take the flow of conciliatory and hostile moves. Is there some principle that governs the redirection of moves from mutually destructive Ds to mutually cooperative Cs? Zartman (2001) has come up with the concept of a "mutually hurting stalemate" for explaining the adversaries' turn to ceasefire and peace negotiations, but it is an idea that has only been applied *ex post* to peace processes because *ex ante* there are no indicators that predict how adversaries define what is mutually hurting. Milosevic lost the Croatian and Bosnian wars, and had economic sanctions imposed on Serbia, and yet he decided to take on NATO over Kosovo. Look at the structure of peace processes when they are engaged by the adversaries in good faith (not simply for the sake of rearming and repositioning their fighters). It is recognized that third-party mediators facilitate peace making and that monitors for verifying the demobilization of the fighting forces and for free and fair post-civil war elections increase the chances for lasting peace. Beyond that, however, the international conflict management experts are divided on many other aspects of peace settlements and implementation: power-sharing governance, justice and amnesty, refugee return and compensation, constitutional design, rebuilding a failed state, economic and social reconstruction, and the part played by external stakeholders in all of these. Moreover, there are disagreements by well-informed and competent observers and students of peace processes on what lessons can be learned from previous cases of peace making and peace building. Instead of optimization or adaptation going about what improves the chances of successful peace, there is a cacophony of voices. Quantitative empirical studies of insurgencies/civil wars/ethno-national conflicts since 1945, or the mid-1950s, or post-cold war, etc. based on complete enumerations of such conflicts have been done repeatedly but do not produce any surprises nor insight into causal mechanisms. One learns from multiple regression analyses that countries that are poor, are non-democratic, have ethnic/religious/language divisions, have a past history of such antagonisms, etc., have a higher chance of experiencing these conflicts compared to countries with the opposed characteristics, and that external assistance to insurgents and terrain conducive to guerilla warfare increase insurgency chances yet further. These are descriptive correlations that do not go beyond common sense and knowledge available from case studies and comparative qualitative analyses. There is much information on duration, intensity, and mode of such fighting (e.g., atrocities and war crimes), and the likelihood that a ceasefire and a peace settlement will break down within two years, or five years, etc. None of this is all that useful for peace-making policy since no one can change the geography of a country and its past history of antagonisms, and because we are just as ignorant about changing a poor, undemocratic, and divided country into a more developed, democratic, and united country as we are on peace making and peace building.

I think one of the major differences between impersonal nature and human affairs is that when you dike a river to control flooding, the waters do not learn how to flood the land in a different fashion, whereas imposing sanctions on a belligerent state does lead to sophisticated evasion of the sanctions, and imposing power sharing with a minority upon an unwilling majority in a peace process can lead to evasion of power sharing during implementation, as the Bosnian Serbs practiced after the Dayton Peace Agreement and the rejectionist Unionists practiced after the Northern Ireland Peace Agreement. The reactive property of interdependent human decision-making presents a formidable challenge for the application of natural science thinking and methods to human affairs.

References

Axelrod, R. (1984) *The Evolution of Cooperation*, Basic Books, New York.

Becker, G. (1981) *A Treatise on the Family*, Harvard University Press, Cambridge, MA.

Lecuyer, B. and Oberschall, A. (1968) Sociology: the early history of social research, in *International Encyclopedia of the Social Sciences*, Macmillan and the Free Press, New York.

Olson, M. Jr. (1968) *The Logic of Collective Action*, Schocken Books, New York.

Oberschall, A. (2007) *Conflict and Peace Building in Divided Societies*, Routledge, London.

Oberschall, A. (1968) The two empirical roots of social theory, in H. Kuklick and E. Long, eds., *Knowledge and Society*, Vol. 6, JAI Press, Greenwich, CT.

Rapoport, A. (1970) *Games, Fights and Debates*, University of Michigan Press, Ann Arbor, MI.

Schelling, T. (1963) *The Strategy of Conflict*, Oxford University Press, New York.

Von Neumann, J. and Morgenstern, O. (1947) *Theory of Games and Economic Behavior*, Princeton University Press, Princeton, NJ.

Woodward, B. (2004) *Plan of Attack*, Simon and Schuster, New York.

Zartman, W. (2001) *Preventive Negotiation: Avoiding Conflict Escalation*, Rowman and Littlefield, Boulder, CO.

Chapter 10
Human Aging and Mortality*

Kenneth G. Manton, Kenneth C. Land, and Eric Stallard

10.1. Introduction

Demographers and biostatisticians share a common interest in the study of human survival and mortality. But the levels of analysis and types of data differ between these two disciplines. The traditional focus of demographers has been on population level analyses with the measurement and mathematical modeling of population processes across the entire human age span—or substantial age segments thereof (e.g., Preston et al. 2001). Even when demographers analyze sample data, the typical focus is on large samples that are representative of the populations from which they are drawn. By comparison, biostatisticians often have to analyze small samples from clinical trials of biomedical procedures or pharmaceutical interventions followed for relatively short periods of time (i.e., generally a few weeks to a few years) in which the intricacies of statistical inference in finite, and possibly non-representative, populations come to the fore (e.g., Lawless 2003).

Given the relatively short time frame of many clinical studies, biostatistical models of survival usually can either treat the sample distributions of certain characteristics of sample members (e.g., age) as fixed or deal with changing characteristics in relatively straightforward ways (e.g., by ignoring period effects). The situation is different when demographers analyze large-sample, long-term (e.g., 50 years in the Framingham study) longitudinal panel studies of human mortality and aging that cover much of the adult human age span. In this case, mortality selection on a static distribution can have a substantial effect on the frequency distributions of characteristics of the remaining sample members after substantial time has elapsed from the initial time point of the study.

The Gaussian random walk model of human mortality and aging (Woodbury and Manton 1977) has proven to be a useful tool for analyzing long-term longitudinal panel data either on (1) physiological parameters, disease incidence, and survival, or on (2) functional disability and survival, where the variables underlying individual risk differences are dynamic and subject to environmental shocks over time. The purpose of this chapter is to describe and substantively interpret the key equations of this model, review some findings from empirical applications of the model, and describe some recent generalizations of the random walk model to deal with correlations between the environmental shocks affecting state variable dynamics. It will be seen that the model can describe changes in human physiological parameters over suitably defined, stochastically perturbed state spaces.

10.2. The Random Walk Model

10.2.1. The Fokker–Planck Diffusion Equation

The random walk model of human mortality and aging has been the subject of many prior publications. Some of the key early publications defining the equations, the model, and computational methods for applying the model are Woodbury and Manton (1977, 1983), Manton and Stallard (1988), Manton et al. (1992), and Yashin and Manton (1997).

The specification of this model assumes that a survival process is operating in a state space for a population of individuals subject to a stochastic environment where only the parameters of the frequency distributions are deterministic. In this state space, under the assumption that the "history" of prior movement in the state space contains no additional information beyond the current state (i.e., the Markov condition), the forward partial differential equation is obtained for the distribution of a population whose movement in the selected state space is determined by the random walk equations. If the initial distribution of the population in the state space is Gaussian, or multivariate normal, then certain assumptions about movement and mortality will operate to preserve normality if the distribution of environmental shocks is also Gaussian. Under the assumption of normality, simultaneous ordinary differential equations can be derived from the forward partial differential equation defining the probability distribution function. Examination of the ordinary simultaneous differential equations shows how parameters for certain models of aging and mortality can be obtained.

To represent the ideas in mathematical terms as simply as possible, consider the following two equations for univariate changes in the state of individuals:

$$dx_i(t) = u(x_i, t)dt + d\xi(x_i, t) \tag{10.1}$$

$$dP(x_i) = -\mu(x_i, t)P(x_i)dt \tag{10.2}$$

Equation (10.1) represents the change for individual i on a selected physiological variable/risk factor x (say, systolic blood pressure), at time t. The movement of the physiological age state, $x_i(t)$, of individual i during time dt is the sum of two effects, one deterministic, $u(x_i, t)$, depending on i's current position, and the other a random walk term, $\xi(x_i, t)$, which is assumed to have independent increments during different time periods. In a number of biological applications, this latter assumption will need to be relaxed.

Equation (10.2) represents the probability of i's survival. The change in the survival probability, $dP(x_i)$, is equal to the probability of death $\mu(x_i, t)dt$ for i at state-space position x and at time t, times the probability of i surviving to reach x, $P(x_i)$.

The random walk model is completed with the specification of the Fokker–Planck equation (Risken 1996) for the change over time in the population distribution $f \equiv f(x, t)$ of a physiological variable/risk factor at time t:

$$\frac{\partial f}{\partial t} = -u\frac{\partial f}{\partial x} - f\frac{\partial u}{\partial x} + \tfrac{1}{2}\sigma_0^2\frac{\partial^2 f}{\partial^2 x} - \mu f \qquad (10.3)$$

The function f(x, t) in Eq. (10.3), where x represents a point in the state space R, where R denotes the set of real numbers, and t is time, defines the frequency distribution of the physiological variable/risk factor x at time t in R.

Equation (10.3) specifies that f(x, t) is the result of four different terms or forces on x at time t in R. The first or advection term involving the primary variables, u(x), represents the drift of the individuals in R. That is, it represents the "average" rate of change of the physiological status of the individuals starting at x. For an individual i, $u \equiv u(x_i)$ is a function only of the physiological status x of i.

At certain points h (rest points), where $u(h) = 0$, this term is zero and we turn attention to the second or divergence term. Divergence, or possibly *convergence*, depending upon whether the probability mass is spread or concentrated by movement in the state space, operates, along with the third term (diffusion), to change the probability distribution. The second term will be, in a homeostatic system, counterbalanced by the effect of the fourth (mortality) term to maintain equilibrium and a nearly constant distribution function. The mortality term is not usually represented in standard physical applications of the Fokker–Planck equation (Risken 1996). If $\partial u(x)/\partial x|_{x=h} < 0$, then this term is positive and the distribution is stable and controlled by homeostatic forces under relatively weak assumptions.

The third term, or *diffusion tensor* (for a multivariate problem), relates to the random-walk component of the motion apart from mean motion or drift. This term represents purely random, exogenous shocks to the system. These movements in the observed physiological space are a result of an organism's position on unmeasured variables. The rate of diffusion will be constant across the entire state space when the physiological variables are appropriately scaled, whereas the rate of divergence (convergence) may be a function of position in the space.

The fourth or *mortality* term refers to the state specific probability of death, i.e., the distribution of "manholes" in the state space as a function of position in the state space. In a Gaussian system, this "selection" or mortality function should be quadratic in form. If there exists a point where advection, divergence, and diffusion compose to produce a stable distribution, and if this point is in a region of low probability of death, it may be characterized as a physiologically homeostatic point. On the other hand, if $\sum(\partial u/\partial x) > 0$ is in some vicinity away from the rest points, positions where there is no drift, then motion is accelerated away from regions of low mortality and out to regions of higher average mortality. This "snowball" effect leads ultimately to higher mortality and systematic removal of "frail" individuals from the population. Thus, the residual population continues to occupy mostly the regions of low mortality displaced away from the highly lethal areas and toward the direction of drift if the force of the mortality gradient vector is weak. The general effect is to keep the variance of the distribution finite by mortality selection of persons in regions of high probability of death. Of course one analytic problem of some interest is the study of how these forces interact over age, especially at advanced ages (e.g., age 95+) where significant changes in dynamics and non-linearities are more likely to be manifest.

The Gaussian random walk model of physiological dynamics and mortality described above originally was developed by Woodbury and Manton (1977, 1983)—in a full vector space form (in contrast to the scalar form illustrated above) in order to accommodate the dynamics of a whole set of *physiological risk factors*, e.g., for following multiple risk factors in a longitudinal panel design such as the Framingham Heart Study. In Manton and Akushevich (2003) the Fokker–Planck equation was generalized to include a "birth" term, i.e., a reverse "hazard" function representing entry of new "persons" to the study population increasing probability mass with some specific set of state-space characteristics. The model was also generalized to accommodate non-Gaussian diffusion processes to describe "lumpy," sparsely populated, high-dimensional state spaces comprised of measures of *disability or functional ability* such as one obtained from data on questions about respondents' limitations with respect to Activities of Daily Living (ADLs) and Instrumental Activities of Daily Living (IADLs) in panel studies such as the National Long-Term Care Surveys (NLTCS); see Manton et al. (1992, 1994). Specifically, the state space, x, was generalized to be a "fuzzy" state space. A fuzzy state-space model may be more robust to model specification error than classical models when used in forecasting. This is because parameters of the state-space distribution and dynamic parameters are simultaneously estimated.

10.2.2. The State-Space and Quadratic Mortality Equations

In conventional applications, the random walk model leads to two systems of equations representing jointly dependent processes. First, systems of autoregressive state-space equations describe linked changes in J state variables—i.e., the specific measures of physiological risk factors or functional ability that are used in the longitudinal study. Denoting these by x_{ijt} (or the J-element vector $\underset{\sim it}{x}$) for individual $i = 1, 2, \ldots, N$, where N denotes sample size on state-space variable $j = 1, 2, \ldots, J$ for time periods $t = 1, 2, \ldots, T$, the state-space equations take the form

$$\underset{\sim it+1}{x} = \underset{\sim 0i}{u} + \underset{\sim 1}{\alpha}\,\mathrm{Age}_{it} + \underset{\sim 2it}{\alpha}\,\underset{\sim it}{x} + \underset{\sim 3}{\alpha}\,\underset{\sim it}{z} + \underset{\sim it}{\varepsilon} \tag{10.4}$$

with $\underset{\sim it}{\mathbf{z}}$ denoting a vector of exogenous variables (e.g., sex, race) for individual i fixed at time t. A case of general interest is to allow for second-order interactions of state variables in Eq. (10.4). It can be shown that such cases of non-linear dynamics may be treated as a Gaussian process where main effects and interactions are both treated as an extended set of state variables (Kulminski et al. 2004).

The second type of jump process is described by a non-negative definite *quadratic mortality or hazard function*:

$$\mu(\underset{\sim it}{x}, \mathrm{Age}_{it}) = \left[\mu_0 + \underset{\sim it}{x}^T \beta + \frac{1}{2} \underset{\sim it}{x}^T \underset{\sim}{B} \underset{\sim it}{x}\right] e^{\theta \mathrm{Age}_{it}} \qquad (10.5)$$

In the force of mortality equation, the term $\mu_0 e^{\theta \mathrm{Age}_{it}}$ represents the conventional "Gompertz" or exponential growth of mortality risk with age. This allows the shape of the quadratic function to change with age.

The Gompertz term is modified by the remainder of the terms interior to the brackets. These contain linear terms (with parameter vector **β**, the coefficients of which describe the linear association of each state-space variable with mortality risk) and a quadratic form (with parameter matrix B the elements of which measure the non-linear association of the state-space variables with mortality risk). Together, the **β** vector and B matrix capture the effects of individual i's physiological, or functional, status in the J variable state-space on i's mortality risk at time t. It is, however, the B-matrix component of Eq. (10.5) that imparts the "quadratic" shape to the hazard function. The linear terms act as a scale "vector" to change the location of the quadratic hazard in the multivariate state space as defined at time/age t.

The substantive rationale for the quadratic hazard function is that, for most physiological variables (e.g., systolic blood pressure), there is an "optimal" level at which mortality risk is minimized. This is illustrated in Fig. 10.1 where the function is centered at a normalized risk factor value of 100. As the physiological variable departs from the optimal value in either direction (lower or higher), however, the mortality risk increases, as is illustrated in the figure. At younger

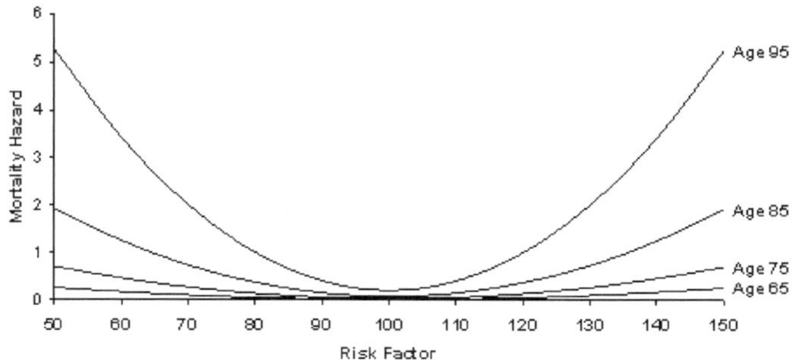

FIGURE 10.1. Illustrative age-specific quadratic hazard function. Univariate quadratic hazard function for single risk factor (X), for a fixed value $\theta = 0.0805$, but at different ages. The quadratic hazard is drawn with the following parameter settings: $\mu(X) = (0.01 + [(x\text{–}x^*)/100]^2 \exp[0.0805(\mathrm{Age}\ 65)]$ and $x^* = 100$

ages (e.g., 65), humans have considerable physiological complexity/redundancy /reserves so the curve is relatively flat. As age advances, however, physiological redundancy tends to decline (i.e., the capacity for self-organization and self-repair declines as environmental "wear" increases the entropy of the system) and the mortality risk for any given level of departure from the optimal value increases more rapidly. In the figure, this is illustrated by the more rapidly increasing quadratic function at ages 75, 85, and 95 than at age 65.

10.3. Findings from Empirical Applications

There have been numerous empirical applications of the Gaussian random walk model. We review findings from two applications.

An important question that has been addressed in prior research is the extent to which controls either for physiological risk factors or for functional status measures, in the quadratic hazard function, can account for the increase of mortality risk with age, as measured by the θ parameter of Eq. (10.5). Using 34-year follow-up data from the Framingham Heart Study, Manton and Stallard (1994) estimated a value of $\theta = 0.1002$ for females. That is, without controls for any state-space variables, mortality risk after age 65 for females was estimated to increase about 10 percent per year of age. This implies a doubling of mortality risk in 6.9 years. Controls for a state space defined by 9 physiological risk factors (diastolic blood pressure, systolic blood pressure, vital capacity, body mass index hematocrit, serum cholesterol, heart rate, blood glucose, smoking, and left ventricular hypertrophy) in the quadratic hazard of Eq. (10.5) reduced the estimated θ for females by 19 percent to 8.1, which corresponds to a doubling of mortality risk every 8.5 years of age. In brief, by keeping one's physiological risk factors at or near their "optimal" levels, application of the random walk model to the Framingham data suggests that the age dependence of mortality risk for females can be reduced by nearly one-fifth. For males in the Framingham study, Manton and Stallard (1994) found a corresponding reduction of about 14 percent.

Similar analyses also have been performed for the age 65 and over population with the NLTCS data. Using the 1982–1984 NLTCS data on females, for example, Manton and Stallard (1994) estimated $\theta = 0.0937$ without controls for any state-space variables. Then, using a state space-of seven fuzzy dimensions of functional ability defined from 27 functional physical performance items in the NLTCS surveys (Manton et al. 1994), they estimated $\theta = 0.0364$. This corresponds to a large reduction in the age dependence of mortality risk after age 65, of 61 percent. That is, with controls for measures of functional status in the $\underset{\sim}{\mathbf{x}}_{it}$ vector, θ is reduced from about 9.4 to 3.6 percent per year. By controlling the functional status of females in a longitudinal panel study, the per year increase in mortality risk declines from over 9 percent to less than 4 percent. Manton and Stallard (1994) report that controlling for income and education as well as functional status further reduces θ to 2.6 percent per year of age. Similar results have been found for males. This implies that the age dependence of mortality risk can be greatly reduced by maintaining functional activity (or by regaining functioning through the use of appropriate rehabilitative services), especially for individuals with higher levels of income and education.

Another application of the random walk model is to make estimates of life expectancies (years of life remaining) at various ages and in various states of health. Demographers and public health researchers have focused attention in recent decades on developing and applying health measures that combine mortality and disability data. Building on work on community and national health measures, and focusing on the elderly population, Katz et al. (1983) operationally defined active life expectancy (ALE) as the period of life free of disability in activities of daily living (ADL). ADLs are personal maintenance tasks performed daily, such as eating, getting in/out of bed, bathing, dressing, toileting, and getting around inside. Freedom from disability in any ADL means the person is able to perform each self-maintenance function without another persons' assistance or by using "special" equipment (assistive devices or a modified "built" environment). Since the Katz et al. (1983) article, the concept of ALE also has been generalized to include not having limitations in instrumental activities of daily living (IADL), which are household maintenance tasks such as cooking, doing the laundry, grocery shopping, traveling, and managing money as well as to physical performance limitations and impairments, disabilities, or social handicaps.

In assessing deviations from the intact or "active" health state, however, there has been controversy in selecting metrics to differentially weight specific physical and cognitive dysfunctions. Obtaining responses directly from affected individuals and whether they perceive that they are chronically functionally limited in an ADL or IADL in the manner described by Katz et al. (1983) is a preferred approach—probably better than only making physical measurements because physical independence at late ages also involves psychological factors, such as the self-perception of health, general morale, and the level of motivation to preserve functions. To move from responses about the ability to perform specific ADLs or IADLs to ALE estimates (sometimes called disability-free life expectancy, DFLE) requires the use of state-dependent life table methods to estimate the average number of years of life remaining free from ADL or IADL impairment at specific ages. The life table methods used for this purpose have evolved from the prevalence-rate method of Sullivan (1971) to double-decrement models (Katz et al. 1983), and to multi-state, or increment–decrement, models (e.g., Rogers et al. 1989; Land et al. 1994). Because of the relative scarcity of national long-term panel studies, however, these models are often applied to synthetic-cohort or "period" data in which general (i.e., non-disability-state-specific) age-specific mortality rates (from vital statistics) and disability rates (measured in health surveys) experienced by a population during a period (e.g., a calendar year) are concatenated across ages to simulate the "experience" of a cohort. The prevalence-rate method remains the most frequently applied technique because of data limitations at the national level (see, e.g, Crimmins et al. 1997).

In addition to a paucity of applications of dynamic ALE concepts to national longitudinal studies, applications of increment–decrement models have often been limited to highly aggregated or "coarse" disability states to define health changes—for example, transitions from not at all disabled to disabled with any ADL or IADL impaired; or transitions from not being disabled to severely disabled where a minimum number of ADL (or IADL) limitations are present. While coarse disability state classifications provide ALE estimates, they do not represent detailed gradations in the levels and types of disability that can be analytically extracted from multiple ADL, IADL, or physical performance responses. In addition, the use of a "not at all disabled/disabled" criterion defined by any ADL or IADL impairment defines a heterogeneous disabled group, containing, for example, individuals with one IADL partly limited (e.g., cooking) as well as individuals completely limited in all ADLs (e.g., bedfast persons).

Manton and Land (2000) addressed these limitations by adapting the random walk model to develop an increment–decrement stochastic process model of transitions among highly-refined functional status profiles interacting with a disability-specific mortality process. Manton and Land (2000) applied the model to data from the 1982, 1984, 1989, and 1994 National Long Term Care Surveys (NLTCSs) to produce ALE estimates for a finely partitioned state space representing individuals in terms of disabilities and disability intensities on multiple (6 or 7) dimensions. These surveys employ a two-stage sample design to focus interviewing resources on detailed assessments of persons with chronic disability (Manton et al. 1993). In the first stage of each NLTCS, persons of age 65+ sampled from Medicare administrative lists are screened for chronic disabilities. If they report at least one chronic (lasting or expected to last 90+ days) disability, or are living in a chronic care institution, they are given either a detailed in-person community or institutional instrument. Because samples are drawn from an administrative list of Medicare enrollees, the follow-up of persons between surveys is nearly 100%. Response rates in all four surveys were 95%.

Manton and Land (2000) made comparisons of their ALE estimates with prevalence-rate-based estimates from synthetic cohort life tables of the US elderly population in 1990 (Crimmins et al. 1997). Results are summarized in Figure 10.2. It can be seen that the ALE estimates of Manton and Land (2000) are 1.8 and 2.6 times larger than the period estimates of Crimmins et al. for males at ages 65 and 85, respectively; 1.6 and 1.9 times higher at these ages for females. The larger multiples observed for males than for females may be due to the fact that males are more likely to recover from some disability states than females. By demographic standards, these differences in ALE estimates are large and have important implications for the population burden of disability among the elderly.

Additional research will be necessary to corroborate the accuracy and reliability of the Manton and Land estimates. Suffice it to say that the more refined state-space descriptions of disability dynamics permitted by the stochastic diffusion life table model they employed have yielded estimates of ALE that

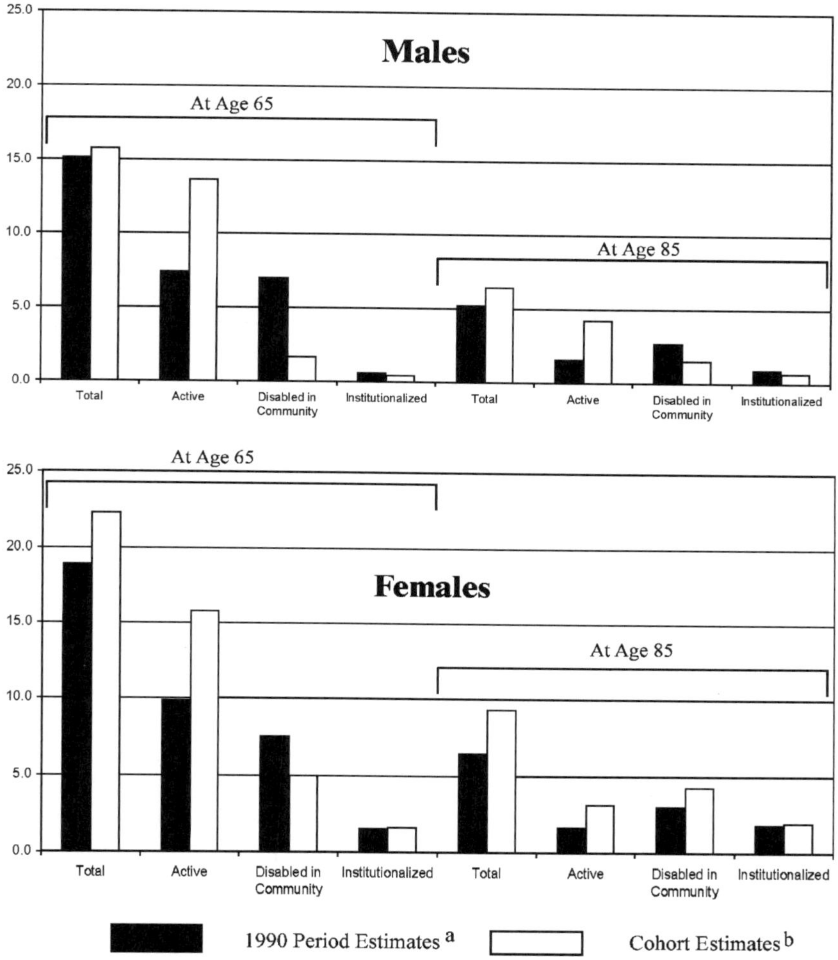

[a] Based on Crimmins et al. (1997). [b] Based on Manton and Land (2000).

Reproduced from Land (2001) with permission of The University of North Carolina Press.

FIGURE 10.2. Comparison of period and completed-cohort estimates of life expectancies, in years, in various health states

appear to be much improved over those produced by the application of the traditional prevalence life table model. If the Manton and Land (2000) ALE estimates continue to be replicated over time as additional updates of National Long Term Care Surveys become available, and if they are further refined to become cohort specific, they will become increasingly valuable social indicators of the health status of the ages 65 and over population in the United States. Furthermore, and what is most important in the context of this chapter, the ALE estimates of Manton and Land are interpretable within the context of a sophisticated mathematical model of human mortality and aging mechanisms that has been developed, applied empirically, and elaborated upon in dozens of research publications over the past two decades.

Manton and Yashin (2000) used both Framingham and NLTCS data to evaluate the magnitude of the effects of diffusion on estimates of mortality and disability changes. Specifically, they evaluated how estimates are perturbed as estimates of diffusion are altered. Manton and Yashin (2000) showed that the effects of inter-measurement diffusion were dependent on specific model assumptions. When the process was updated monthly by interpolating between measurements observed at 5-year intervals, life expectancy was higher than when the process was updated annually under the assumption that disability was constant over each 5-year measurement interval (i.e., 20.8 years at age 65 when using interpolated values versus 20.1 years when using constant values; or 3.5% higher). This was because (1) the monthly model allowed stochasticity to more frequently "shuffle" the deck so that high transition rate groups were less prevalent at each specific age and (2) the parameters of the monthly model were based on interpolated data, rather than the constant data of the annual model.

Another comparison was conducted by formulating a discrete state, discrete time Markov process based on assuming constant disability rates over each 5-year interval (like the annual model), but with the additional restriction of no within-group heterogeneity among the "fuzzy-set" groups formed from the underlying Grade of Membership analysis. With these assumptions, the frail and institutional population prevalences were higher than in the annual model, which in turn were higher than in the monthly model (Manton and Yashin 2000, p. 138). These comparisons indicated that the updating of disability states within the 5-year measurement intervals and the appropriate representation of within-group heterogeneity have potentially important effects on disability-specific life expectancy estimates.

Manton and Yashin (2000, p. 140) also showed how an unobserved variable could significantly affect estimates. Specifically, they re-estimated the process conditional on education level and found that this reduced the effect of diffusion compared to the model where education was treated as unobserved.

Other analyses have evaluated the assumption that the diffusion matrix was constant (Akushevich et al. 2005). This hypothesis was rejected because the force of homeostasis tended to weaken faster with age than mortality selection operated to remove the individuals with the highest rate of loss of stability.

10.4. Extensions of the Random Walk Model

Innovative population models, such as the random walk model, can be seen as generalizations of classical population models. Classical models date back to Gompertz (1825), who specified a two-parameter analytic expression for mortality $\mu_0(t) = \mu_G(t) = \alpha \exp \theta t$. This model still applies today for adult mortality in most human populations from about age 25 to age 85. Other long-standing mortality population models include the two-parameter Weibull model, $\mu_0(t) = \mu_W(t) = \alpha t^{m-1}$, which was applied to models of carcinogenesis (Armitage and Doll 1954). Strehler and Mildvan (1960) developed a theoretical model to describe how the combined effects of individual aging and environmental stresses produced the Gompertz mortality curve. Sacher and Trucco (1962) gave an early description of a stochastic mechanism of aging and mortality. These models predated the development of a series of models based on the demographic conception of frailty (Vaupel et al. 1979), i.e., frailty models, correlated frailty models, debilitation models, repair capacity models (see the review by Yashin et al. 2000).

A stochastic process generalization of the random walk model was explicitly formulated by Yashin and Manton (1997). This model uses the same assumptions about dynamics of covariates and the form (quadratic) of the hazard function (i.e., mortality or incidence functions) as discussed above. Specifically, in this approach the system of stochastic differential equations where the dynamics are specified as (W_t is a Wiener process),

$$dx(t) = [a_0(t) + a_1(t)x(t)]dt + a_2(t)dW_t \tag{10.6}$$

and mortality is specified as,

$$\mu[x(t), t] = \mu_0(t) + 2b(t)x(t) + x^*(t)B(t)x(t) \tag{10.7}$$

Parameters of this model are estimated using the likelihood,

$$L = \prod_{i=1}^{N} \hat{\mu}[\tau_i, \hat{x}(\tau_i)]^{\delta_i} \exp\left(-\int_0^{\tau_i} \hat{u}[u, \hat{x}_i(u)]du\right) \times \prod_{j=1}^{k_i} f[x_i(t_j)|\hat{x}i(t_{j-1})] \tag{10.8}$$

where $\hat{\mu}(\hat{x}(t), t) = m^*(t)B(t)m(t) + 2b(t)m(t) + tr(B(t)\gamma(t)) + \mu_0(t)$ has the sense of a right-continuous mortality rate. The vector of covariate means m(t) and the covariance matrix $\gamma(t)$ are defined by systems of ordinary differential equations evaluated at intervals (t_j, t_{j+1})

$$dm(t)/dt = a_0(t) + [a_1(t) - 2b(t)]m(t) - 2\gamma(t)B(t)m(t), \quad m(t_j) = \hat{x}(t_j) \tag{10.9}$$

and

$$d\gamma(t)/dt = a_1(t)\gamma(t) + \gamma(t)a_1^*(t) + a_2(t)a_2^*(t) - 2\gamma(t)B(t)\gamma(t), \quad \gamma(t_j) = 0 \tag{10.10}$$

where both equations represent both dynamic and mortality selection effects. All parameters are defined within the joint likelihood function in Eq. (10.8), so only one numerical procedure is needed to estimate all parameters simultaneously. It is not necessary to use any ad hoc procedures to fill in missing data values because the model generated values m(t) are effectively used in Eq. (10.8) for them. Furthermore, the projections are obtained as a solution of differential equations, so the intervals between times of assessment (e.g., surveys) may be variable.

Recent generalizations of the random walk model examine the use of Fokker–Planck equations where diffusion is "anomalous" or non-Gaussian. In the equations above, we represented non-Gaussian effects in a limited way by conditioning on variables reflecting unobserved or unobservable factors, e.g., by introducing age dependence in the quadratic mortality function or by generalizing the state space to have higher-order interactions. This implies an alteration or transformation of the state-space topology so that Gaussian assumptions continue to hold. Using pairwise interactions as state variables means that the fourth-order moments of the original state variables may be significant—a non-Gaussian distribution function possibly with significant skewness and kurtosis (Kulminski et al. 2004).

In fact non-Gaussian Fokker–Planck equations can be generated in other ways to represent a range of different non-Gaussian stochastic processes. One is to include instrumental variables in the drift function to represent the mean effects of systems at different levels of biological organizations (Shiino 2003). Another is to have the parameters of the Fokker–Planck equations for a system affected by unobserved environmental variables. Such variables could cause correlations in diffusion effects, so that the force of diffusion is no longer Gaussian (Frank, 2005). In this case, entropy is also re-defined so that there are interactions in environmental conditions. Entropy for this more general "non-extensive" case is sometimes called "Tsallis" entropy. In these non-Gaussian applications, the distribution of disability status is no longer random. In models of neural function, for example, there may be a hierarchical ordering of absorbing energy states if the state space is assumed to have p-adic properties (Manton et al. 2004).

10.5. Conclusions

According to the *constructal law* (Bejan 2000, 2005; Bejan and Lorente 2005), a system not in equilibrium will, over time, generate paths that allow "currents" (probability density in the state space) to flow from a point to points with the easiest access and least resistance. Typical applications of constructal theory involve an optimization of an objective function subject to constraints.

We have described herein the mathematics of the random walk model of human aging and mortality, some results from empirical applications of this model, and some of its generalizations. This model explicitly represents the "flows" of persons with specific values of physiological variables and/or physical functioning disabilities within the context of a continuously operating risk of mortality. The random walk model was not explicitly developed with a view toward the articulation of an objective function that humans optimize (consciously or not), subject to constraints. However, the components of the Fokker–Planck equations could be used to define an objective function. The most obvious candidate is the quadratic hazard function. This could lead to a quadratic optimization problem where, if survival is increased as a desirable state, lethal events could be given weight (or costs) for an individual. If, instead, convex scores (Manton et al. 1994) are used to represent disability, and use of exogenous factors z in Eq. (10.4) as control variables, with interactions (x • z) also defined, then z could be manipulated to optimize the population distribution of disability scores x. We believe that further pursuit of the intersection of constructal law ideas with those of the random walk model could be a fruitful avenue of further exploration and analysis of human aging and health dynamics, using these or similar approaches to define an appropriate objective function.

Acknowledgements The research in this chapter was supported by grants from the National Institute on Aging (Grants No. R01-AG01159 and P01-AG17937).

References

Akushevich, I., Kulminski, A. and Manton, K (2005) Life tables with covariates: Dynamic model for nonlinear analysis of longitudinal data. *Math. Popul. Stud.* **12**(2), 51–80.

Armitage, P. and Doll R. (1954) The age distribution of cancer and a multistage theory of carcinogenesis. *Br. J. Cancer* **8**, 1–12.

Bejan, A. (2000) *Shape and Structure, from Engineering to Nature*. Cambridge University Press, Cambridge, UK.

Bejan, A. (2005) The constructal law of organization in nature: Tree-shaped flows and body size. *J. Exper. Biol.* **208**(9), 1677–1686.

Bejan, A. and Lorente, S. (2005) *La Loi Constructale*. L'Harmattan, Paris.

Crimmins, E. M., Saito, Y. and Ingegneri, D. (1997) Trends in disability-free life expectancy in the United States, 1970–90. *Popul. Dev. Rev.* **23**, 555–572.

Frank, T. D. (2005) *Nonlinear Fokker-Planck Equations: Fundamentals and Applications*. Springer, New York, XII, 407 p. 86.

Gompertz, B. (1825) On the nature of the function expressive of the law of human mortality, and on a new mode of determining the value of life contingencies. *Phil. Trans. R. Soc. Lond.* **115**, 513–583.

Katz, S., Branch, L. G., Branson, M. H., Papsidero, J. A., Beck, J. C. and Greer, D. S. (1983) Active life expectancy. *N. Engl. J. Med.* **309**, 1218–1224.

Kulminski, A., Akushevich, I. and Manton, K. (2004) Modeling nonlinear effects in longitudinal survival data: Implications for the physiological dynamics of biological systems. *Front. Biosci.* **9**, 481–493.

Land, Kenneth C. (2001) Models and Indicators. *Soc. Forces* **80**, 381–410.

Land, K. C., Guralnik, J. M. and Blazer, D. G. (1994) Estimating increment-decrement life tables with multiple covariates from panel data: The case of active life expectancy. *Demography* **31**, 297–319.

Lawless, J. F. (2003) *Statistical Models and Methods for Lifetime Data*. Wiley, New York.

Manton, K. G. and Akushevich, I. (2003) State variable methods for demographic analysis: A mathematical theory of physiological regeneration and aging. *Nonlinear Phenomena in Complex Systems*, **6**(3), 717–727.

Manton, K. G., Corder, L. S. and Stallard, E. (1993) Estimates of change in chronic disability and institutional incidence and prevalence rates in the U.S. elderly population from the 1982, 1984, and 1989 National Long Term Care Survey. *J. Gerontology: Soc. Sci.* **47**(4), S153–S166.

Manton, K. G. and Land, K. C. (2000) Active life expectancy estimates for the U.S. elderly population: A multidimensional continuous mixture model of functional change applied to completed cohorts, 1982–1996, *Demography* **37**, 253–266.

Manton K. G. and Stallard E. (1988) *Chronic Disease Risk Modelling: Measurement and Evaluation of the Risks of Chronic Disease Processes.* In the Griffin Series of the Biomathematics of Diseases. Charles Griffin Limited, London, UK.

Manton, K. G. and Stallard E. (1994) Medical demography: Interaction of disability dynamics and mortality in L. G. Martin and S. H. Preston (eds.), *Demography of Aging*. Washington, DC: National Academy Press, pp. 217–178.

Manton, K. G., Stallard E. and Singer, B. H. (1992) Projecting the future size and health status of the U.S. elderly population. *Int. J. Forecasting* **8**, 433–458.

Manton, K. G., Volovyk, S. and Kulminski, A. (2004) ROS effects on neurodegeneration in Alzheimer's disease and related disorders: On environmental stresses of ionizing radiation. *Current Alzheimer Research* (Lahiri DK, Ed.) **1**(4), 277–293.

Manton, K. G., Woodbury, M. A. and Tolley, H. D. (1994) *Statistical Applications Using Fuzzy Sets.* Wiley, New York, p. 312.

Manton, K. G. and Yashin A. I. (2000) *Mechanisms of Aging and Mortality: Searches for New Paradigms.* Monographs on Population Aging, 7, Odense University Press, Odense, Denmark.

Preston, S. H., Heuveline, P. and Guillot, M. (2001) *Demography: Measuring and Modeling Population Processes*. Blackwell, Malden, MA.

Risken, H. (1996) *The Fokker-Planck Equation: Methods of Solutions and Applications* (2nd Edition). Springer, New York.

Rogers, A., Rogers, R. G. and Branch, L. G. (1989) A multistate analysis of active life expectancy. *Public Health Rep.* **104**, 222–226.

Sacher, G. A. and Trucco, E. (1962) The stochastic theory of mortality. *Ann. NY Acad. Sci.* **96**, 985–1007.

Shiino, M. (2003) Stability analysis of mean-field-type nonlinear Fokker-Planck equations associated with a generalized entropy and its application to the self-gravitating system. *Phys. Rev. E* **67**, 056118-1–056118-16.

Strehler, B. L. and Mildvan, A. S. (1960) General theory of mortality and aging. *Science* **132**, 14–21.

Sullivan, D. F. (1971) A single index of mortality and morbidity. *HSMHA Health Rep.* **86**, 347–354.

Vaupel, J. W., Manton, K.G. and Stallard, E. (1979) The impact of heterogeneity in individual frailty on the dynamics of mortality. *Demography* **16**(3), 439–454.

Woodbury, M. A. and Manton, K. G. (1977) A random-walk model of human mortality and aging. *Theor. Popul. Biology* **11**, 37–48.

Woodbury, M. A. and Manton, K. G. (1983) A theoretical model of the physiological dynamics of circulatory disease in human populations. *Human Biology* **55**, 417–441.

Yashin, A. I. and Manton, K. G. (1997) Effects of unobserved and partially observed covariate processes on system failure: A review of models and estimation strategies. *Stat. Sci.* **12**(1), 20–34.

Yashin, A. I., Iachine, I. A. and Begun, A. S. (2000) Mortality modeling: A review. *Math. Popul. Stud.* **8**(4), 305–332.

Chapter 11
Statistical Mechanical Models for Social Systems

Carter T. Butts

11.1. Summary

In recent years, researchers in the area of constructal theory have sought to apply principles from the modeling of engineered systems to problems throughout the sciences. This chapter provides an application of statistical mechanical models to social systems arising from the assignment of objects (e.g., persons, households, or organizations) to locations (e.g., occupations, residences, or building sites) under the influence of exogenous covariates. Two illustrative applications (occupational stratification and residential settlement patterns) are presented, and simulation is employed to show the behavior of the location system model in each case. Formal analogies between thermodynamic and statistical interpretations of model elements are discussed, as is the compatibility of the location system model with the assumption of stochastic optimization behavior on the part of individual agents.

11.2. Introduction

At their core, developments in the growing body of literature known as "constructal theory" (Bejan 1997, 2000; Bejan and Lorente 2004; Bejan and Marden 2006; Reis and Bejan 2006) reflect an effort to apply principles from the modeling of engineered systems to problems throughout the sciences. In this respect, constructal theory falls within the tradition of researchers such as Zipf (1949) and Calder (1984), who have sought to capture the behavior of a diverse array of systems via a combination of physical constraints and optimization processes. Within the social sciences, constrained optimization has been central to research on choice theoretic models (the dominant paradigm in economics) and has had a strong influence on the literatures in organization theory (particularly organizational design) and human judgment and decision making. While optimization-based models have not always proven correct, their simplicity and generalizability continue to attract interest from researchers in many fields. In contrast, physical constraints – or even models borne of physical processes – have seen only intermittent integration into social science research. This is unfortunate, given both the reality of physical limits on social organization and the potential applicability of certain physical models to social systems.* Here, we describe one modeling framework which incorporates both elements to capture the behavior of high-dimensional social systems whose components exhibit complex dependence. Although inspired by related models within social network analysis, this framework can be shown to admit a physical interpretation, thereby facilitating the application of insights from other fields (particularly statistical mechanics) to a broad class of social phenomena. It is hoped that the results shown here will serve to encourage further development of social models which capitalize on analogies with physical processes.

11.2.1. Precursors Within Social Network Analysis

A fundamental problem facing researchers in the social network field has been the need to model systems whose elements depend upon one another in non-trivial ways. For instance, in modeling directed relations (i.e., those with distinct senders and receivers), it will generally be the case that ties (or "edges," in the language of graph theory) sharing the same endpoints will depend upon one another. This form of dependence (called *dyadic dependence*) was combined with the notion of heterogeneity in rates of tie formation to form the first family of what would eventually be called *exponential random graph models*, the p_1 family (Holland and Leinhardt 1981). While dyadic dependence was relatively simple in nature, the models created to cope with it were rapidly expanded into more complex cases. For instance, Frank and Strauss (1986) famously considered graph processes in which two edges are dependent if they share any endpoint; this lead to the *Markov graphs*, whose properties are far more intricate than those of processes exhibiting only dyadic dependence. Extensions to other, still more extensive forms of dependence followed (e.g., Strauss and Ikeda 1990; Wasserman and Pattison 1996; Pattison and Wasserman 1999; Robins et al. 1999; Pattison and Robins 2002), along with corresponding innovations in simulation and inferential methods (Crouch et al. 1998; Snijders 2002; Hunter and Handcock 2006).

For our present purposes, it is important to emphasize that these developments did not arise from efforts within the social network field alone. Rather, they resulted from an interdisciplinary synthesis in which insights from areas such as spatial statistics (e.g., Besag 1974, 1975) and statistical physics (Strauss 1986; Swendsen and Wang 1987), as well as innovations in computing technology and simulation methods (Geyer and

* Recognition of this connection is at least as old as the founding of sociology, which was initially envisioned as leading to a form of "social physics" (Quetelet 1835, Comte 1854). Although primarily invoked for rhetorical purposes, this stance foreshadowed later developments such as the work of Coleman (1964), who drew extensively on physical models.

Thompson 1992; Gamerman 1997), were leveraged to formulate and solve problems within the social sciences. In the best tradition of such "borrowing," scientific advances were made neither by ignoring external developments nor by blindly importing models from other disciplines (nor still by an "invasion" of scientists from these disciplines into the network field). Progress resulted instead from the recognition of structural similarities between problems in network research and problems in other substantive domains, followed by the adaptation and translation of models from these fields into the network context. This approach has allowed network researchers to avoid many of the pitfalls identified by Fararo (1984) associated with the importation of mathematically attractive models with poor empirical motivation. It may also serve as a useful example to be emulated by workers in areas such as constructal theory, who seek to apply their ideas to novel substantive domains.

While the focus of this chapter is not on network analysis, the approach taken here owes much to the exponential family modeling tradition described above. It also draws heavily on the statistical mechanical framework to which that tradition is closely related. More specifically, this chapter presents a family of models for social phenomena which can be described in terms of the arrangement of various (possibly related) objects with respect to a set of (again, possibly related) locations. This family is designed so as to leverage the large literature on the stochastic modeling of systems with non-trivial dependence structures. It is also constructed so as to be applicable across a wide range of substantive contexts; to scale well to large social systems; to be readily simulated; to be specifiable in terms of directly measurable properties; and to support likelihood-based inference using (fairly) standard methods. Although this model family is not obviously "constructal" in the sense of Bejan (1997, 2000), it does incorporate elements of (local) optimization and physical constraint. Thus, it may be of some interest to those working in the area of constructal theory per se.

The structure of the remainder of the chapter is as follows. After a brief comment on notation, we present the core formalism of the chapter (the generalized location system). Given this, we turn to a discussion of modeling location systems, including both conceptual and computational issues. Finally, we illustrate the use of the location system model to examine two classes of processes (occupational stratification and residential settlement patterns), before concluding with a brief discussion.

11.2.2. Notation

We here outline some general notation, which will be used in the material which follows. A graph, G, is defined as G = (V, E), where V is a set of vertices and E is a set of edges on V. When applied to sets, $|\cdot|$ represents cardinality; thus $|V|$ is the number of vertices (or order) of G. In some cases (particularly when dealing with valued graphs), it will be useful to represent graphs in adjacency matrix form, where the adjacency matrix X for graph G is defined as a $|V| \times |V|$ matrix such that X_{ij} is the value of the (i, j) edge in G. By convention, $X_{ij} = 0$

if G contains no (i, j) edge. A tuple of graphs $(G_1, \ldots, G_n)$ on common vertex set V may be similarly represented by a $n \times |V| \times |V|$ adjacency array, X, such that X_{ij} is the adjacency matrix for G_i .

When referring to a random variable, X, we denote the probability of a particular event x by $\Pr(X = x)$. More generically, $\Pr(X)$ refers to the probability mass function of X (where X is discrete). Expectation is denoted by the operator **E**, with subscripts used to designate conditioning where necessary. Thus, the parametric pmf $\Pr(X|\theta)$ leads to the corresponding expectation $\mathbf{E}_\theta(X)$. (Likewise for variance, written $\mathrm{Var}_\theta(X)$.) When discussing sequences of realizations of a random variable X, parenthetical superscript notation is used to designate particular draws (e.g., $(x^{(1)}, \ldots, x^{(n)})$). Distributional equivalence is denoted by $\sim$ (read: "is distributed as"), so $X \sim Y$ implies that X is distributed as Y. For convenience, this notation may also be extended to pmfs, such that $X \sim f$ (for random variable X and pmf f) should be understood to mean that X is distributed as a random variable with pmf f.

11.3. Generalized Location Systems

Our focus here is on what we shall call *generalized location systems*, which represent the allocation of arbitrary entities (e.g., persons, objects, organizations) to "locations" (e.g., physical regions, jobs, social roles). While our intent is to maintain a high level of generality, we will limit ourselves to systems for which both entities and locations are discrete and countable, and for which it is meaningful to treat the properties of entities and locations as relatively stable (at least for purposes of analysis). Relaxation of these constraints is possible, but will not be discussed here; even with these limitations, however, the present framework still allows for a great deal of flexibility.

We begin by positing a system of n identifiable *objects*, $O = (o_1, \ldots, o_n)$, each of which may reside in exactly one of m identifiable *locations*, $L = (l_1, \ldots, l_m)$. The state of this system at any given time is represented by a *configuration vector*, $\ell \in \{1, \ldots, m\}^n$, which is defined such that $\ell_i = j$ iff o_1 resides at location l_j. Depending on the system in question, not all hypothetical configuration vectors are physically realizable; the set of all such realizable vectors is said to be the set of *accessible configurations*, and is denoted $\mathbb{C}$. $\mathbb{C}$ may be parameterized in a number of ways, perhaps the most important of which being in terms of *occupancy constraints*. We define the *occupancy function* of a location system as

$$\mathbf{P}(x, \ell) = \sum_{i=1}^{n} I(\ell_i = x), \tag{11.1}$$

where I is the standard indicator function. The vectors of maximum and minimum occupancies for a given location system are composed of the maximum/minimum values of the occupancy function for each state under $\mathbb{C}$ (respectively). That is, we require that $P_i^- \leq \mathbf{P}(i, \ell) \leq P_i^+$ for all $i \in 1, \ldots, m$, $\ell \in \mathbb{C}$, where P^-, P^+ are the minimum and maximum occupancy vectors. If $P_i^- = P_i^+ = 1 \quad \forall \quad i \in 1, \ldots, m$, then it follows that ℓ is a permutation vector on $1, \ldots, n$, in which case we must have $m = n$ for non-empty $\mathbb{C}$. This is an important special case, particularly in organizational contexts (White 1970). By contrast, it is frequently the case in geographical contexts (e.g., settlement) that $P_i^- = 0$ and $P_i^+ > n \quad \forall \quad i \in 1, \ldots, m$, in which case occupancy is effectively unconstrained.

In addition to configurations and labels, objects and locations typically possess other properties of scientific interest. We refer to these as *features*, with F_O being the set of object features and F_L being the set of location features. While we do not (initially) place constraints on the feature sets, it is worth highlighting two feature types which are of special interest. Feature vectors provide ways of assigning numerical values to individual objects or locations, e.g., age, average rent level, or wage rate. Adjacency matrices can also serve as important features, encoding dyadic relationships among objects or locations. Examples of such relationships can include travel distance, marital ties, or demographic similarity. Because relational features allow for coupling of objects or locations, they play a central role in the modeling of complex social processes (as we shall see).

To draw the above together, we define a generalized location system by the tuple $(L, O, \mathbb{C}, F_L, F_O)$. The state of the system is given by ℓ, which will be of primary modeling interest. Various specifications of $\mathbb{C}$ are possible, but particular emphasis is placed on occupancy constraints, which specify the range of populations which each location can support. With these elements, it is possible to model a wide range of social systems, and it is to this problem that we now turn.

11.4. Modeling Location Systems

Given the definition of a generalized location system, we now present a stochastic model for its equilibrium state. In particular, we assume that – given a set of accessible configurations, $\mathbb{C}$– the system will be found to occupy any particular configuration, ℓ, with some specified probability. Our primary interest is in the modeling of these equilibrium probabilities, although some dynamic extensions are possible.

Given the above, we first define the set indicator function

$$I_{\mathbb{C}}(\ell) = \begin{cases} 1 & \text{if } \ell \in \mathbb{C} \\ 0 & \text{otherwise} \end{cases} \tag{11.2}$$

The equilibrium probability of observing a given configuration can then be written as

$$\Pr(S = \ell) = I_{\mathbb{C}}(\ell) \frac{\exp(P(\ell))}{\sum_{\ell' \in \mathbb{C}} \exp(P(\ell'))} \tag{11.3}$$

where S is the random state, and P is a quantity called the *social potential* (defined below). The sum

$$Z(P, \mathbb{C}) = \sum_{\ell' \in \mathbb{C}} \exp(P(\ell')) \tag{11.4}$$

is the *normalizing factor* for the location model, a quantity which corresponds directly to the partition function of statistical mechanics (Kittel and Kroemer 1980). Equation (11.3) defines a discrete exponential family on the elements of $\mathbb{C}$, and is complete in the sense that any pmf on $\mathbb{C}$ can be written in the form of Eq. (11.3). This completeness is an important benefit of the discrete exponential family framework, but there are other benefits as well. For instance, models of this type have been widely explored in both physics and mathematical statistics (see, e.g., Barndorff-Nielsen 1978; Brown 1986), facilitating the cross-application of existing knowledge to new modeling problems. Also significant is the fact that, for appropriate parameterizations of P, a number of well-known results allow for likelihood-based inference of model parameters from empirical data (Johansen 1979). These advantages may be contrasted with, for instance, those of intellective agent-based approaches, which (while dynamically flexible) frequently exhibit poorly understood behavior, and which rarely admit a principled theory of inference.

While Eq. (11.3) can represent any distribution on $\mathbb{C}$, its scientific utility clearly lies in identifying a theoretically appropriate specification of P. Intuitively, the social potential for any given configuration is equal to its log-probability, up to an additive constant. Thus, the location system is more likely to be found in areas of high potential, and/or (in a dynamic context) to spend more time in such states. While any number of forms for P could be proposed, we here work with a constrained family which incorporates several features of known substantive importance for a variety of social systems. This family of potential functions is introduced in the following section.

11.4.1. A Family of Social Potentials

As noted above, we seek a family of functions $P : \mathbb{C} \mapsto \mathbb{R}$ such that $\Pr(S = \ell) \propto \exp(P(\ell))$. This family should incorporate as wide a range of substantively meaningful effects as possible; since it is not reasonable to expect effects to be identical in every situation, the family should be *parameterized* so as to allow differential weighting of effects. Ideally, the social potential family should also be easily computed, and its structure easily interpreted.

An obvious initial solution to this problem is to construct P from a linear combination of deterministic functions of F_L and F_O, which then act as sufficient statistics for the resulting distribution. Employing such a potential function within Eq. (11.3) leads to a regular exponential family on $\mathbb{C}$ (Johansen 1979), which has a number of useful statistical implications. The so-called "curved" exponential family models (which are formed by allowing P to be a non-linear function of statistics and parameters) (Efron 1975) are also possible, and may be useful in certain cases (e.g., where one must enforce a functional relationship among large numbers of parameters; see Hunter and Handcock 2006, for a network-related example). Here, we restrict ourselves to the linear case.

Even if a linear form is supposed, however, we are still left with the question of which effects should be included in the social potential. By definition, these effects must be parameterized as functions of the location and object features. Further, both location and object features (as we are using the term) can include both attributes (features of the individual location or object per se) and relations (features of object or location sets). Here, we will limit ourselves to relations which are dyadic (i.e., defined on pairs) and single-mode (i.e., which do not mix objects and locations). Thus, our effects should be functions of feature vectors, and/or (possibly valued) graphs.

While this constraint still admits a wide range of possibilities, we can further focus our attention by noting that the purpose of P is ultimately to control the assignment of objects to locations. This suggests immediately that the effects of greatest substantive importance will be those which draw objects toward or away from particular locations. Table 11.1 provides one categorization of such effects by feature type. In the first (upper left) cell, we find effects which express direct attraction or repulsion between particular objects and locations, based on their attributes. In the second (upper right) cell are effects which express a tendency for objects linked through connected locations to be particularly similar or distinct. (Spatial autocorrelation is a classic example of such an effect.) The converse family of effects is found in the third (lower left) cell; these effects represent a tendency for objects to be connected to other objects with similar (or different) locations. Homophily in career choice—where careers are interpreted as "locations"—serves as an example of a location homogeneity effect. Finally, in the fourth (lower right) cell we have effects based on the tendency of location relations to align (or disalign) with object relations. Propinquity, for example, is a tendency for adjacent objects to reside in nearby locations.

Taken together, these four categories of effects combine to form the social potential. Under the assumption of linear decomposability, we thus posit four sub-potentials (one for each category) such that

$$P(\ell) = P_\alpha(\ell) + P_\beta(\ell) + P_\gamma(\ell) + P_\delta(\ell)\,. \tag{11.5}$$

TABLE 11.1. Elements of the social potential

	Location attributes	Location relations
Object Attributes	Attraction/Repulsion Effects	Object Homogeneity/ Heterogeneity Effects (through Locations)
Object Relations	Location Homogeneity/Heterogeneity Effects (through Objects)	Alignment Effects

We now consider each of these functions in turn. The first class of effects which must be represented in any practical location system are global attraction/repulsion—also called "push/pull"—effects. Residential locations, potential firm sites, occupations, and the like have features which make them generally likely to attract or repel certain objects (be they persons, organizations, or other entities). Such effects are naturally modeled via product-moments of attributes. Let $Q \in \mathbb{R}^{m \times a}$, $X \in \mathbb{R}^{n \times a}$ be exogenous features reflecting location and object attributes (respectively), and let $\alpha \in \mathbb{R}^a$ be a parameter vector. Then we may define P_α as

$$P_\alpha(\ell) = \sum_{i=1}^{a} \alpha_i t_i^\alpha(\ell) \tag{11.6}$$

$$= \sum_{i=1}^{a} \alpha_i \sum_{j=1}^{n} Q_{\ell_j i} X_{ji}, \tag{11.7}$$

where t^α is a vector of sufficient statistics.

A second class of effects concerns object homogeneity/heterogeneity—that is, the conditional tendency for associated locations to be occupied by objects with similar (or different) features. Let $Y \in \mathbb{R}^{n \times b}$ be a matrix of object attributes, $B \in \mathbb{R}^{b \times m \times m}$ be an adjacency array on the location set, and $\beta \in \mathbb{R}^b$ a parameter vector. Then we define the object homogeneity/heterogeneity potential by

$$P_\beta(\ell) = \sum_{i=1}^{b} \beta_i t_i^\beta(\ell) \tag{11.8}$$

$$= \sum_{i=1}^{b} \beta_i \sum_{j=1}^{n} \sum_{k=1}^{n} B_{i\ell_j\ell_k} \left|Y_{ji} - Y_{ki}\right|, \tag{11.9}$$

where, as before, t^β is a vector of sufficient statistics. It should be noted that the form of t^β is closely related to Geary's C, a widely used index of spatial autocorrelation (Cliff and Ord 1973). t^β is based on absolute rather than squared differences, and is not normalized in the same manner as C, but its behavior is qualitatively similar in many respects.

The parallel case to P_β is P_γ, which models the effect location homogeneity or heterogeneity through objects. Let $R \in R^{m \times c}$ be a matrix of location features, $A \in \mathbb{R}^{c \times n \times n}$ be an adjacency array on the object set, and $\gamma \in \mathbb{R}^c$ be a parameter vector. Then P_γ is defined as follows:

$$P_\gamma(\ell) = \sum_{i=1}^{c} \gamma_i t_i^\gamma(\ell) \tag{11.10}$$

$$= \sum_{i=1}^{c} \gamma_i \sum_{j=1}^{n} \sum_{k=1}^{n} A_{ijk} \left|R_{\ell_j i} - R_{\ell_k i}\right|, \tag{11.11}$$

As implied by the above, t^γ is the vector of sufficient statistics for location homogeneity. t^γ is at core similar to t^β, save in that the role of object and location are reversed: absolute differences are now taken with respect to *location* features, and are evaluated with respect to the connections between the objects occupying said locations.

The final element of the social potential is the alignment potential, P_δ, which expresses tendencies toward alignment or disalignment of object and location relations. Given object and location adjacency arrays $W \in \mathbb{R}^{d\times n\times n}$ and $D \in \mathbb{R}^{d\times m\times m}$ (respectively) and parameter vector $\delta \in \mathbb{R}^d$, the alignment potential is given by

$$P_\delta(\ell) = \sum_{i=1}^{d} \delta_i t_i^\delta(\ell) \tag{11.12}$$

$$= \sum_{i=1}^{d} \delta_i \sum_{j=1}^{n} \sum_{k=1}^{n} W_{ijk} D_{i\ell_j\ell_k}, \tag{11.13}$$

where, as in the prior cases, t^δ represents the vector of sufficient statistics. The form chosen for t^δ is Hubert's Gamma, which is the standard matrix cross-product moment (see Hubert 1987, for a range of applications).

It should be noted that all four effect classes can actually be written in terms of matrix cross-product moment statistics on suitably transformed adjacency arrays, and hence only P_δ is formally required to express P. Although formally equivalent to that shown above, this parameterization obscures the substantive interpretation of matrix/vector effects outlined in Table 11.1, and requires pre-processing of raw adjacency data; for this reason, we will continue to treat the sub-potentials as distinct in the treatment which follows. Given this parameterization, we may complete our development by substituting the quantities of Eqs. (11.7–11.13) into Eq. (11.5) which gives us

$$P_{(\ell)} = \sum_{i=1}^{a} \alpha_i t_i^\alpha(\ell) + \sum_{i=1}^{b} \beta_i t_i^\beta(\ell) + \sum_{i=1}^{c} \gamma_i t_i^\gamma(\ell) + \sum_{i=1}^{d} \delta_i t_i^\delta(\ell) \tag{11.14}$$

in terms of sufficient statistics, or

$$\begin{aligned} P(\ell) = &\sum_{i=1}^{a} \alpha_i \sum_{j=1}^{n} Q_{\ell_j i} X_{ji} + \sum_{i=1}^{b} \beta_i \sum_{j=1}^{n} \sum_{k=1}^{n} B_{i\ell_j\ell_k} \left|Y_{ji} - Y_{ki}\right| \\ &+ \sum_{i=1}^{c} \gamma_i \sum_{j=1}^{n} \sum_{k=1}^{n} A_{ijk} \left|R_{\ell_j i} - R_{\ell_k i}\right| + \sum_{i=1}^{d} \delta_i \sum_{j=1}^{n} \sum_{k=1}^{n} W_{ijk} D_{i\ell_j\ell_k} \end{aligned} \tag{11.15}$$

in terms of the underlying covariates. Together with Eq. (11.3), Eq. (11.15) specifies a regular exponential family of models for the generalized location system. As we have seen, this family allows for the independent specification of attraction/repulsion, heterogeneity/homogeneity, and alignment effects. Although motivated on purely social grounds, it is noteworthy that this model is fundamentally statistical mechanical in nature. Given the broader focus of this book on the application of physical modeling strategies to a wide range of substantive areas, we now consider this connection in greater detail.

11.4.2. Thermodynamic Properties of the Location System Model

We have already seen that the stochastic location system model of Eq. (11.3) can be viewed as directly analogous to a standard class of statistical mechanical models. This fact allows us to employ some useful results from the physics literature (see, e.g., Kittel and Kroemer 1980) to elucidate several aspects of model behavior. (Interestingly, many of these results have parallels within the statistical literature, and can be derived in other ways; see, e.g., Barndorff-Nielsen (1978). See also Strauss (1986) for a similar discussion in the context of exponential random graph models.)

As noted above, the normalizing factor $Z(P, \mathbb{C})$ is directly analogous to the *partition function* of statistical mechanics. The quantity $F = -\ln Z(P, \mathbb{C})$, in turn, corresponds to the *free energy* of the location system. In a classical statistical mechanical system, the probability of observing the system in microstate j is given by

$$p_j = \frac{\exp(-\varepsilon_j/\tau)}{Z}, \tag{11.16}$$

where ε_j is the microstate energy of j and τ is the temperature. Thus, the log-probability of microstate j is a linear function of the free and microstate energies:

$$\ln p_j = F - \frac{\varepsilon_j}{\tau}. \tag{11.17}$$

Returning to Eq. (11.3), it is immediately apparent that the social potential P plays the role of $-\varepsilon/\tau$. Indeed, inspecting Eq. (11.14) reveals an even closer correspondence: the realizations of the sufficient statistics associated with the elements of t^α, t^β, t^γ, and t^δ are similar to microstate energies, and the corresponding parameters (α, β, γ, and δ) can be thought of as vectors of inverse temperatures. More precisely, each sufficient statistic is analogous to the energy function (or Hamiltonian) associated with a particular "mode" of ℓ, just as the total microstate energy of a particle system might combine contributions from translational, rotational, and/or vibrational modes. The "energy" associated with a particular microstate, ℓ, in each mode is given by the value of the sufficient statistic for that microstate (i.e., $t_i^\theta(\ell)$). As in the physical case, the log-probability of observing a particular realization of the location system can be expressed as a "free energy" minus a linear combination of microstate "energies" whose coefficients correspond to "inverse temperatures." While one does not conventionally encounter multiple temperatures in a physical system (although a close examination of parameters such as the chemical potential shows them to act as de facto temperature modifiers), we will find that this metaphor is useful in understanding the behavior of the location system. This point was foreshadowed by Mayhew et al. (1995), who invoked an "energy distribution principle" in describing the occurrence of naturally forming groups. The present model implements a notion of precisely this sort, for more general social systems. As exponential family models also have the property of maximizing entropy conditional on their parameters and sufficient statistics, these location system models can also be thought of as a family of *baseline models* in the sense of Mayhew (1984a).

In addition to providing insight into system behavior, the above relations are also helpful in deriving other characteristics. For instance, the average microstate energy of the system described by Eq. (11.16) is given by $dF/d\theta$, where $\theta=\tau^{-1}$. It follows for our purposes that

$$E_{(\alpha,\beta,\gamma,\delta)}t_i^{\theta}=\frac{-d\ln Z(P,\mathbb{C})}{d\theta_i}, \tag{11.18}$$

where θ_i represents any parameter of the system. Thus, expectations for arbitrary sufficient statistics can be obtained through the partition function. Second moments may be obtained in a similar manner: the Hessian matrix $\frac{-d^2F}{d\theta^2}$ yields the variance-covariance matrix for all sufficient statistics in the system. (In the physical case, this corresponds to the *energy fluctuation*, or the *variance in energy.*)

Moments of sufficient statistics are useful for a variety of purposes, but other statistical mechanical properties of the location system may also be of value. For instance, the "heat capacity" of the system for parameter θ_i is given by $\mathrm{Var}(t_i^{\theta}(\ell))\theta_i^2=\left.\frac{-d^2F}{d\theta^2}\right|_{ii}\theta_i^2$. In the physical case, heat capacity reflects the capacity of a system to store energy (in the sense of the change in energy per unit temperature). Here, heat capacity for parameter θ_i reflects the sensitivity of the corresponding statistic t_i^{θ} to changes in the "temperature" $1/\theta_i$. For instance, if θ_i corresponds to a attraction parameter between income and gender, then heat capacity can be used to parameterize the income consequences of a weakening (or strengthening) of the attractive tendency within the larger system.

Using arguments similar to the above, it is possible to derive analogs to various other thermodynamic properties such as pressure and entropy (the latter also obtainable through information-theoretic arguments). While one must always be careful in interpreting such quantities, they may nevertheless provide interesting and useful ways of describing the properties of location systems. We will see some of the interpretational value of thermodynamic analogy below, when we consider some sample applications of the location system model; before proceeding to this, however, we turn to the question of how location system behavior may be simulated.

11.4.3. Simulation

For purposes of both prediction and inference, it is necessary to simulate the behavior of the location system model for arbitrary covariates and parameter values. Generally, it is not possible to take draws from the location system model directly due to the large size of $\mathbb{C}$: except in very special cases, the computational complexity of calculating $Z(P, \mathbb{C})$ is prohibitive, and hence the associated probability distribution cannot be normalized. Despite this limitation, approximate samples from the location system model may be readily obtained by means of a Metropolis algorithm. Given that numerous accessible references on the Metropolis algorithm and other Markov chain Monte Carlo methods are currently available (see, e.g., Gamerman 1997; Gilks et al. 1996; Gelman et al. 1995), we will focus here on issues which are specific to the model at hand. Fortunately, the location system model is not especially difficult to simulate, although certain measures are necessary to ensure scalability for large systems.

To review, a Metropolis algorithm proceeds in the following general manner (see Gilks et al. 1996, for further details). Let S be the (random) system state. We begin with some initial state $\ell^{(0)} \in \mathbb{C}$, and propose moving to a candidate state $\ell^{(1)}$ which is generally chosen so as to be in a neighborhood of $\ell^{(0)}$. (Some additional constraints (e.g., detailed balance) apply to the candidate distribution, but these do not affect the results given here.) The candidate state is then "accepted" with probability $\min\left(1, \frac{\Pr(S=\ell^{(1)}|P,\mathbb{C})}{\Pr(S=\ell^{(0)}|P,\mathbb{C})}\right)$. If accepted, the candidate becomes our new base state, and we repeat the process for $\ell^{(2)}$. If rejected, $\ell^{(1)}$ is replaced by a copy of $\ell^{(0)}$, and again the process is repeated. This process constitutes a Markov chain whose equilibrium distribution (under certain fairly broad conditions) converges to the target distribution (here, $\Pr(S|P, \mathbb{C})$). It is noteworthy that this process requires only that the target distribution be computable up to a constant factor; this feature makes Metropolis algorithms (and related MCMC techniques) very attractive to those working with exponential family models (e.g., Strauss 1986; Snijders 2002; Butts 2006).

11.4.3.1. The Location System Model as a Constrained Optimization Process

In addition to the application of analogies from physical systems, a central element of constructal theory is constrained optimization. In that regard it is interesting to note that the equilibrium behavior of the location system model can be shown to emerge from a choice process in which individual agents (here, our "objects") act to stochastically maximize a utility function, subject to constrained options. In particular, let $u(\ell)$ be the utility of configuration ℓ for each agent, and let us imagine that opportunities for agents to change location arrive at random times. When such an opportunity arises, the agent in question is able to choose between moving to a specified alternative location and remaining in place. If the move in question is utility increasing (i.e., if $u(\ell') > u(\ell)$ for a move leading to location vector ℓ'), then the agent relocates. Otherwise, we presume that the agent has some chance of moving regardless, corresponding to $\exp\left[u(\ell') - u(\ell)\right]$. This can be understood as a form of bounded rationality, in which agents occasionally overestimate the value of new locations, or as arising from unobserved heterogeneity in agent preferences. Under fairly mild conditions regarding the distribution of movement opportunities (most critically, each agent must have a non-zero probability of having the opportunity to move to any given location through some move sequence in finite time), this process forms a Markov chain whose equilibrium distribution is proportional to u; indeed, it is a special case of the Metropolis algorithm, described above. Given this, it follows that the equilibrium behavior of such a system can be described by the location system model in the case for which u = P. While this is not the only dynamic system which gives rise to this equilibrium distribution, it is nevertheless sufficient to show that the location system model can arise from a process of constrained stochastic optimization. This constitutes another affinity with the constructal perspective, albeit an attenuated one (since the optimization involved is only approximate).

11.5. Illustrative Applications

The location system model can be employed to represent a wide range of social systems. This breadth of potential applications is illustrated here by means of two simple examples, one involving economic inequality and another involving residential segregation. Although both examples shown here are stylized for purposes of exposition, they do serve to demonstrate some of the phenomena which can be captured by the location system model.

11.5.1. Job Segregation, Discrimination, and Inequality

Our first application employs the location system to model occupational stratification. We begin by positing a stylized "microeconomy" of 100 workers (objects), who are matched with 100 distinct jobs (locations) on a 1:1 basis. The population of workers is taken to consist of equal numbers of men and women, who are allocated at random into heterosexual couples such that all individuals have exactly one partner. To represent other individual features which may affect labor market performance, we also rank the workers on a single dimension of "human capital," with ranks assigned randomly in alternating fashion by gender. (Thus, males and females have effectively identical human capital distributions, and human capital ranks are uncorrelated within couples.) Like workers, jobs vary in features which may make them more or less desirable; here, we assign each job a "wage" (expressed in rank order), and group jobs into ten contiguous occupational categories. Thus, the top ten jobs (by wage) are in the first category, the next ten are in the second category, etc. While this setting is greatly simplified, it nevertheless allows us to explore basic interactions between occupational segregation, household effects, and factors such as discrimination. Elaborations such as hierarchical job categories, distinct unemployed states, additional job or worker attributes, and relaxations of 1:1 matching, could easily be employed to model more complex settings.

To examine the behavior of the location system model under different assignment regimes, we simulate model draws across a range of parameter values.

For instance, the first panel of Fig. 11.1 shows the mean male/female wage gap (i.e., the difference between mean male and female wages) under a model incorporating human capital and discrimination effects. Both effects are implemented as α parameters, and thus reflect general sorting tendencies; in particular, higher values of α_1 (discrimination) reflect stronger tendencies to place males in high-wage occupations, while higher values of α_2 (human capital) reflect stronger tendencies to place workers with high levels of human capital in high-wage occupations. At $\alpha_i = 0$, the corresponding tendency is fully inactive; $\alpha = (0, 0)$ is thus a random mixing model. Negative values of α_1 indicate discrimination in favor of females (i.e., a tendency to place males in low-wage occupations); since human capital is unlikely to have a negative effect on earnings, negative α_2 values are not considered. Within Fig. 11.1, each plotted circle corresponds to the outcome of a simulation at the corresponding coordinates. Circle area indicates the magnitude of the observed effect (mean gap for the left-hand panel, variance in wage gap for the right-hand panel) and shading indicates the mean direction of the gap in question (dark favoring males and light favoring females). For the simulation 500 coordinate pairs were chosen, and 1,000 MCMC draws taken at each pair (thinned from a full sample of 200,000 per pair, with a burn-in period of 100,000 draws). To speed convergence, all chains were run in parallel using a coupled MCMC scheme based on Whittington (2000), with state exchanges among randomly selected chain pairs occurring every 25 iterations. Coordinates were placed using a two-dimensional Halton sequence (Press et al. 1992), as this was found to produce more rapid convergence for the coupled MCMC sampler than a uniform grid (but covers the space more evenly than would a pseudo-random sample). As shown by the first panel of Fig. 11.1, increasing discrimination in favor of males ($\alpha_1 > 0$) or females ($\alpha_1 < 0$) generates a corresponding increase in the wage gap. This effect is persistent, but attenuates in the presence of strong human capital effects; since human capital is here uncorrelated with gender, selection on this dimension tends to "dampen out" the effects of discrimination. This is particularly clear in the second panel of Fig. 11.1, which shows that wage gap variance diminishes rapidly as the merit effect climbs. Thus, both the stratified and unstratified states arising in the upper portion of the parameter space are appreciably lower in variance than the unstratified region close to the origin, an effect which is not captured by the mean gap alone.

An interesting counterpoint to the purely inhibitory effect of additional α effects (here, human capital) on stratification is the impact of occupational segregation. To capture the latter effect, we replace the second α effect with a β effect expressing the tendency for jobs within the same occupational category to be more (or less) heterogeneous with respect to their gender composition. Negative β effects act to inhibit heterogeneity, and hence model (in this context) the effect of occupational segregation. Positive β values, by contrast, imply supra-random mixing (as might be produced, for instance, by an affirmative action policy). The results from simulations varying both effects are shown in Fig. 11.2 (simulations for Figs. 11.2 and 11.3 were performed in the same manner as those of Fig. 11.1). While discrimination continues to have its usual effect,

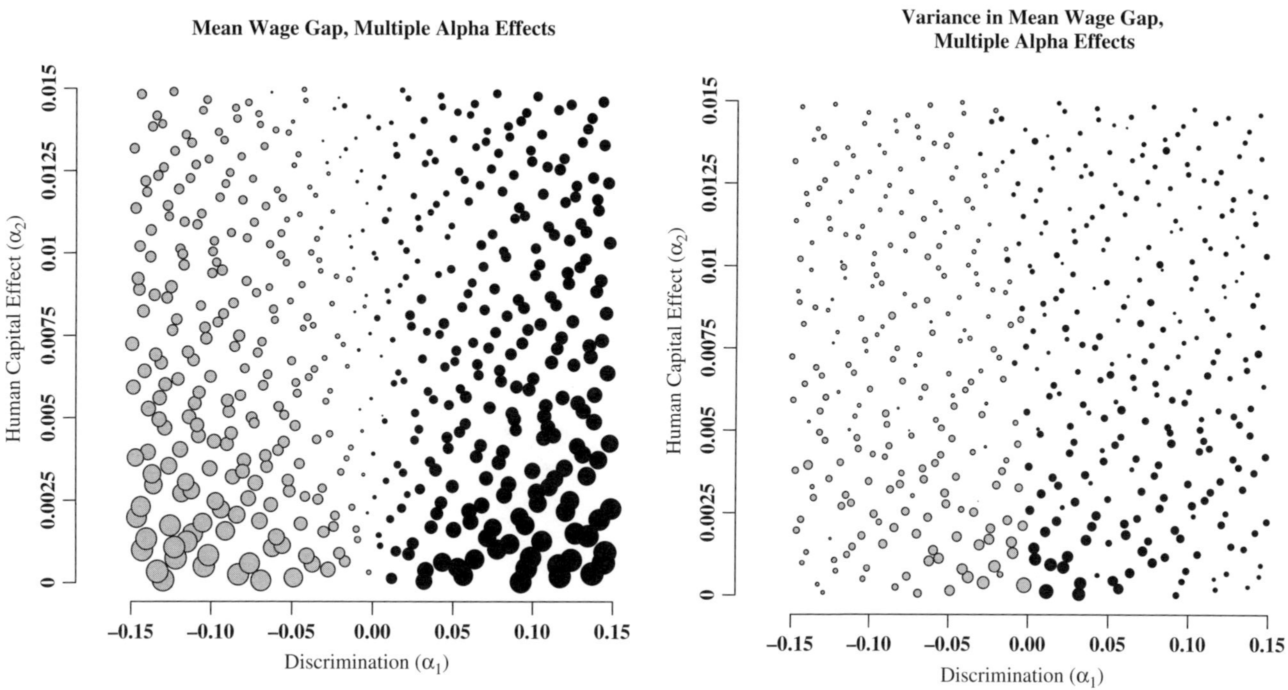

FIGURE 11.1. Gender difference in mean wage rank, discrimination, and human capital effects

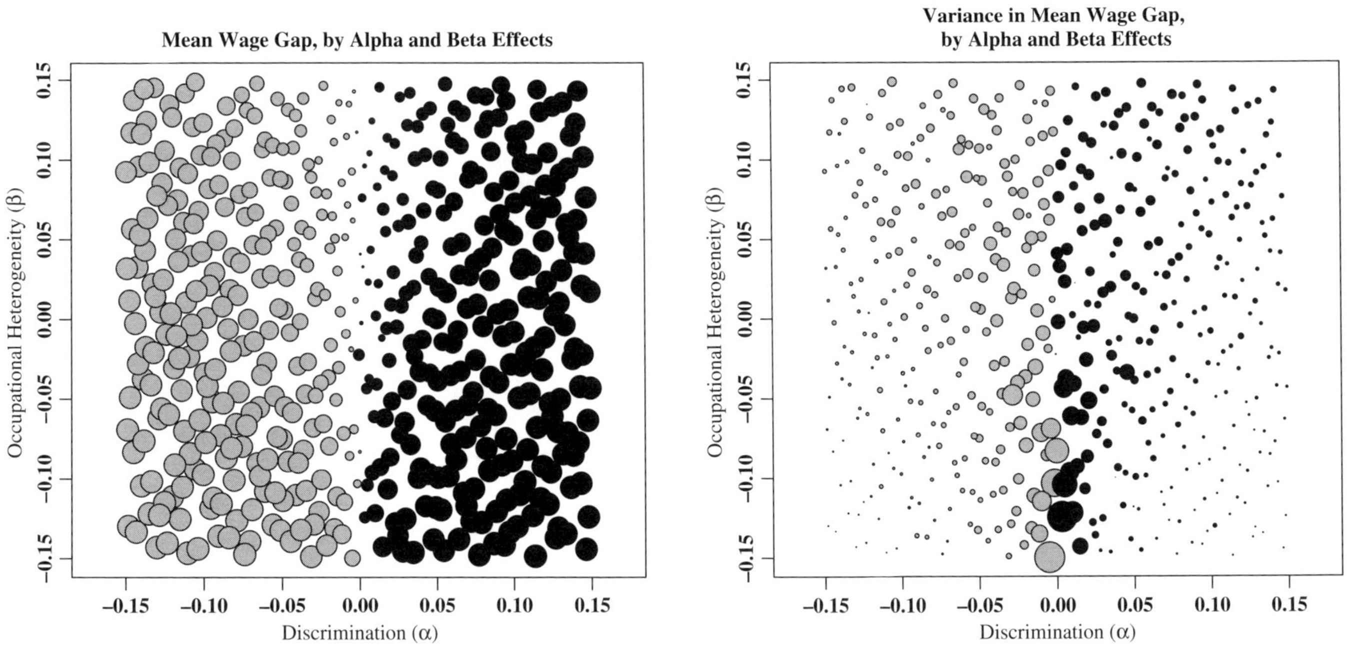

FIGURE 11.2. Gender difference in mean wage rank, discrimination, and segregation effects

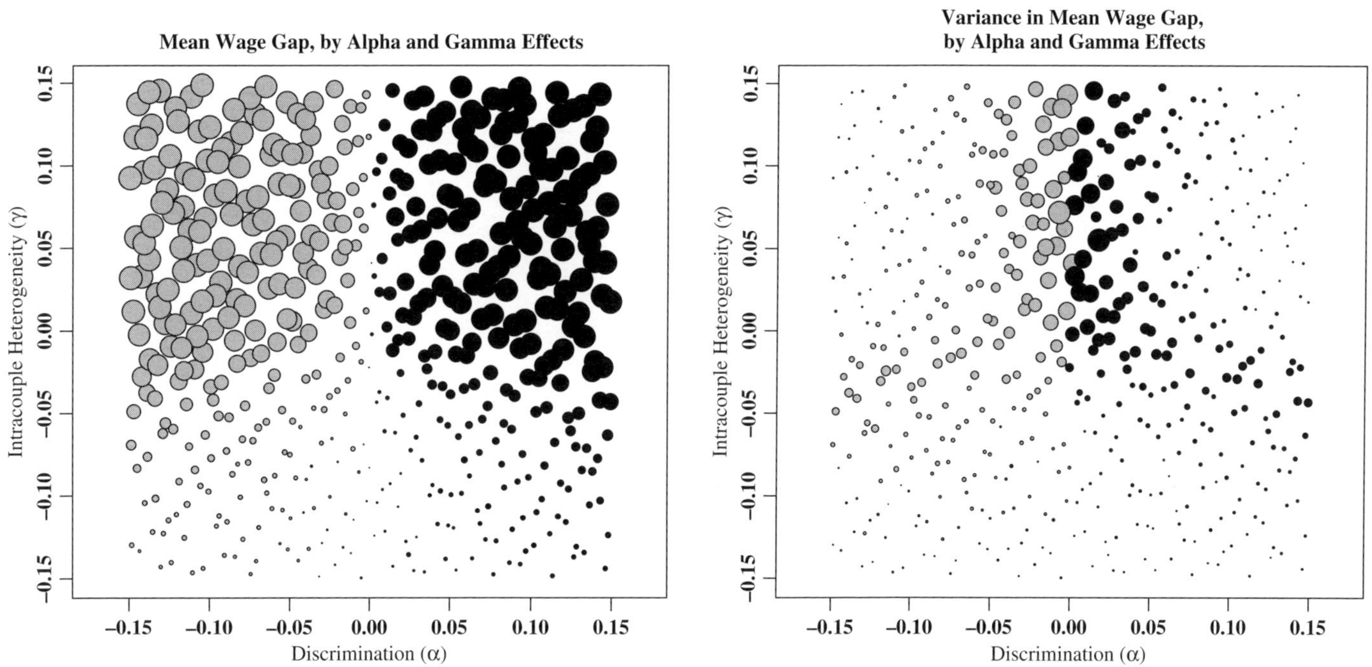

FIGURE 11.3. Gender difference in mean wage rank, discrimination, and couple-level homogeneity effects

the role of segregation is more complex. As the first panel of Fig. 11.2 shows, strong segregation effects substantially increase the rate of transition from one stratification regime (e.g., male dominant) to the other (e.g., female dominant); where segregation is strong, there is an almost immediate phase transition at $\alpha = 0$ from one regime to the other. A mixing regime (where the wage gap is of small magnitude) continues to exist near the origin, and enlarges slightly for $\beta > 0$. This confirms the intuition that supra-random mixing does inhibit convergence to a highly stratified regime, albeit weakly. The explanation for both effects lies in the way in which segregation effects alter the allocation of men and women to occupational categories. When segregation is strong ($\beta \ll 0$), each category tends to be occupied exclusively by members of a single gender; in the presence of even a weak discrimination effect, high-ranking categories will then tend to become the exclusive province of the dominant group, while the subordinated group will be similarly concentrated almost exclusively into the low-ranking blocks. By contrast, when $\beta \gg 0$, it becomes nearly impossible for any occupational category to be gender exclusive. Thus, there must be some members of the subordinate group in the high-ranking blocks, and some members of the dominant group in the low-ranking blocks. This makes extreme stratification much more difficult to achieve, hence the inhibitory effect on the wage gap. Interestingly, this same phenomenon implies a non-monotonic interaction with discrimination on the variance of the wage gap. When $\beta \ll 0$ and $\alpha \approx 0$, occupational categories are highly segregated with no general tendency for any particular category to be dominated by males or females. If, by chance, it happens that the males wind up with the high-end categories, then the wage gap will be large (and positive); these categories could be as easily dominated by females, however, in which case the gap will be negative but also of large magnitude. Thus, segregation in the absence of discrimination should act to greatly increase the variance of the wage gap, without impacting the mean. On the other hand, we have already seen that high segregation in the presence of discrimination results in a highly stratified regime, in which variance should be low. This divergence should disappear in the supra-random mixing case, since the net impact of this effect is to push the job allocation process toward uniformity. As it happens, we see all three of these phenomena in the second panel of Fig. 11.2, which shows the variance of the wage gap across the parameter space. Compared with the $\beta = 0$ baseline, $\beta \ll 0$, $\alpha \approx 0$ shows highly elevated variance, falling almost immediately to low levels as α increases in magnitude. By contrast, variance is much less sensitive to α where $\beta \gg 0$, tending to remain moderate even at more extreme α values.

Just as one can consider the effect of homogeneity or heterogeneity with respect to persons within the same occupation, one can also (via γ effects) consider forces toward or away from heterogeneity with respect to couples. Mechanisms such as social influence (Freidkin 1998), homophily on unobserved characteristics (McPherson et al. 2001), and diffusion of opportunity for social ties (Calvo-Armengol and Jackson 2004) can potentially lead to a net tendency toward similarity of within-couple wage rates. By contrast, incentives for specialization in home versus market production (Becker 1991), normative pressures for intensive parenting (Jones and Brayfield 1997), and the like can lead to high levels of intra-couple wage heterogeneity. To explore these effects, we replace the β effect used to model segregation in Fig. 11.2 with a γ effect for intra-couple wage heterogeneity, and simulate draws from the location system model. As shown in Fig. 11.3, the results are striking: while positive intra-couple heterogeneity effects slightly encourage convergence to a stratified regime, even modestly negative values dampen stratification altogether. How can this be? The secret lies in the observation that the absolute value of the male/female wage gap must be less than or equal to the mean of the absolute intra-couple wage differences. As a result, intra-couple wage heterogeneity acts as a "throttle" on the wage gap: force it to diminish (by setting $\gamma < 0$), and stratification must likewise decrease. This effect similarly reduces the variance of the wage gap (see Fig. 11.3, panel 2), resulting in a "homogeneous mixing regime" in which stratification is uniformly minimal. By contrast, high intra-couple heterogeneity requires one member of each couple to have a much higher wage rank than the other; like segregation, this inflates variance where discrimination is low, but reduces it where discrimination is high. Between, there exists a thin band of entropic mixing, where the various forces essentially cancel each other out.

While these simulations only hint at what is possible when using the location system to model occupational stratification, the effects they suggest are nevertheless interesting and non-obvious. Particularly striking is the relative power of couple-level heterogeneity effects in suppressing labor market discrimination, a result which suggests a stronger connection between processes such as mate selection and marital bargaining with macro-level stratification than might be supposed. The exacerbation of discrimination effects by discrimination is less surprising, but no less important, along with the somewhat weaker inhibiting effect of active desegregation. These phenomena highlight the importance of capturing dependencies among both individuals and among jobs when modeling stratification in labor market settings. Such effects can be readily parameterized using the location system, thereby facilitating a more complete theoretical and treatment of wage inequality within the occupational system.

11.5.2. Settlement Patterns and Residential Segregation

Another problem of long-standing interest to social scientists in many fields has been the role of segregation within residential settlement processes (Schelling 1969; Bourne 1981; Massey and Denton 1992; Zhang 2004). Here, we illustrate the use of the location model on a stylized settlement system involving 1,000 households (objects) allocated to regions on a uniform 20-by-20 spatial grid (locations). Unlike the job allocation system described above, this system places no occupancy constraints on each cell; however, "soft" constraints may be implemented via density dependence effects. For purposes of demonstration, each household is assigned a random "income" (drawn independently from a log-normal distribution with parameters 10 and 1.5) and an "ethnicity" (drawn from two types, with 500 households belonging to each type). Households are tied to one another via social ties, here modeled simply as a Bernoulli graph with mean degree of 1.5. Regions, for their part, relate to one another via their spatial location. Here, we will make use of both Euclidean distances between regional centroids and Queen's contiguity (for purposes of segregation). Each region is also assigned a location on a "rent" gradient, which scales with the inverse square of centroid distance from the center of the grid.

With these building blocks, a number of mechanisms can be explored. Several examples of configurations resulting from such mechanisms are shown in Fig. 11.4. Each panel shows the 400 regions comprising the location set, with household positions indicated by circles. (Within-cell positions are jittered for clarity.) Household ethnicity is indicated by color, and network ties are shown via edges. While each configuration corresponds to a single draw from the location model, a burn-in sample of 100,000 draws was taken (and discarded) prior to

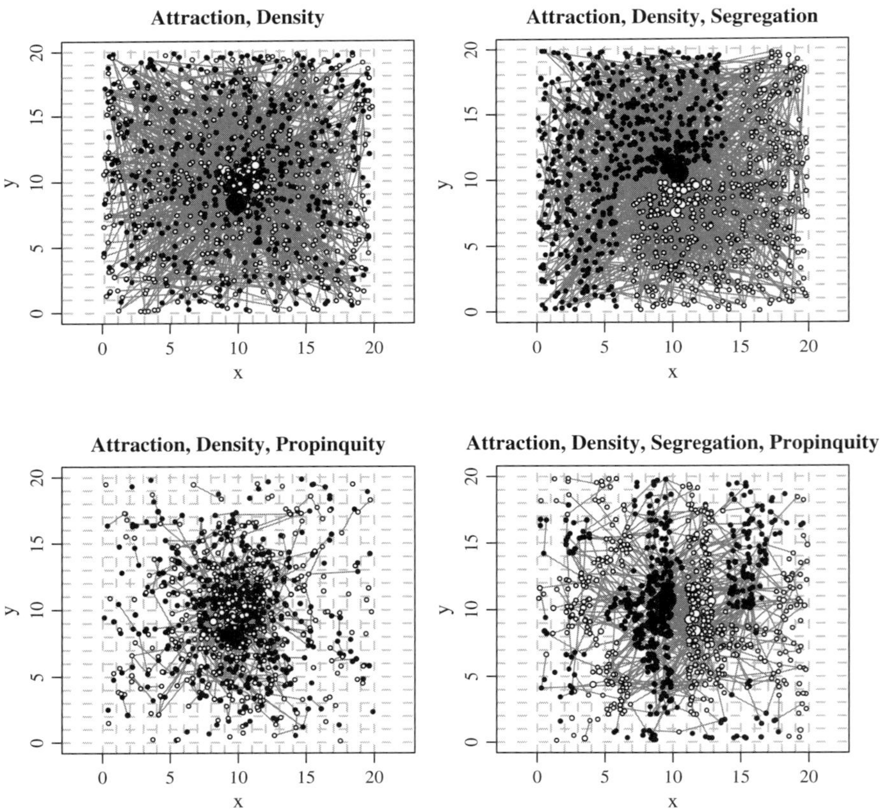

FIGURE 11.4. Location model draws, spatial model.

sampling. Configurations shown here are typical of model behavior for these covariates and parameter values.

The panels of Fig. 11.4 nicely illustrate a number of model behaviors. In the first (upper left) panel, a model has been fit with an attraction parameter based on an interaction between rent level and household income ($\alpha = 0.0001$), balanced by a negative density dependence parameter $\delta = -0.01$. (Density is modeled by an alignment statistic between an identity matrix (for locations) and a matrix of ones (for objects); the associated product moment is the sum of squared location occupancies.) Although the former effect tends to pull all households toward the center of the grid, the density avoidance effect tends to prevent "clumping." As a result, high-income households are preferentially clustered in high-rent areas, with lower-income households displaced to outlying areas. Note that without segregation or propinquity effects, neither ethnic nor social clustering are present; this would not be the case if ties were formed homophilously, and/or if ethnicity was correlated with income. Clustering can also be induced directly, of course, as is shown in the upper right panel of Fig. 11.4. Here, we have added an object homogeneity effect for ethnicity through Queen's contiguity of regions ($\beta = -0.5$), which tends to allocate households to regions so as to reduce local heterogeneity. As can be seen, this induces strong ethnic clustering within the location system; while high-income households are still preferentially attracted to high-rent areas, this sorting is not strong enough to overcome segregation effects. Another interesting feature of the resulting configurations is the nearly empty "buffer" territory which lies between ethnic clusters. These buffer regions arise as a side-effect of the contiguity rule, which tends to discourage direct contact between clusters. As this suggests, the neighborhood over which segregation effects operate can have a substantial impact on the nature of the clustering which results. This would seem to indicate an important direction for empirical research.

A rather different sort of clustering is generated by adding a propinquity effect to the original attraction and density model. Propinquity is here implemented as an alignment effect between the inter-household network and the Euclidean distance between household locations ($\delta = -1$). As one might anticipate, the primary effect of propinquity (shown in the lower-left panel of Fig. 11.4) is to pull members of the giant component together. Since many of these members also happen to be strongly attracted to high-rent regions, the net effect is greater population density in the area immediately surrounding the urban core. Another interesting effect, however, involves households on the periphery: since propinquity draws socially connected households into the core, peripheral households are disproportionately those with few ties and/or which belong to smaller components. The model thus predicts an association between social isolation and geographical isolation. Ironically, this situation is somewhat attenuated by the reintroduction of a residential segregation effect (lower-right panel). While there is still a tendency for social isolates to be forced into the geographical periphery, the consolidation of ethnic clusters limits this somewhat. Because ties are uncorrelated with ethnicity, propinquity also acts to break the settlement pattern into somewhat smaller, "band-like" clusters with interethnic ties spanning the inter-cluster buffer zones. (One would not expect to observe this effect in most empirical settings, however, due to the strong ethnic homophily of most social ties (McPherson et al. 2001).)

Broader information on the behavior of the settlement model can be obtained by simulating draws from across the parameter space, as with Figs. 11.1–11.3. (Simulations were performed in the same manner as those for the occupation model, but were thinned from 300,000 rather than 200,000 draws per chain.) Figure 11.5 shows the mean local heterogeneity statistic (t^{β}) for ethnicity by Queen's contiguity as a function of segregation (β) and propinquity (δ) parameters. Each circle within the two panels reflects the mean or variance of the realized heterogeneity statistic (respectively), with circle shading indicating the corresponding mean or variance in the realized alignment statistic (t^{δ}). As the figure indicates, a clear phase transition occurs at $\beta = 0$, as one transitions sharply from a segregated, low-variance regime to a heterogeneous, high-variance regime. Interestingly, the realized level of propinquity varies greatly only in the upper left-hand quadrant of the parameter space (an environment combining segregative tendencies with high δ values). Propinquity has no substantial effect on heterogeneity in this case, demonstrating that there exist some structural effects which are only weakly coupled.

Unlike propinquity, population density effects interact much more strongly with segregation. Figure 11.6 shows mean/variance in the realized population density (circle area) and heterogeneity (circle shading) statistics as a function of their associated parameters. Unsurprisingly, pressure toward density quickly tips the system into a highly concentrated population regime; somewhat more surprisingly, however, the variance of this state is much higher than the diffuse regime. This reflects the fact that pressures toward population density tend to lead to a rapid collapse into local clusters, which change only unevenly across draws: thus, there is rather more variability here in realized density than there is in the case where households are forced to spread thinly across the landscape. The low-density regime also tends to support high levels of homogeneity, while very high densities tend to inhibit it somewhat; interestingly, however, this inhibition appears to occur (in many parts of the parameter space) via a series of sudden phase transitions, rather than a gradual shift (the exception being the lower right-hand quadrant, in which heterogeneity pressure gradually overcomes resistance toward concentration). Thus, pressure toward or away from segregation can enhance or inhibit the concentration of population into small areas, and vice versa, and this process can occur very suddenly when on the border between regimes.

As Schelling long ago noted, even mild tendencies toward local segregation can result in residential segregation at larger scales (Schelling 1969). While the location system model certainly bears this out, the model also suggests that factors such as population density and inter-household ties can interact with segregation in non-trivial ways. Using the location system framework, such interactions are easy to examine, and the strength of the relevant parameters can

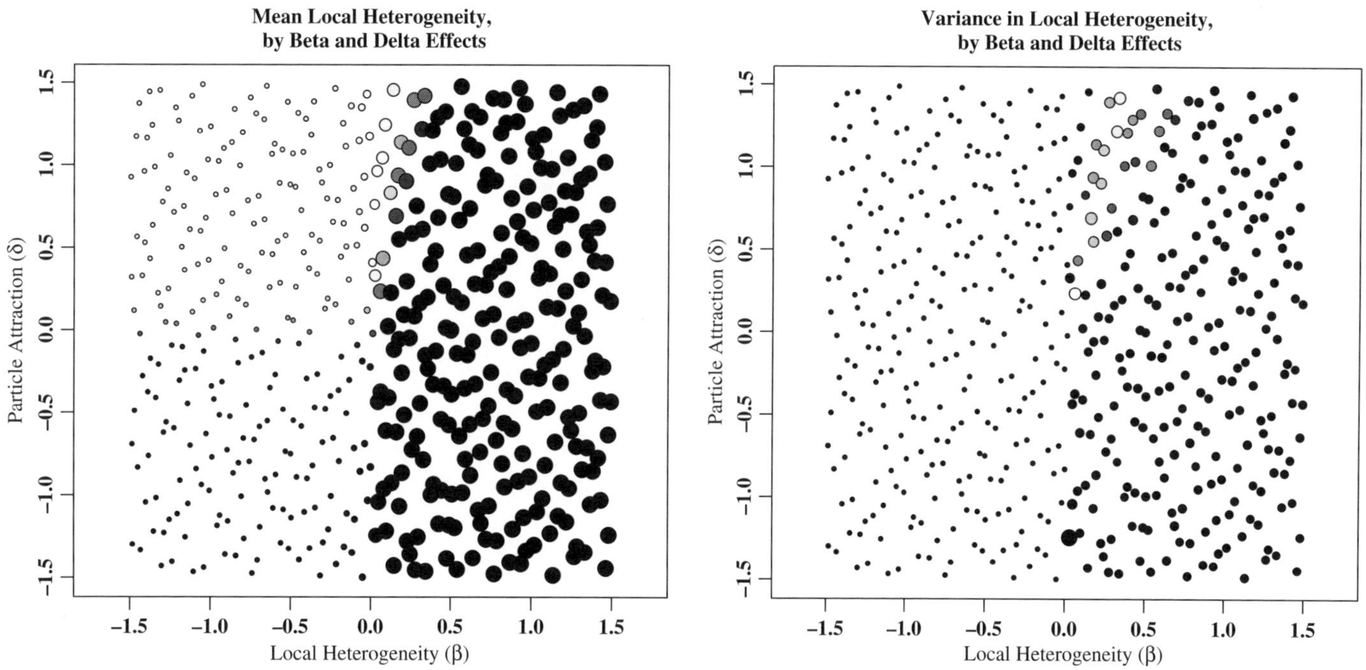

FIGURE 11.5. Mean segregation statistic, by segregation and propinquity effects

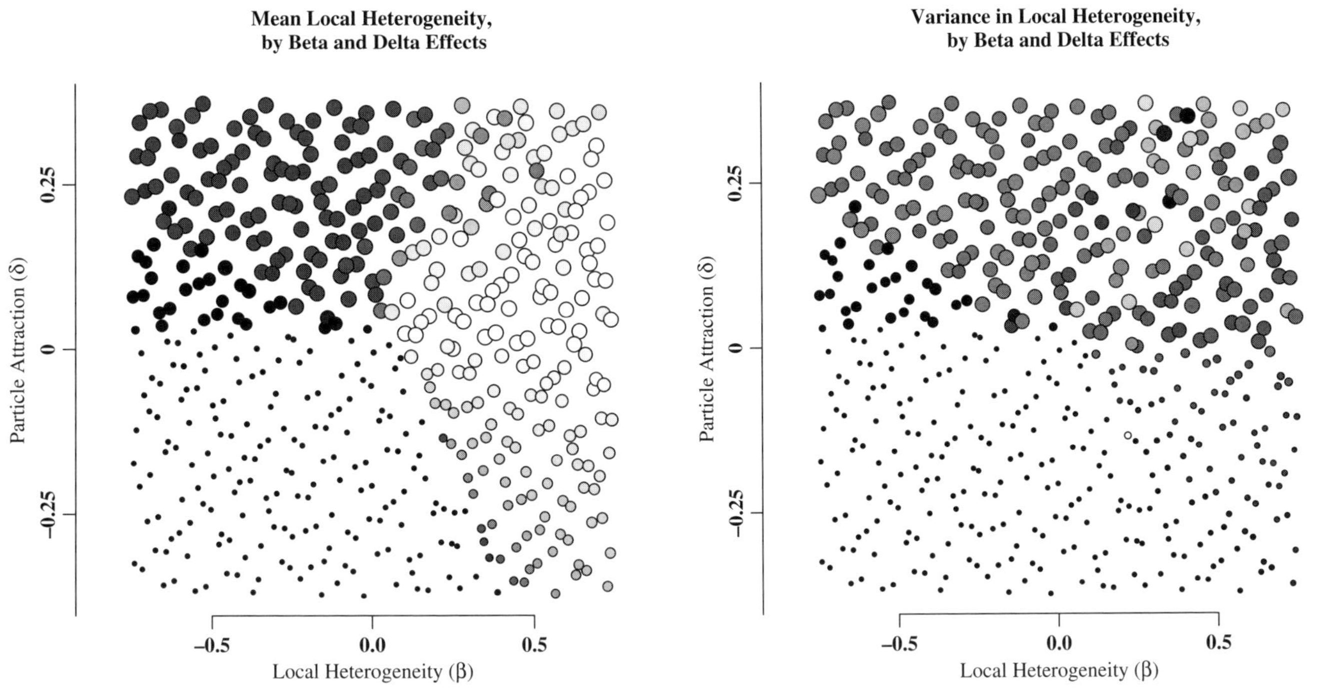

FIGURE 11.6. Mean concentration statistic, by density and segregation effects

be readily estimated from census or other data sources. It is also a simple matter to introduce objects of other types (e.g., firms) which relate to households and to each other in distinct ways (as represented through additional covariates). In an era in which geographical data is increasingly available, such capabilities create the opportunity for numerous lines of research.

11.6. Conclusions

In the foregoing, we have used a stochastic modeling framework (the generalized location system) to illustrate the applicability of physical models to a broad class of social systems. While the location system has antecedents in many fields (including spatial statistics and social networks), its strong formal connection with statistical mechanics is of particular relevance for researchers in areas such as constructal theory, who seek to identify productive ways of integrating physical principles into the social sciences. The applicability of the location system to problems such as occupational stratification and residential settlement patterns highlights not only the versatility of the model, but also the extent to which many apparently disparate social phenomena have strong underlying commonalities. Recognizing and exploiting those commonalities may allow us not only to cross-apply findings between the physical and social sciences, but also to leverage knowledge across different problems within the social sciences themselves.

Acknowledgment The author would like to thank Mark Handcock, Miruna Petrescu-Prahova, John Skvoretz, Miller McPherson, Garry Robins, and Pip Pattison for their comments regarding this work. This research was supported in part by NIH award 5 R01 DA012831-05.

References

Barndorff-Nielsen, O. (1978) *Information and Exponential Families in Statistical Theory.* Wiley, New York.

Becker, G. S. (1991) *A Treatise on the Family*, expanded edition, Harvard University Press, Cambridge, MA.

Bejan, A. (1997) *Advanced Engineering Thermodynamics*, second edition, Wiley, New York.

Bejan, A. (2000) *Shape and Structure: from Engineering to Nature* Cambridge University Press, Cambridge, UK.

Bejan, A. and Lorente, S. (2004) The constructal law and the thermodynamics of flow systems with configuration. *Int. J. Heat Mass Transfer* **47**, 3203–3214.

Bejan, A. and Marden, J. H. (2006) Unifying constructal theory for scale effects in running, swimming and flying. *J. Exp. Biol.* **209**, 238–248.

Besag, J. (1974) Spatial interaction and the statistical analysis of lattice systems, *J. Royal Statistical Society*, Series B **36**, 192–236.

Besag, J. (1975) Statistical analysis of non-lattice data, *The Statistician* **24**, 179–195.

Bourne, L. (1981) *The Geography of Housing* Winston, New York.

Brown, L. D. (1986) *Fundamentals of Statistical Exponential Families, with Applications in Statistical Decision Theory*, Institute of Mathematical Statistics Hayward, CA.

Butts, C. T. (2006) Permutation models for relational data. *Sociological Methodology*, in press.

Calder, W. A. (1984) *Size, Function, and Life History*, Harvard University Press, Cambridge, MA.

Calvo-Armengol, A. and Jackson, M. O. (2004) The effects of social networks on employment and inequality, *American Economic Review* **94**, 426–454.

Cliff, A. D. and Ord, J. K. (1973) *Spatial Autocorrelation*, Pion, London.

Coleman, J. S. (1964) *Introduction to Mathematical Sociology*, Free Press, New York.

Comte, A. (1854) *The Positive Philosophy*, volume 2. Appleton New York.

Crouch, B., Wasserman, S. and Trachtenburg, F. (1998) Markov chain Monte Carlo maximum likelihood estimation for p* social network models, Paper presented at the XVIII Int. Sunbelt Social Network Conference, Sitges, Spain.

Efron, B. (1975) Defining the curvature of a statistical problem (with application to second order efficiency) (with Discussion), *Annals of Statistics* **3**, 1189–1242.

Fararo, T. J. (1984) Critique and comment: catastrophe analysis of the Simon-Homans model, *Behavioral Science* **29**, 212–216.

Frank, O. and Strauss, D. (1986) Markov graphs, *J. American Statistical Association* **81**, 832–842.

Freidkin, N. (1998) *A Structural Theory of Social Influence*, Cambridge University Press Cambridge, UK.

Gamerman, D. (1997) *Markov Chain Monte Carlo: Stochastic Simulation for Bayesian Inference*, Chapman and Hall, London.

Gelman, A., Carlin, J. B., Stern, H. S. and Rubin, D. B. (1995) *Bayesian Data Analysis*. Chapman and Hall, London.

Geyer, C. J. and Thompson, E. A. (1992) Constrained Monte Carlo maximum likelihood calculations (with Discussion), *J. Royal Statistical Society*, Series C **54**, 657–699.

Gilks, W. R., Richardson, S. and Spiegelhalter, D. J. (eds.) (1996) *Markov Chain Monte Carlo in Practice*. Chapman and Hall, London.

Holland, P. W. and Leinhardt. S. (1981) An Exponential Family of Probability Distributions for Directed Graphs (with Discussion), *J. American Statistical Association* **76**, 33–50.

Hubert, L. J. (1987) *Assignment Methods in Combinatorial Data Analysis*, Marcel Dekker, New York.

Hunter, D. R. and Handcock, M. S. (2006) Inference in curved exponential family models for networks, *J. Computational and Graphical Statistics* forthcoming.

Johansen, S. (1979) *Introduction to the Theory of Regular Exponential Families*, University of Copenhagen, Copenhagen.

Jones, R. and Brayfield, A. (1997) Life's greatest joy?: European attitudes toward the centrality of children, *Social Forces* **75**, 1239–1269.

Kittel, C. and Kroemer, H. (1980) *Thermal Physics*, second edition, Freeman, New York.

Massey, D. S. and Denton, N. A. (1992) *American Apartheid: Segregation and the Making of the Underclass* Harvard University Press, Cambridge, MA.

Mayhew, B. H. (1984a) Baseline models of sociological phenomena,. *J. Mathematical Sociology* **9**, 259–281.

Mayhew, B. H. (1984b) Chance and necessity in sociological theory, *J. Mathematical Sociology* **9**, 305–339.

Mayhew, B. H., McPherson, J. M., Rotolo, M. and Smith-Lovin, L. (1995) Sex and race homogeneity in naturally occurring groups, *Social Forces* **74**, 15–52.

McPherson, J. M., Smith-Lovin, L. and Cook, J. M. (2001) Birds of a feather: homophily in social networks, *Annual Review of Sociology* **27**, 415–444.

Pattison, P. and Robins, G. (2002) Neighborhood-based models for social networks, *Sociological Methodology* **32**, 301–337.

Pattison, P. and Wasserman, S. (1999) Logit models and logistic regressions for social networks, II. Multivariate relations, *British J. Mathematical Statistical Psychology* **52**, 169–193.

Press, W. H., Teukolsky, S. A., Vetterling, W. T. and Flannery, B. P. (1992) *Numerical Recipes: The Art of Scientific Computing,* second edition, Cambridge University Press, Cambridge, UK.

Quetelet, A. (1835) *Sur L'Homme et Sur Developpement de Ses Facultes, ou Essai de Physique Sociale*, Bachelier Paris.

Reis, A. H. and Bejan, A. (2006) Constructal theory of global circulation and climate, *Int. J. Heat Mass Transfer* **49**, 1857–1875.

Robins, G., Pattison, P. and Wasserman, S. (1999) Logit models and logistic regressions for social networks, III. Valued relations, *Psychometrika* **64**, 371–394.

Schelling, T. C. (1969) Models of segregation, *American Economic Review***59**, 483–493.

Snijders, T. A. B. (2002) Markov chain Monte Carlo estimation of exponential random graph models, *J. Social Structure* **3**,http://www.cmu.edu/joss/content/articles/volume3/Snijders.pdf

Strauss, D. (1986) On a general class of models for interaction, *SIAM Review* **28**, 513–527.

Strauss, D. and Ikeda, M. (1990) Pseudolikelihood estimation for social networks, *J. American Statistical Association* **85**, 204–212.

Swendsen, R. G. and Wang, J. S. (1987) Non-universal Critical dynamics in Monte Carlo simulation, *Physical Review Letters* **58**, 86–88.

Wasserman, S. and Pattison, P. (1996) Logit models and logistic regressions for social networks: I. An introduction to Markov graphs and p*, *Psychometrika* **60**, 401–426.

White, H. C. (1970) *Chains of Opportunity: System Models of Mobility in Organizations*, Harvard University Press, Cambridge, MA.

Whittington, S. G. (2000) MCMC methods in statistical mechanics: Avoiding quasi-ergodic problems. In *Monte Carlo Methods*, edited by N. Madras, vol. 26 of *Fields Institute Communications*, pp. 131–141, American Mathematical Society. Providence, RI.

Zhang, J. (2004) A dynamic model of residential segregation, *J. Mathematical Sociology* **28**, 147–170.

Zipf, G. K. (1949) *Human Behavior and the Principle of Least Effort*, Hafner, New York.

Chapter 12
Discrete Exponential Family Models for Ethnic Residential Segregation

Miruna Petrescu-Prahova

12.1. Introduction

Ethnic residential segregation has been a visible and salient aspect of urban life in the US, especially after the country experienced massive waves of immigration during the 19th and early 20th century. Empirical studies conducted at the beginning of the 20th century noted the existence of ethnic neighborhoods in metropolitan areas with large immigrant populations, such as Chicago and New York (Thomas 1921). Post-1965 immigrants, although hailing from different origins than their predecessors, have exhibited the same tendency to form ethnic communities in which institutions and services are tailored to the characteristic needs of the ethnic groups (Zhou 1992; Foner 2000; Waldinger 2001). Apart from ethnic neighborhoods that formed as a result of immigration, cities in the US are home to a large African-American population, which is, and has consistently been, residentially segregated from the native-born white population (Taeuber and Taeuber 1965; Massey and Denton 1993; Gottdiener and Hutchinson 2000).

Previous residential segregation studies have sought to identify the factors that determine residential settlement patterns (Clark 1992; Zubrinsky Charles 2001; Wilson and Hammer 2001; Alba and Nee 2003). They suggest several factors, which can be classified into three main categories: physical characteristics of the urban environment, individual and aggregate socioeconomic characteristics, and individual preferences for neighborhood composition. Important strides have recently been made in the direction of studying the interactions among these factors by researchers using agent-based and cellular automata models (Epstein and Axtell 1996; Mare and Bruch 2003; Benenson 2004; Zhang 2005; Fossett 2006), based on early work by Schelling (1969, 1971) and Sakoda (1971). Starting from Schelling's initial result that the outcome of a multitude of interrelated individual choices, where unorganized segregation is concerned, is a complex system with collective results that bear no close relation to individual behaviors each taken by themselves (Schelling 1969), agent-based model research has furthered our understanding of how factors such as neighborhood composition preferences and socioeconomic characteristics influence spatial residential patterns. Fossett (2006) shows that

> Ethnic preferences and social distance dynamics can, when combined with status preferences, status dynamics, and demographic and urban-structural settings common in American cities, produce highly stable patterns of multi-group segregation and hypersegregation (i.e., high levels of ethnic segregation on multiple dimensions) of minority populations (p. 185).

However, these results have largely been based on simulations of "toy" worlds, and the efforts to extend the analyses to real cases have been hampered by a lack of inferential tools to connect theoretical models with extant data.

In this study we use a novel statistical framework based on discrete exponential family models, which bridges this "inferential gap," allowing the researcher both to simulate simple scenarios in order to understand basic mechanisms, and to make inference based on existing data in order to identify mechanisms in real settings. Here we present results based on simulations of a simple scenario that will allow us to enhance our understanding of the behavior of the model and to build intuitions that will guide empirical data analysis, which makes the subject of future research.

12.2. Potential Determinants of Ethnic Residential Segregation

The physical characteristics of the urban environment are a set of factors that were many times emphasized by the classic urban sociologists of the Chicago School but are often overlooked in more recent studies. Modern cities have certain man-made features, which are intrinsic to their structure and to some extent independent of their resident population, as well as natural features, all of which may be conducive to certain patterns of land use (McKenzie 1924; Hawley 1950). Fixed infrastructure (e.g., roads, factories), the spatial distribution of land available for residential use (as opposed to economic use), and the number of housing units, combined with natural barriers such as rivers or hills, can influence settlement patterns, since locations which present spatially isolated clusters of housing units may be more prone to segregation than locations with minimal barriers between units. [(One of the expressions through which the urban vernacular has captured this situation is "the wrong side of the tracks," which reflects the fact that the borders of segregated neighborhoods are determined by such barriers as railroad tracks (Massey and Denton 1993)]. Foner (2000) notes that in the early years of the Jewish and Italian influx into New York, most immigrants settled in the downtown neighborhoods situated below Fourteenth Street, which ensured that they were living close to the sources of jobs – docks, warehouses, factories, and business streets (p. 39). They were able to move out of these neighborhoods only after the infrastructure of public transportation, roads, and bridges eased the access to new destinations such as Harlem, Brooklyn, and Queens. However, even in the extremely densely populated area below Fourteenth Street, Italians and Jews were rarely close neighbors. The grid structure of the streets provided the barriers, and "most blocks were heavily dominated, if not exclusively populated, by one or the other immigrant group" (Foner 2000, p. 41).

Another set of factors are individual and aggregate socioeconomic characteristics, especially personal income and rent levels. The relationship between rent and personal income is a hard constraint on residential choice, especially for low-income households. As a consequence, households with comparable incomes seek locations with similar and affordable rent levels and consequently cluster together in certain parts of the metropolis (Hawley 1950). If, in addition, we take into account the fact that poverty disproportionately affects members of the minority ethnic groups, we have the premises of ethnic residential segregation through income levels alone (Clark 1986b; Gottdiener and Hutchinson 2000). On the other hand, settlement patterns of ethnic groups in urban areas are determined partly by social networks of kinship, friendship, and co-ethnicity. To a large extent, these networks offer support to new immigrants, who are unfamiliar with American society and frequently lacking proficiency in English. This leads to geographic concentration of ethnic or even national origin groups (Thomas 1921; MacDonald and MacDonald 1970; Massey et al. 1998; Menjivar 2000). This phenomenon is not restricted to immigrants, however; human geography studies suggest that internal migrants also make settlement decisions based on the geographic location of friends and relatives (Clark 1986a).

One of the most influential theories for the interpretation of ethnic population distribution across metropolitan space is the spatial assimilation framework, developed by Massey (1985), on the basis of the work of members of the Chicago School such as Robert Park and Louis Wirth. According to this framework, which is related to the normative view of immigrant assimilation in the host societies (as presented by Gordon 1964), immigrant groups initially settle in enclaves located in the inner city, mainly in economically disadvantaged areas. As their members experience social mobility and acculturation, they usually leave these areas and move to "better" neighborhoods, namely areas that do not have such a high concentration of ethnic minorities, leading to a reduction in residential segregation levels.

The underlying assumptions of the framework are that neighborhood location and housing are largely determined by market processes and that individuals are motivated to improve their residential status once they have acculturated and made some socioeconomic gains. In this context, residential exposure to the majority group is hypothesized to improve as a result of gains in an ethnic family's socioeconomic standing, acculturation (as measured, for instance, by its members' proficiency in speaking English), and generational status or, in the case of first generation immigrants, length of residence in the country of destination.

Residence in the suburbs is also taken into account in the model because it is seen as a sign of enhanced residential assimilation. A series of studies of spatial assimilation for some of the main metropolitan regions, summarized by Alba and Nee (2003), focus especially on the median household income of the census tract of residence and the percent of non-Hispanic whites, the majority group, among residents, as indicators of spatial assimilation. For Asians and Hispanics, the most powerful determinant of living in a high income, high percent white neighborhood is their own socioeconomic position: the greater their income and the higher their educational status, the larger, for instance, the percentage of non-Hispanic whites in the population of the neighborhood where they reside.

The spatial assimilation framework does not apply, however, to African-American communities and to immigrant groups that have mixed African ancestry (Haitians, West Indians), because of racial discrimination by the white population (Massey and Denton 1993). Apart from this shortcoming, the spatial assimilation model, which was built primarily on the experience of the mainly Southern and Eastern European immigrant flows in the early 20th century, fails to account for the experience of new immigrant groups. Responding to these concerns, Portes and Zhou (1993) propose the theory of segmented assimilation, prompted by various research that showed different assimilation outcomes for ethnic groups in the post-1965 wave, which stands in contrast with the classic view of immigrant assimilation as a straight-line process. One of the assimilation trajectories is characterized by upward social and economic mobility in the context of the preservation of ethnic identity and culture, and strong ties with the ethnic community. The achievement of social mobility is no longer linked with the exit from the ethnic community—especially for those groups that have financial capital when they arrive in the US—and remaining in the ethnic community represents a choice rather than a constraint for members of some national-origin groups such as Cubans (South et al. 2005).

The final set of factors suggested by previous research as a potential determinant of ethnic residential segregation are individual preferences for neighborhood composition (Clark 1992; Zubrinsky Charles 2001), which can vary according to the reference combination of ethnic groups. One of the first factors emphasized is the preference for homogeneity, which can be understood either as a desire to be close to co-ethnics (homophily) or a desire to be apart from ethnic "others" (xenophobia). This type of preference is mostly exhibited by the non-Hispanic white population, who prefers neighborhoods that are 70 percent or more white, when viewed as combinations of non-Hispanic white and black households (Clark 1992). In contrast, blacks appear to want a sizable population of coethnics and substantial integration at the same time, leading to a preference for 50/50 neighborhoods (Zubrinsky Charles 2001). Hispanics tend to approximate the preferences of blacks, when the reference composition is Hispanic/non-Hispanic white, but approach a preference for neighborhoods that are 75 percent Hispanic when the potential neighbors are black. In turn, Asian respondents are much more open to integration with non-Hispanic whites than with other groups and find integration with blacks least appealing, while at the same time showing strong preferences for co-ethnic neighbors (Zubrinsky Charles 2001).

Apart from influencing personal residential choices, neighborhood composition preferences are important because they can lead to discrimination in the housing market, for instance through restrictive covenants signed by neighborhood associations, which limited the choices available to minority groups and led to the creation of segregated neighborhoods (Massey and Denton 1993). Although some of these extreme, formally implemented measures are now illegal, personal discrimination by real-estate agents is harder to identify and eradicate, and its global effects are not well known (Clark 1992).

Despite the wealth of empirical studies that analyze the potential determinants of residential segregation, very few of them have attempted to compare the relative impact of these factors in generating residential settlement patterns, or to explicitly identify the manner in which such patterns emerge from the interaction of these elements in field settings (Clark 1986b). It is to the latter issue that we hope to contribute with this study.

12.3. Research Methodology

The assumption on which the present approach is built is that at any point in time, we can interpret the spatial residential pattern as an equilibrium state of a system of households and areal units, in which households are located in the areal units. However, this system contains various kinds of dependencies: people are tied to one another by kin or friendship relations, and geographic locations are related by virtue of being contiguous or being a certain distance apart from one another. As such, a traditional regression framework is not going to be very reliable in explaining outcomes, and it will fail to represent the complex dependencies within the system. One area of sociology that has seen tremendous advances toward developing stochastic models for social systems with complex dependence structures is social network analysis, where researchers have drawn on earlier results in other scientific fields such as spatial statistics and statistical physics (Robins and Pattison 2005; Butts 2005). Building further on these developments, Butts (2005) has proposed "a family of models for social phenomena which can be described in terms of the arrangement of various (possibly related) objects with respect to a set of (again, possibly related) locations" (p. 2). These "generalized location systems" can be used to characterize a range of social processes such as occupational segregation, stratification, and settlement patterns. In the case of residential settlement patterns, households represent objects, areal units such as census tracts or block groups represent locations, and we model the probability of observing a particular assignment (i.e., the observed distribution of households across areal units) as resulting from the interaction of factors such as availability of housing, wealth, and preferences for neighborhood composition.

The advantages of this framework are that it can be readily simulated, allowing for the testing of simple scenarios, it is specifiable in terms of directly measurable properties, and supports likelihood-based inference (using Markov Chain Monte Carlo methods). Another set of characteristics that recommends the use of this framework for the study of residential settlement patterns is the ability to include as covariates a range of factors such as population density, inter-household ties, individual preferences and areal unit characteristics, and examine the effect of their interactions in determining residential patterns.

The generalized location system model is defined as a stochastic model for the equilibrium state of a generalized location system, which represents the assignment of objects (persons, organizations, etc.) to locations (places, jobs, etc.). Given a set of possible configurations (C), the system will be found to occupy any particular configuration with some specified probability. The equilibrium probability of observing a given configuration can be written as

$$\Pr(S=l) = I_C(l)\frac{\exp(P(l))}{\sum_{l' \in C}\exp(P(l'))} \tag{12.1}$$

where S is the random state, l is a particular configuration, and P is the quantity we are most interested in, the *social potential*. In this model, the location system is more likely to be found in areas of high potential (which, in turn, are areas of high probability). We need, therefore, to specify a functional form for the social potential that allows us to incorporate as many substantively meaningful effects as possible.

To start with, we can take into account the fact that both objects and locations have features, which can be attributes or relations among objects or among locations. A simple two-by-two table (Table 12.1) gives us a range of effects that we may include in the social potential function:

We now consider these four classes of effects and some examples, without paying attention, at this moment, to their functional form. After reviewing these we present the functional form of the social potential as a linear combination of the four types of effects.

Attraction/repulsion (frequently called push/pull) effects are based on object and location attributes. Locations (neighborhoods, for example) have attributes that make them attractive (or undesirable) to objects (e.g., households) with particular attributes. High-income neighborhoods attract individuals with high income, and at the same time repel individuals with low incomes. Another

TABLE 12.1. Elements in the social potential function

	Location attributes	Location relations
Object attributes	Attraction/repulsion effects	Object homogeneity/ Heterogeneity effects
Object relations	Location homogeneity/ Heterogeneity effects	Alignment effects

important case of this type of effect is discrimination. In this framework, discrimination may be understood as a conditional tendency for households with certain features to be placed in (or denied access to) certain locations.

The second category of effects deals with object homogeneity/heterogeneity based on location relations. In other words, this effect captures the tendency for associated locations to be occupied by objects with similar (or different) features. Xenophobia effects can be understood in this framework as the tendency for people of the same race or ethnic origin to reside in contiguous neighborhoods, based on their desire to reduce heterogeneity. This then leads to the formation of clusters of areas with high percentages of people from that group.

Effects of location homogeneity/heterogeneity through relations of objects capture the tendency for locations that are similar to be occupied by people who are associated in some way. An example of such effects is recruitment by entrepreneurs through networks of immigrants. The result is that similar types of jobs (supermarket assistants, for instance) are occupied by people from the same family or community. It is slightly more difficult to interpret this type of effect when locations are geographical units, and it will not be included in the simulations presented below.

Finally, alignment effects express the tendencies for objects that are related to occupy locations that are related in their turn. The first example is propinquity, the tendency for people who are linked (through kinship or friendship) to reside in neighboring locations. This category of effects is the most flexible, since this function uses matrices as inputs and many mechanisms can be expressed in terms of products of matrices (for instance, density avoidance or homophily).

The social potential is constructed as a linear function of these effects, and has the following expression:

$$P(l) = \sum_{i=1}^{a} \alpha_i t_i^{\alpha}(l) + \sum_{i=1}^{b} \beta_i t_i^{\beta}(l) + \sum_{i=1}^{c} \gamma_i t_i^{\gamma}(l) + \sum_{i=1}^{d} \delta_i t_i^{\delta}(l) \tag{12.2}$$

where α, β, γ, and δ are the model parameter vectors, and t^{α}, t^{β}, t^{γ}, and t^{δ} are vectors of sufficient statistics.

We can also express the social potential in terms of the underlying covariates as

$$\begin{aligned} P(l) = & \sum_{i=1}^{a} \alpha_i \sum_{j=1}^{n} \mathbf{Q}_{l_i i} \mathbf{X}_{ji} + \sum_{i=1}^{b} \beta_i \sum_{j=1}^{n} \sum_{k=1}^{n} \mathbf{B}_{il_j l_k} \left| \mathbf{Y}_{ji} - \mathbf{Y}_{ki} \right| \\ & + \sum_{i=1}^{c} \gamma_i \sum_{j=1}^{n} \sum_{k=1}^{n} \mathbf{A}_{ijk} \left| \mathbf{R}_{l_j i} - \mathbf{R}_{l_k i} \right| + \sum_{i=1}^{d} \delta_i \sum_{j=1}^{n} \sum_{k=1}^{n} \mathbf{W}_{ijk} \mathbf{D}_{il_j l_k} \end{aligned} \tag{12.3}$$

where **X** and **Y** are vectors of object (e.g., household) attributes, **Q** and **R** are vectors of areal unit (e.g., census tract) attributes, **B** and **D** are arrays of areal unit relation adjacency matrices, and **A** and **W** are arrays of household relation adjacency matrices. By specifying these parameters in a simulated scenario, we can obtain assignments of households to locations that illustrate what the spatial patterns would be like if these particular mechanisms/effects were at play.

12.4. Simulation Results

The simulation of a simple scenario with a small population and few effects that are added successively allows us to better understand the behavior of the model and observe how the assignment of households to areal units (which from now on will be referred to as "neighborhoods") is affected by the incorporation of new effects. This scenario includes the following effects:

- Attraction: based on household income and neighborhood rent
- Xenophobia: object homogeneity effect for ethnicity based on the contiguity matrix of the neighborhoods
- Homophily: alignment effect between ethnicity and neighborhood contiguity, interpreted as a preference for being close to similar alters, without any preference toward members of the other group. In a scenario with two groups it can have two forms:

    - Single homophily, where only the members of one of the groups prefer to be close to similar alters
    - Double homophily, where members of both groups prefer to be close to similar alters

This effect can be expressed as

$$\delta \sum_{k=1}^{n} \sum_{k=1}^{n} \mathbf{W}_{jk} \mathbf{D}_{l_j l_k} \tag{12.4}$$

where, for single homophily

$$\mathbf{W}_{jk} = \begin{cases} 1 \;\; \mathit{if} \quad \mathbf{Y}_j = \mathbf{Y}_k = 1(\mathit{or}\ 0) \\ 0 \qquad\qquad\qquad\qquad \mathit{otherwise} \end{cases} \tag{12.5}$$

for double homophily

$$\mathbf{W}_{jk} = \begin{cases} 1 \;\; \mathit{if} \quad \mathbf{Y}_j = \mathbf{Y}_k \\ 0 \qquad\quad \mathit{otherwise} \end{cases} \tag{12.6}$$

and where **D** is a neighborhood contiguity matrix and **Y** is a vector of household characteristics, in this case, ethnicity

- Density: alignment effect based on total population counts in neighborhoods, prevents clustering in any one neighborhood, acting as an occupancy constraint

- Propinquity: alignment effect between the inter-household network and the matrix of Euclidean distances between neighborhood centroids.

Our principal focus in this analysis is on homophily and xenophobia effects and so comparisons are drawn mainly between model specifications that do and do not include these effects.

In order to characterize and compare the assignments we use residential segregation indices. Researchers concerned with identifying and measuring residential segregation have developed a series of indices that reflect different ways of conceptualizing segregation (Duncan and Duncan 1955; Lieberson 1981; Massey and Denton 1988; Grannis 2002). Massey and Denton's (1988) classic analysis of segregation indices identifies 20 measures, classified according to five key dimensions of segregation: evenness (the differential distribution of the population), exposure (referring to potential contact between members of different groups), concentration (the relative amount of physical space occupied by groups), centralization (indicating the degree to which a group is located near the center of the city), and clustering (the degree to which minority group members live in contiguous areas). All of these indices measure the degree to which two or more groups live separately from one another, and their calculation is based on a division of the urban area into "neighborhoods," which most often are Census tracts, and the percentages of the various group populations in the total and neighborhood population. In the present analysis we use two of these indices, the dissimilarity index and the spatial proximity index.

The dissimilarity index, D, is the most widely used evenness index, and one of the most widely used segregation indices overall. It measures departure from the even distribution of minority and majority population across areal units, and can be interpreted as the percentage of a group's population that would have to change residence for each neighborhood to have the same percentage of that group as the urban area overall. For example, a value for D of 0.6 in an area where the minority group represents 20 percent of the whole population would mean that 60 percent of the members of the minority group population would have to move in order for all neighborhoods in the area to have a 20-percent minority population. The index ranges from 0 (complete integration) to 1 (complete segregation), and its formula is

$$D = \frac{\sum_{i=1}^{n} (t_i \, |p_i - P|)}{2TP(1-P)} \tag{12.7}$$

where n is the number of neighborhoods (or tracts) in the urban area, T is the total population of the area, t_i is the total population of neighborhood i, P is the proportion minority in the total population, and p_i is the proportion minority in population in area i.

Although they are based on proportions of minority/majority population in clearly defined neighborhoods, most residential segregation indices do not take into account the location of these spatial units of measurement relative to each other, thus ignoring important aspects of segregation such as the geographic distance between two group concentrations (White 1983; Massey and Denton 1988; Grannis 2002). Clustering indices address this shortcoming and measure “the extent to which areal units inhabited by minority members adjoin one another, or cluster, in space” (Massey and Denton 1988, p. 293). The spatial proximity index (*SP*) is a clustering index proposed by White (1986), which calculates the average of intragroup proximities for the minority and majority populations, weighted by the proportions each group represents of the total population.

$$SP = \frac{XP_{xx} + YP_{yy}}{TP_{tt}} \tag{12.8}$$

where

$$P_{xx} = \sum_{i=1}^{n}\sum_{j=1}^{n}\left[\frac{x_i x_j c_{ij}}{X^2}\right] \tag{12.9}$$

$$P_{yy} = \sum_{i=1}^{n}\sum_{j=1}^{n}\left[\frac{y_i y_j c_{ij}}{Y^2}\right] \tag{12.10}$$

$$P_{tt} = \sum_{i=1}^{n}\sum_{j=1}^{n}\left[\frac{t_i t_j c_{ij}}{T^2}\right] \tag{12.11}$$

and x_i is the minority population of neighborhood i, y_i is the majority population of neighborhood i, t_i is the total population of neighborhood i, X is the total minority population in the urban area, Y is the total majority population, T is the total population, and c_{ij} has a value of 1 if neighborhoods i and j are contiguous, and 0 otherwise (i.e., c_{ij} is the value of the *ij*-th cell in the contiguity matrix).

Spatial proximity equals 1 if there is no differential clustering between minority and majority group members. It is greater than 1 when members of each group live nearer to one another than to members of the other group, and is less than 1 if minority and majority members live nearer to members of the other group than to members of their own group.

We begin by specifying the covariates and parameter values used in this simulation scenario.[1] A number of 1000 households are allocated to 400 neighborhoods, represented by squares in a 20×20 grid. Each household has one of two types of ethnicity, which is randomly assigned in equal proportions (500 households belong to each type), and is given a random income (drawn independently from a log-normal distribution with parameters 10 and 1.5). Households are tied by social ties (kin or friendship, for example), which are modeled as a Bernoulli graph with mean degree of 1.5 (i.e., a graph in which each edge is an

[1] For details on the simulation process, which is based on the Metropolis algorithm, see Butts (2005).

independent Bernoulli trial with probability approximately 0.0015). Each neighborhood is assigned a rent value which scales with the inverse of the distance between its centroid and the center of the grid. Neighborhoods have equal area and relationships among them are expressed in terms of either Euclidean distance between centroids or Queen's contiguity (i.e., two neighborhoods are considered contiguous if they share a border or a point).

The parameter values used in this analysis are listed in Table 12.2. They are constant across model specifications and have been selected to provide the best illustration of the effect they quantify.

Figures 12.1 through 12.11 illustrate simulated draws from various specifications of the model. For each model specification the figures correspond to one Metropolis draw, which was sampled after a burn-in sample of 100,000 draws was taken and discarded. Households are represented by filled or empty circles (according to their ethnicity), whose diameter scales with income levels (the bigger the diameter, the larger the household income). Network ties among households are represented by dotted lines, neighborhood boundaries are given by the black lines of the grid, and within-neighborhood household positions are jittered to prevent overlap. All models include attraction and density effects and we build on this base by adding various combinations of the other effects. Values of the dissimilarity and spatial proximity indices for configurations determined by each model's specification are listed in Table 12.3.

The first set of configurations we analyze is illustrated in Figs. 12.1 through 12.4. We begin with the model specification that includes only attraction and density effects (Fig. 12.1), and then add, in turn, the xenophobia (Fig. 12.2), single homophily (Fig. 12.3), and double homophily effects (Fig. 12.4). When only attraction and density are present, evenness (as measured by D) has moderate levels: 47 percent of the population would have to move in order for all the neighborhoods in the grid to have the 50/50 distribution that characterizes the total population ($D = 0.47$). Clustering, on the other hand (as measured by SP, which in this case equals 1.04), is almost non-existent, with the exception, perhaps, of a tendency for higher income households of both ethnicities to congregate close to the center of the grid.

TABLE 12.2. Parameter values in the simulation scenario

Parameter	Effect	Value
α	Attraction	0.00075
β	Xenophobia	−0.
δ_1	Single homophily	0.05
	Double homophily	0.03
δ_2	Propinquity	−1
δ_3	Density	−0.01

TABLE 12.3. Residential segregation index values for different model specifications

Effects included in the model	Dissimilarity index	Spatial proximity index
Attraction, density	0.47	1.04
Attraction, density, xenophobia	1.00	1.99
Attraction, density, single homophily	0.89	1.70
Attraction, density, double homophily	0.31	1.03
Attraction, density, xenophobia, single homophily	1.00	1.98
Attraction, density, xenophobia, double homophily	1.00	2.00
Attraction, density, xenophobia, propinquity	0.98	1.97
Attraction, density, single homophily, propinquity	0.58	1.21
Attraction, density, double homophily, propinquity	0.33	1.05
Attraction, density, xenophobia, single homophily, propinquity	1.00	1.98
Attraction, density, xenophobia, double homophily, propinquity	0.87	1.87

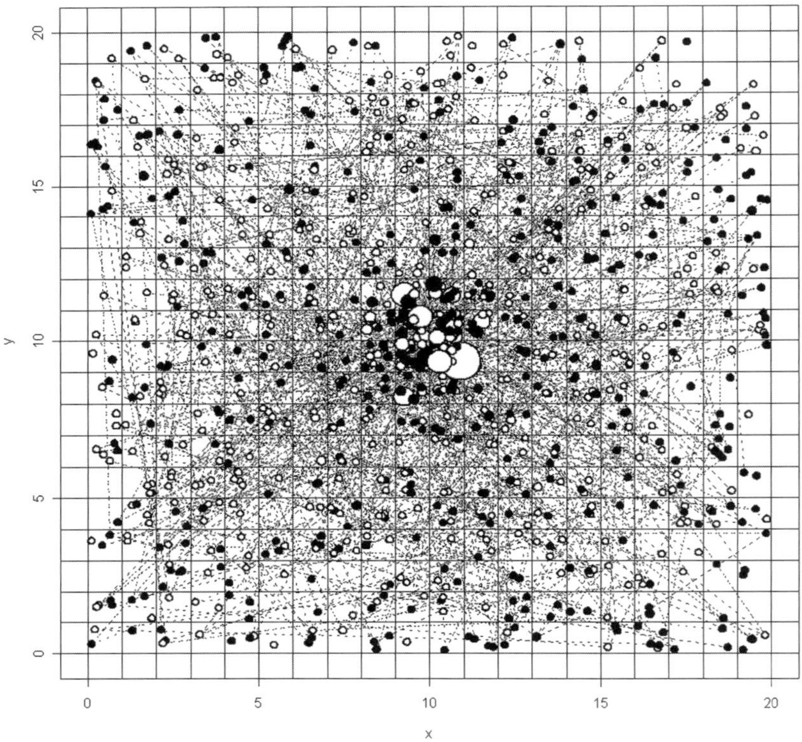

FIGURE 12.1. Attraction, Density Effects

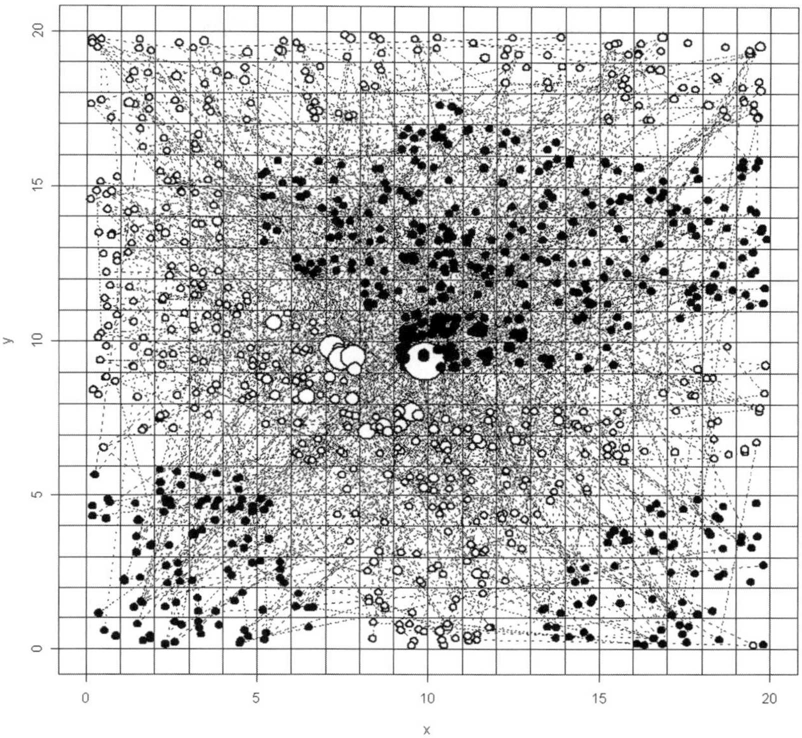

FIGURE 12.2. Attraction, Density, Xenophobia Effects

When we turn to the model specification in which xenophobia is added (Fig. 12.2), we observe an assignment of households to neighborhoods that is highly, even completely, segregated, as measured by both indices ($D = 1.00, SP = 1.99$). (For this scenario with two groups of equal size, the maximum value of SP is 2). The areas occupied by the two groups are separated by an almost empty band, due to the fact that xenophobia is based on the neighborhood contiguity matrix and therefore direct contact between the two groups is discouraged. (In contrast, using Euclidean distance in this case would push the two groups as far apart as possible, in diagonal corners of the grid.)

Adding a single homophily effect to the initial attraction and density model generates a configuration that is less segregated than the one that includes xenophobia, but still has relatively high values on both indices ($D = 0.89$, $SP = 1.70$). By adding the single homophily effect (which in this case refers to an above chance tendency for filled-circle households to be found close to one another), a cluster of filled-circle households is formed around the center of the grid. This has two consequences. First, there are now many neighborhoods for which the filled-circle/empty-circle ratio departs from 50/50, leading to a high value for D. Second, since SP measures clustering directly, its value increases relative to

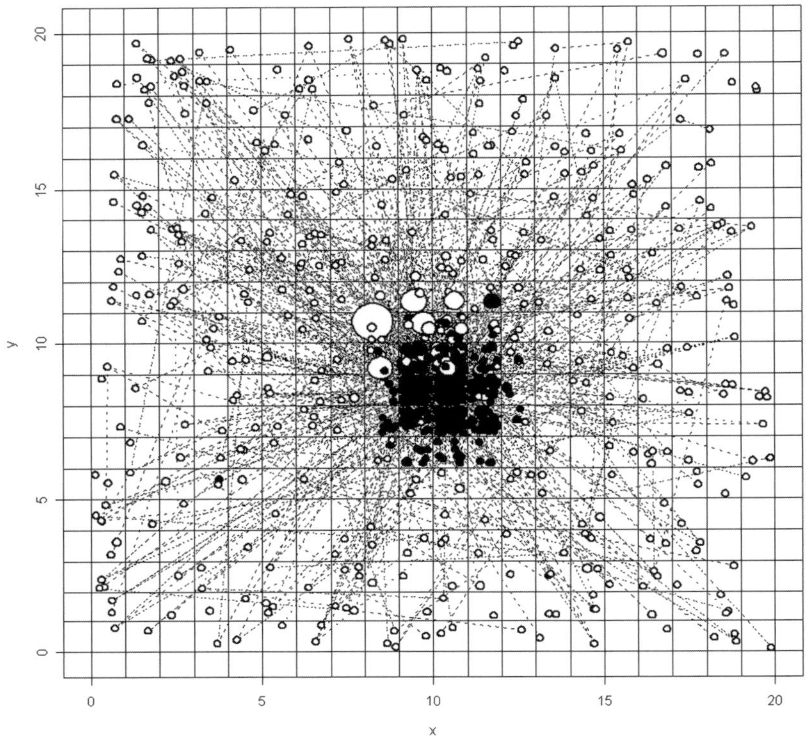

FIGURE 12.3. Attraction, Density, Single Homophily Effects

the one obtained in the attraction and density assignment, but as empty-circle households are still mixed with filled-circle ones in some neighborhoods, it does not reach its maximum value as in the previous case that included attraction, density, and xenophobia effects.

The double homophily effect that we add last leads to an even less segregated configuration ($D = 0.31$, $SP = 1.03$). Both groups are clustered around the center of the grid, and as they occupy roughly the same area, very low levels of segregation are present.

The purpose of analyzing the next group of configurations is to enhance our understanding of the manner in which the simultaneous presence of xenophobia and homophily effects influences the assignment of households to neighborhoods. As can be gleaned from Fig. 12.5 (xenophobia and single homophily, $D = 1.00$, $SP = 1.98$) and Fig. 12.6 (xenophobia and double homophily, $D = 1.00$, $SP = 2.00$), the presence of xenophobia *and* homophily at the same time leads to (almost) complete segregation. These configurations and index values stand in stark contrast with the two configurations in Figs. 12.3 and 12.4, in which homophily effects were present just by themselves. This result shows that

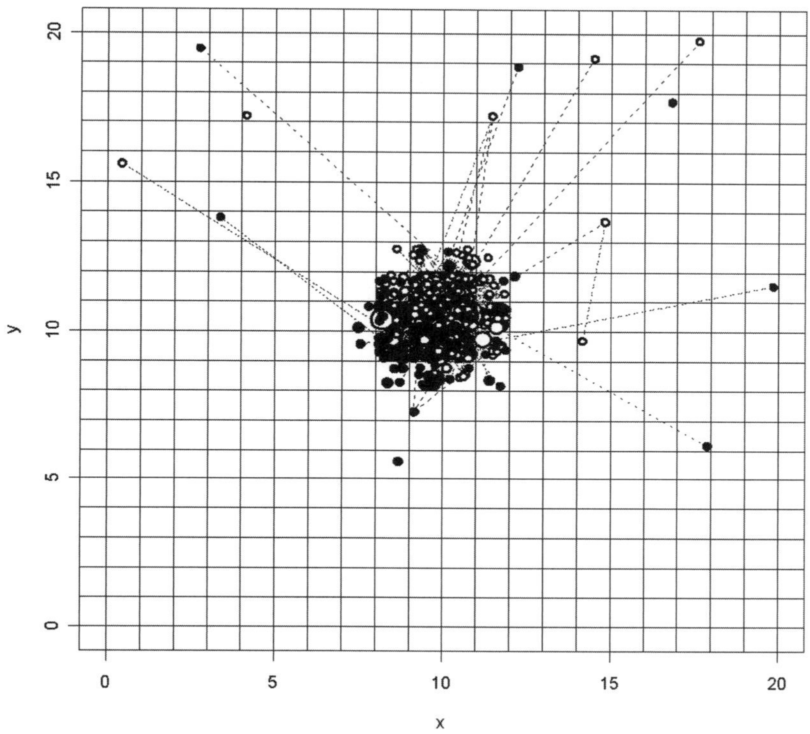

FIGURE 12.4. Attraction, Density, Double Homophily Effects

the presence of homophily, not accompanied by xenophobia, is not sufficient to produce high levels of segregation, especially in the case of double homophily.

In the third set of configuration (Figs. 12.7–12.9) we focus on the consequences of adding the propinquity effect to the model. As we noted above, propinquity is an alignment effect which implies that households that are linked via social network ties tend to be found in neighborhoods that are close to each other. In this case, "closeness" is determined by Euclidean distance between neighborhood centroids rather than by contiguity.

When propinquity is added to the attraction, density, and xenophobia model, households belonging to the two groups cluster into ethnically homogeneous bands separated by empty regions. The big areas occupied by the two groups in the previous configuration determined by the attraction, density, and xenophobia effects is broken into smaller bands that are formed so that households that are tied can be found in neighborhoods that are close to each other in Euclidean space. However, the two groups remain highly segregated along ethnic lines ($D = 0.98$, $SP = 1.97$).

By comparing the three configurations in this set, we see again that the model that includes xenophobia leads to the most segregated configuration

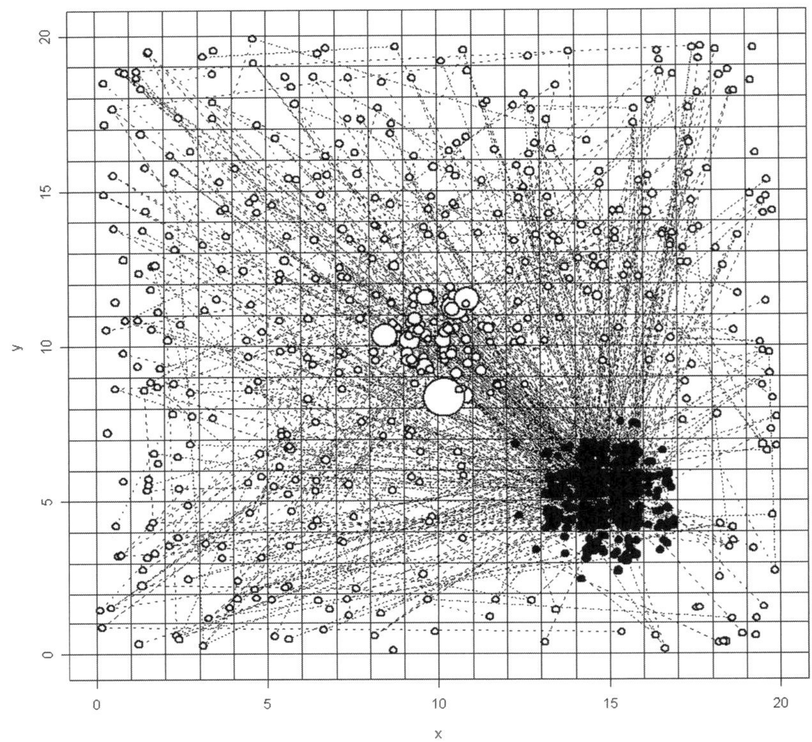

FIGURE 12.5. Attraction, Density, Xenophobia, Single Homophily Effects

(D = 0.98,SP = 1.97, compared with $D = 0.58$, $SP = 1.21$ for single homophily and $D = 0.33$, $SP = 1.05$ for double homophily). An interesting consequence of adding the propinquity effect, for all three cases, is the fact that isolates and lone dyads now appear on the periphery of the grid; the combination of higher income and bigger number of ties has pulled the other households toward the center. This effect is more apparent in the model that includes a single homophily effect, since empty-circle households do not exhibit the tendency to be close to ethnically similar alters, thus suggesting a connection between low income and social and geographic isolation.

The last set, which comprises Figs. 12.10 and 12.11, presents draws from two models that include all effects we have considered so far. In these cases the main "structural signatures" observed so far for each of the effects are present: the buffer zone characteristic for xenophobia separates the areas occupied by the two groups, filled-circle households are clustered together, while empty-circle ones are either scattered (single homophily) or clustered (double homophily), and tighter clusters as well as isolates and lone dyads are present due to the propinquity effect. Both configurations are highly segregated, with segregation index values of $D = 1.00$, $SP = 1.98$, and $D = 0.87$, $SP = 1.87$, respectively.

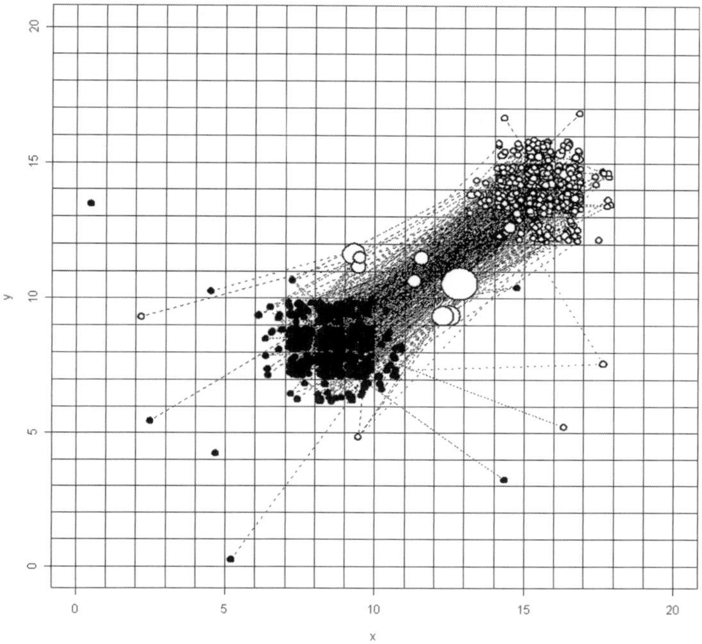

FIGURE 12.6. Attraction, Density, Xenophobia, Double Homophily Effects

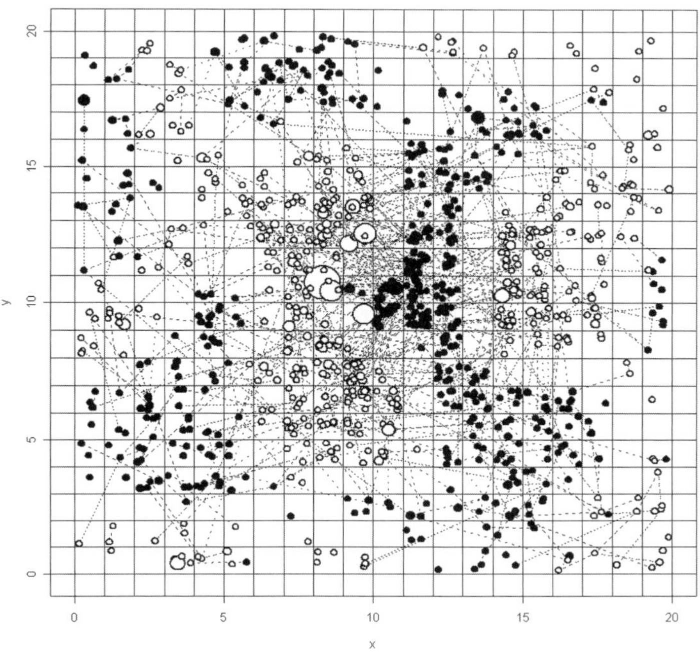

FIGURE 12.7. Attraction, Density, Xenophobia, Propinquity Effects

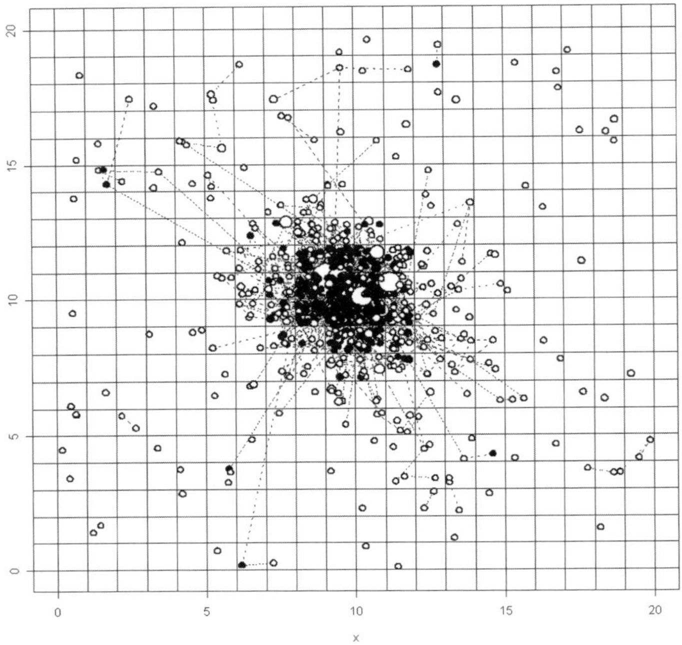

FIGURE 12.8. Attraction, Density, Single Homophily, Propinquity Effects

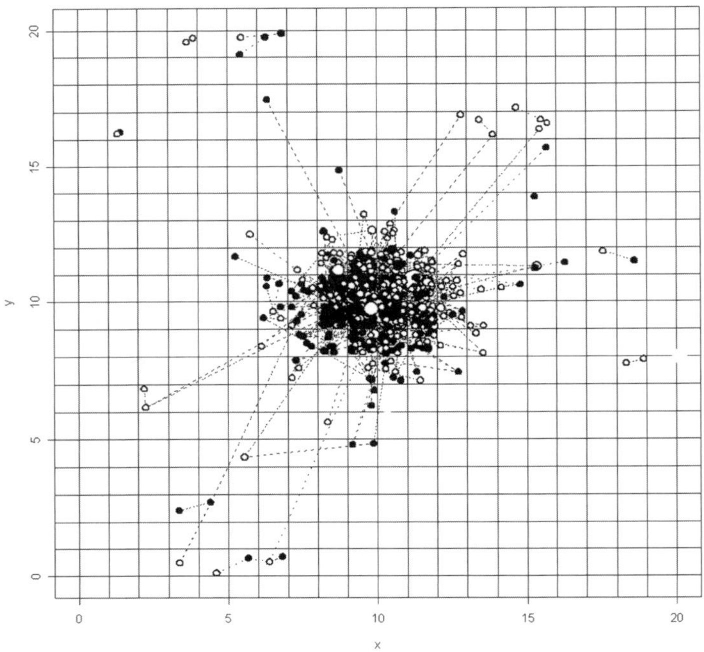

FIGURE 12.9. Attraction, Density, Double Homophily, Propinquity Effects

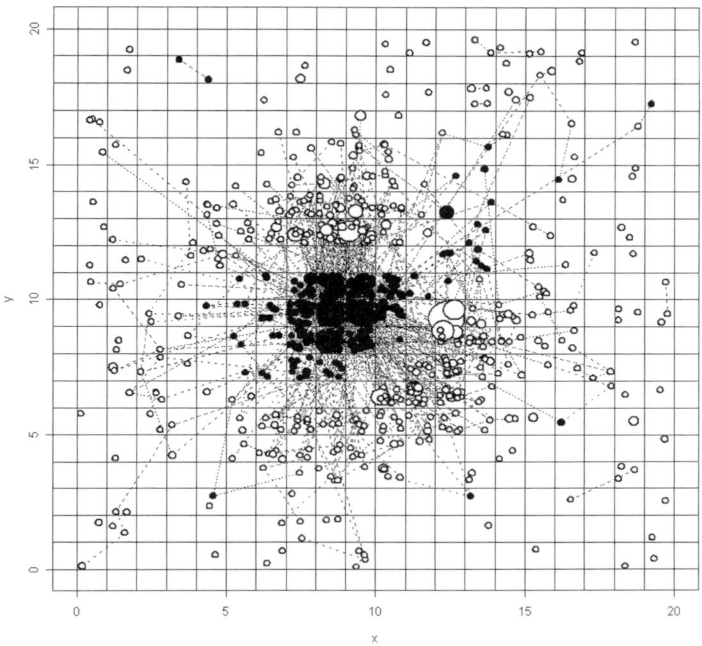

FIGURE 12.10. Attraction, Density, Xenophobia, Single Homophily, Propinquity Effects

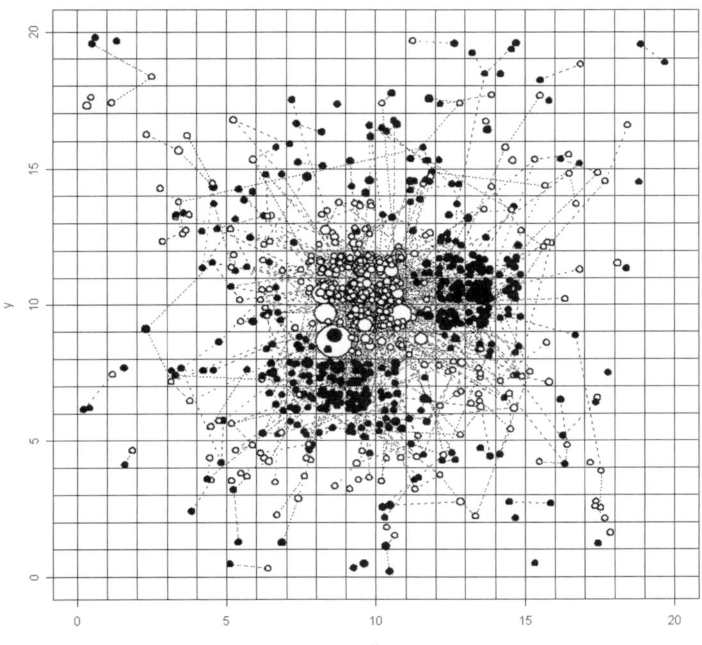

FIGURE 12.11. Attraction, Density, Xenophobia, Double Homophily, Propinquity Effects

12.5. Conclusion

The model proposed in this study differs from agent-based and cellular automata models based on ethnic preferences most importantly because it allows researchers to differentiate between the tendency to be close to similar alters (homophily) and the tendency to be far from dissimilar alters (xenophobia), in contrast with the percentage/threshold approach based on the ethnic composition of the neighborhood employed by previous studies. Several important conclusions can be drawn from the analysis of the simulation results presented here:

- Homophily and xenophobia are distinct processes: models that include a xenophobia effect always lead to segregated configurations, while those including homophily only do so under certain conditions.
- Even within homophily, distinguishing between single and double homophily can provide useful insights: models that include a single homophily effect sometimes lead to moderately segregated configurations, while those including a double homophily effect almost always do not.
- Homophily and xenophobia have different structural signatures in terms of spatial patterns of residential settlement, and interact in different and non-trivial ways with other effects.

We must emphasize here that these conclusions are based on the particular covariates and parameters used in simulating the model. Further research in which multiple covariate and parameter values are employed will help improve our understanding of model behavior and residential segregation processes.

References

Alba R, Nee V (2003) Remaking the American mainstream: Assimilation and contemporary America. Harvard University Press, Cambridge.

Benenson I (2004) Agent-based modeling: From individual residential choice to urban residential dynamics. In Goodchild MF, Janelle DG (eds.) Spatially Integrated Social Science. Oxford University Press, New York, pp. 67–94.

Butts CT (2005) Building inferentially tractable models of complex social systems: A generalized location framework. Institute of Mathematical and Behavioral Sciences, Technical Report MBS 05/08. University of California, Irvine.

Clark WAV (1986a). Human Migration. Sage Publications, Beverly Hills, CA.

Clark WAV (1986b) Residential segregation in American cities: A review and interpretation. Population Research and Policy Review 5:95–127.

Clark WAV (1992) Residential preferences and residential choices in a multiethnic context. Demography 29:451–466.

Denton NA, Massey DS (1989) Racial identity among Caribbean Hispanics: The effect of double minority status on residential segregation. American Sociological Review 54:790–808.

Duncan OD, Duncan B (1955) A methodological analysis of segregation indices. American Sociological Review 20:210–217.

Epstein, Joshua M. and Robert Axtell. 1996. Growing artificial societies: Social science from the bottom up The Brookings Institution Press, Washington, D.C.
Foner N (2000) From Ellis Island to JFK: New York's two great waves of immigration. Yale University Press.
Fossett M (2006) Ethnic preferences, social distance dynamics, and residential segregation: Theoretical explorations using simulation analysis. Journal of Mathematical Sociology 30:185–274.
Frey WH, Farley R (1994) Latino, Asian, and Black segregation in U.S. metropolitan areas: Are multi-ethnic metros different? Demography 33:35–50.
Gordon MM (1964) Assimilation in American life. Oxford University Press, New York.
Gottdiener M, Hutchinson R (2000) The new urban sociology. McGraw-Hill, Boston.
Grannis R (2002) Discussion: Segregation indices and their functional inputs. Sociological Methodology 32:69–84.
Hawley AH (1950) Human ecology: A theory of community structure. Ronald, New York.
Iceland J, Weinberg D, Steinmetz E (2002) Racial and ethnic residential segregation in the United States: 1980–2000. U.S. Census Bureau, Washington, DC.
Lieberson S (1981) An asymmetrical approach to segregation. In Peach C, Robinson V, Smith S (eds), Ethnic Segregation in Cities. Croom-Helm, pp. 61–82.
MacDonald JS, MacDonald L (1974) Chain migration, ethnic neighborhood formation, and social networks. In Tilly C (ed) An urban world. Little and Brown, Boston, pp. 226–235.
Mare R, Bruch E (2003) Spatial inequality, neighborhood mobility, and residential segregation. California Center for Population Research Working Paper No. 003-03. University of California, Los Angeles.
Massey DS (1985) Ethnic residential segregation: A theoretical synthesis and empirical review. Sociology and Social Research 69:315–350.
Massey DS, Denton NA (1985) Spatial assimilation as a socioeconomic outcome. American Sociological Review 50:94–106.
Massey DS, Denton NA (1987) Trends in the residential segregation of Blacks, Hispanics, and Asians: 1970–1980. American Sociological Review, 94:802–825.
Massey DS, Denton NA (1988). The dimensions of residential segregation. Social Forces 67:281–315.
Massey DS, Denton NA (1993). American Apartheid: Segregation and the making of the underclass. Harvard University Press, Cambridge, MA.
Massey DS., White MJ and Phua V (1996) The dimensions of segregation revisited. Sociological Methods and Research 25:172–206.
McKenzie R (1924) The ecological approach to the study of urban community. Reprinted in Short JF (ed.) (1971). The social fabric of the metropolis: Contributions of the Chicago school of urban sociology. University of Chicago Press, Chicago, pp. 17–32.
Menjivar C (2000) Fragmented ties: Salvadoran immigrant networks in America. University of California Press, Berkeley, CA.
Portes A, Zhou M (1993) The new second generation: Segmented assimilation and its variants among post-1965 immigrant youth. Annals of the American Academy of Political and Social Science 530:74–98.
Robins G, Pattison P (2005) Interdependencies and social processes: Dependence graphs and generalized dependence structures. In Carrington PJ, Scott J and Wasserman S (eds) Models and methods in social network analysis, Cambridge University Press, Cambridge, MA, pp. 192–214.

Sakoda JM (1971) The checkerboard model of social interaction. Journal of Mathematical Sociology 1:119–132.
Schelling TC (1969) Models of segregation. American Economic Review 59:483–493.
Schelling TC (1971) Dynamic models of segregation. Journal of Mathematical Sociology 1:143–186.
South SJ, Crowder K, Chavez E (2005) Geographic mobility and spatial assimilation among U.S. Latino immigrants. International Migration Review 39:577–607.
Tauber KE, Tauber AF (1965) Negroes in cities: Residential segregation and neighborhood change. Aldine, Chicago.
Thomas WI (1921) The immigrant community. Reprinted in Short JF (ed.) (1971). The social fabric of the metropolis: Contributions of the Chicago school of urban sociology. University of Chicago Press, Chicago, pp. 120–130.
Waldinger R (ed.) (2001). Strangers at the gates: New immigrants in urban America. University of California Press, Berkeley.
Waters MC (1999) Black identities: West Indian immigrant dreams and American realities. Harvard University Press, Cambridge, MA.
White MJ (1983) The measurement of residential segregation. American Journal of Sociology 88:1008–1019.
White MJ (1986) Segregation and diversity: Measures in population distribution. Population Index 52:198–221.
Wilson FD, Hammer RB (2001) The causes and consequences of racial residential segregation. In O'Connor A, Tilly C, Bobo L (eds) Urban inequality in the United States: Evidence from four cities. Russell Sage Foundation, New York, pp. 272–303.
Zhang J (2004) A dynamic model of residential segregation. Journal of Mathematical Sociology 28:147–170.
Zhou M (1992) Chinatown: The socioeconomic potential of an urban enclave. Temple University Press, Philadelphia, PA.
Zubrinsky Charles C (2001). Processes of racial residential segregation. In O'Connor A, Tilly C, Bobo L (eds) Urban inequality in the United States: Evidence from four cities. Russell Sage Foundation, New York, pp. 217–271.

Chapter 13
Corporate Interlock

Lorien Jasny

13.1. Abstract

An important test of the theory put forth in this book is its explanatory power across disciplines and medium. In this section, I will examine a social system for the presence of structures expected under constructal theory. There is an expansive literature in social network analysis of interlocking directorates of corporate boards of directors, and using this medium to comment on the social implications of constructal theory seems a logical step—this analysis will examine a well-studied area of social science but through the new lens of constructal theory.

13.2. Introduction

Scientists in many different fields pursue unifying theories of action. Physicists attempt to discern the rules governing the motion of particles, while sociologists try to find patterns in the dynamics of social groups. The interaction between theories in both fields, although limited, is not new. In his 1949 work *Human Behavior and the Principle of Least Effort: an Introduction to Human Ecology*, George Zipf employed a notion in physics called the "Principle of Least Action" in the analysis of linguistics. The Principle of Least Action states that physical systems act to minimize "exertion," often illustrated in physical examples as kinetic energy. It was first introduced in the study of light reflection by Pierre-Louis Moreau de Maupertuis (Maupertuis 1744), and used by other mathematicians (Euler, Lagrange, and others) as a different view and approach to problems of motion than Newtonian mechanics. Newtonian mechanics relates force to mass, whereas a principle of least action view will model the system through energy expenditure, and derive the trajectories of lowest "cost." Although both approaches arrive at the same solution, the principle of least action provides an alternative orientation and framework from which to understand physics. Replace particles with human actors, and the principle of least action becomes a framework for the understanding of human interaction, and is the basis for many theories of human nature. Zipf's work on word patterns derives from a view of linguistics as a system in which the energy used to produce and express thoughts can be minimized (Zipf 1949). Other notable extensions of the principle of least effort into the social sciences include the schools of economics built upon utility theory and the related concept of Pareto efficiency. These theories differ on application and breadth, but the common thread is that they assume cognitive individuals will act to minimize overall expenditure of energy while maximizing payoff. Zipf's Principle of Least Effort posits a macro-approach in time: actions that appear to contradict the principle of least effort will, over time, work to create an averaged effect of least effort (Zipf 1949). This amendment permits an individual to take the harder path if he or she perceives better payoff in the long run rather than immediate expenditure. Utilitarianism looks at that aggregation of individuals in society, arguing that economic policies, while sometimes harmful for a specific individual, should work for the maximal utility of the whole. Thus, in the same manner that a physical system settles on an equilibrium point of minimal expended energy, these theories of social behavior argue that patterns of aggregate human response may operate on a similar mechanism as kinetic energy in the Principle of Least Effort.

In recent work, Adrian Bejan (2000) has made a contribution to the exploration of the Least Action Principle in other scientific disciplines by marshaling many examples (thermodynamics, aerodynamics, etc.; see Bejan 2000), in support of a general model of structure he labels "Constructal Theory." Bejan's formulation of his Constructal Law is that "for a finite-size system to persist in time (to live), it must evolve in such a way that it provides easier access to the imposed currents that flow through it" (Bejan 2005). The framing of this statement is intended for specific cases where the "currents" are easily identified: a touted comparison are the examples of heating and cooling systems in electronics and the organization of oxygen flow in the human lung. Bejan argues that the lung has developed over the course of human evolution to maximize the dispersal of necessary "current," and this structural layout is reflected in the optimization of the cooling problem in modern electronics (Bejan 2000). The solution to these optimization problems is not new; however, Bejan's unification of the structures found in natural and man-made systems under the label of Constructal geometry is his own addition to this body of work. The derivation from the Constructal Law to a specific geometry follows a specific logic:

> The origin of the generation of geometric form rests in the balancing (or distributing) of the various flow resistances through the system. A real system owes its irreversibility to several mechanisms, most notably the flow of fluid, heat and electricity. The effort to improve the performance of an entire system rests on the ability to minimize all its internal flow resistances, together and simultaneously, in an integrative manner (Bejan, 2000).

Bejan concludes that the optimal, and frequently found, geometry derived from the constructal law is a tree-shaped network (for proof, see Bejan 2005). In the examples given, both cooling units and the human lung rely on tree-shaped networks of coolant or veins for the efficient diffusion of coolant or oxygen through the system, thus employing the specified geometry. Bejan argues this specific network architecture maximizes the flow potential, necessary in both engineered and natural systems. This formulation provides a testable hypothesis for the Constructal Law (Bejan 2005). He argues that this framing of evolutionary process includes a crucial element missing from theories of natural selection: a system is "fittest" or most efficient when it complies with the Constructal Law, thus provides a method for prediction not only of animals, but all natural systems (Bejan 2005).

Bejan argues that this principle has obvious purchase in economic systems under the heading of maximized utility, but its extension into the social realm is so far untested. Thus, the impetus for this current research is to again take a principle from physics and test its application in the social sciences. The specific hypothesis of the presence of tree-shaped networks presents a simple test to see if in one particular case, that of corporate interlocks, the empirical data supports the Constructal Law. If we could include social networks under the umbrella of the Constructal Law, social scientists would gain a powerful platform from which to incorporate other methods from the physical sciences for the investigation of social sciences. The converse is equally interesting: if the networks in question do not follow the principles maximized under the Constructal Law, this investigation can provide some understanding of the dynamics at work in a socially constructed network adverse to the principles found in the other networks Bejan has investigated. These particular examinations of the corporate interlock networks will help to illuminate the structure and dynamics of these specific cases over time. The broader impact is a test for the possible inclusion of the principle of least action in social networks.

13.3. Corporate Interlocks

Corporate interlocks have been a focus of study and debate since their inception in the early 20th century. When one individual sits on the board of directors of two different corporations, those two companies are interlocked through their shared board member. This connection between corporations facilitates cooperation and communication between the two companies. The complete set of interlocks on a collection of firms can be thought of as a network, in which the firms are nodes, and two nodes are joined by an edge if only if they have an interlock. Consider Fig. 13.1 which shows a simple network of 14 firms. Here, the company Halliburton is the central node, with lines drawn to nodes labeled A–F representing Halliburton's six board members. In addition to belonging to Halliburton's board, these individuals also sit on the boards of several other companies. The outer nodes that A–F are also connected to thus represent the companies that Halliburton is interlocked with through their respective board members. The actual networks analyzed smoothes over the board member nodes, only displaying the links between companies. Thus, the figure below is two mode representation of part of a corporate interlock network.

The range and advantage provided by corporate interlocks is at the heart of a debate over the structure of power in the economic market. This relationship

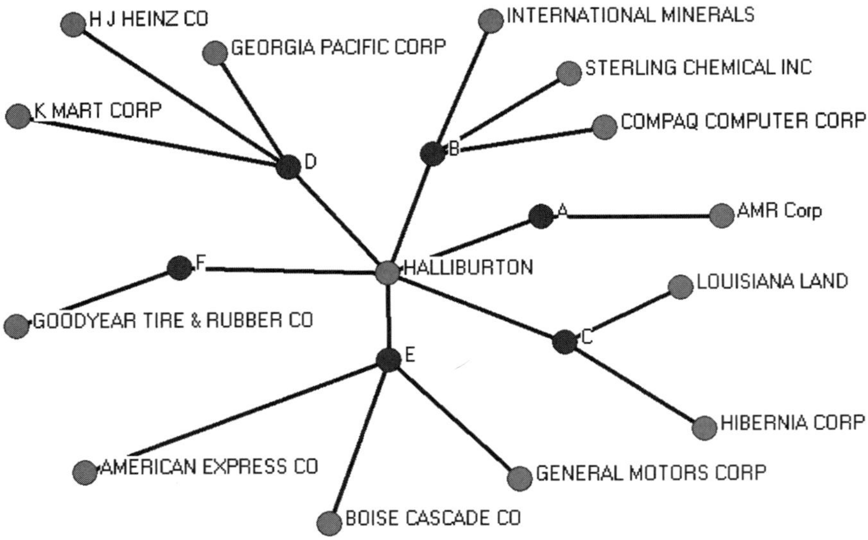

FIGURE 13.1. Sociogram depicting a corporate interlock

formed the basis for the C. Wright Mill's argument in 1956 that a "power elite" existed in American business, composed of these board members who exercised unsurpassed power through their connections. According to Mills, these corporate officers created a new group with even greater ability to forward their class interests than any faction previously observed. The response to Mills' conception of a group of board members controlling the actions of the majority of the American corporate system was the pluralist notion that corporations are often at odds with each other in a capitalist system. Due to the forces of the free market, while two companies are aligned on issue A, they oppose each other on issue B, regardless of the views of any particular board member (Neustadtl, Scott, and Clawson 1991). This division, imposed regularly in a capitalist system, should prevent any conspiratorial machinations on the part of the board members that are heavily interlocked. This argument is continually rehashed through the work of other scholars as interlocks rise and fall in popularity, but the dispute underlies the importance of understanding the nature of inter-corporate ties.

The network literature on corporate interlocks uses a different framework. Rather than discussing the implications of an elite interlocked core outside of the network of organizations, these authors investigate the structural importance of interlock ties themselves and the consequences for the involved corporations. In this perspective, having a board member take a position on another board, subsequently forming an interlock, is viewed as a strategic move by an organization to gain control over their environment. For the individual, this opportunity involves an expenditure of time and responsibility with an increase in money and prestige. The company, too, accepts benefits and losses as a result of this tie: the board member has less time to devote to one particular organization, and may forward the thinking of another corporation in the board room. However, the presence of interlocks indicates some benefit to the corporation since the number has persisted and increased over the 20th century. Researchers theorize that the benefits of interlocks to a company take the form of strengthened ties to other companies that produce goods they depend on, and more knowledge of the decisions other companies are making and problems they are facing (Allen 1974; Mizruchi 1982). Two important studies examine interlock connections as the basis for the diffusion of practices and information throughout the corporate network (Powell 1996; Davis 1991). Both authors use interlocks as one measure of how innovation spreads through the networks of different organizations as board members report to other boards about new ideas and practices (Powell 1996; Davis 1991).

Most of the studies of corporate interlocks examine the relationships from the point of view of specific organizations. Fewer authors consider the corporate interlocks from the perspective of the whole network. One notable exception is Mark Mizruchi, who explains in Theories of the Modern Corporation that "the corporation must be viewed as an element of an interorganizational system, in which no one corporation can be understood without locating its position within the system" (Mizruchi 1982). This comment adds another dimension to the interlock relationship; not only are the direct interlocks significant to a company's operation itself, but Mizruchi argues that a firm's position within the interlock network is crucial to its performance (Mizruchi 1987). Thus, some understanding of the characteristics of the network itself is necessary for understanding how information and innovation diffuse through the corporate world. From this view, the network of corporate interlocks is a prime target for scrutiny under the premise of Constructal Law. Has the system of corporate interlocks evolved to best transmit information through the channels of interlocks across the system of companies? Given that there is some cost to the creation of a tie, but more benefits gained through inclusion in the network, can we detect structure in accordance with the Constructal Law? Constructal theory provides some interesting and testable hypotheses for such a system. This approach is not without precedence in organizational theory: in a study of organizational learning, Krackhardt explained "links are not without costs in a social system. They take time and resources to maintain" (Krackhardt 1994). Krackhardt developed his notion of graph efficiency to measure precisely how often a network deviates from a strict hierarchy, pictorially a tree shape with information or "current" flowing down the hierarchical ladder or branches of the tree. Under Constructal Law, the tree structure is considered optimal. Krackhardt efficiency measures how often a network deviates from a strict hierarchy, a tree, and therefore characterizes how dense the network is beyond that barely needed to keep the social group even indirectly connected to one another. Network formation is the result of an optimization problem; the question is, whether it fits into those categorized by a Constructal geometry. The system of corporate interlocks certainly has some limitations, but with those understood, this case provides an interesting testing ground for Constructal Theory in the social sciences. An ideal system, in the Constructal sense, would settle on the most efficient format that kept organizations in the loop, afforded the board members enough positions, and did not overly cost the organizations included. The analysis at hand will test if the network of corporate interlocks is organized under a Constructal geometry to maximize the flow of this information to the different corporations.

13.4. Data

The data collected in this study are from the Yoo, Davis, and Baker study of corporate interlocks (Davis et al. 2001). The sample of companies investigated was drawn from the Fortune 1000 list for 1990 and 2000. The 1990 dataset contains 726 nodes, corporations, and 3315 links, interlocks between corporations created by shared board members. The 2000 dataset has 916 nodes and 3799 interlocks between the corporations. These are the empirical networks that herein are compared to simulations. The simulated networks are generated as random graphs with the same sizes and number of edges as each of the empirical networks examined. One hundred random networks are generated for each simulation to generate a range of each value, and this was done using the rgnm function in R written by Professor Carter Butts (2006).

13.5. Methodology

Many studies of interlocking directorates search a network for the most centralized nodes and compare those results against indicators of performance or some other independent variable (see Mizruchi and Galaskiewicz 1993 for a review). The purpose of this chapter, rather, is not to compare nodes within the networks, but the structures of the networks as systems themselves; this involves first comparing all the data to randomly generated networks and then comparing the 1990 data to 2000 data. Thus, the statistics to be examined are those drawn from Social Network Analysis that yield measures of the network as a whole. If the results vary greatly from the randomized networks, then the empirical data is organized along some rules that result in patterns different from those that appear randomly. The degree distribution will first classify the type of network. Many standard network patterns confine how many other nodes one is connected to, and this is the degree distribution. Returning to Fig. 13.1, Halliburton has a degree of 13, as it is connected to 13 other companies. The path distances will describe how closely tied the corporations are, and how that length compares to the networks generated through simulation. Again, in Fig. 13.1, Halliburton is at a distance of 1 from each company that it is interlocked with. Two other companies depicted are at a distance of two, since they can reach each other through Halliburton in two steps. If the path length is shorter than that which appears in simulation, then the companies are more closely connected. Longer paths would show a looser, more spread out network where some corporations are further than most others. These relations impact how currents could flow through such a network, and how cohesive the network of corporations is. Third, the triad counts look at the types of relations that predominate in the network, establishing a much more in-depth view of the actual relationships formed between the interlocks in the network. For background on the use of triads in network analysis, see Faust 2006. Finally, Krackhardt efficiency is the last statistic presented and is the most direct translation of Constructal theory into social network methodology. This statistic measures how much the network patterns deviate from a formal hierarchy, or the tree structures found in those systems that support Constructal theory.

To test if the networks have developed "efficiently," I compare each network to randomly simulated networks of the same size and edge distribution. The results determine to what extent the empirical networks deviate from what result could happen if the interlocks occurred at random. The simulated network is constructed given the same number of nodes and edges as in the empirical network. This ensures that the same number of links exists in each trial, so no difference in structure can be the result of differing parameters. I replicated the simulated networks one hundred times and averaged the values for a mean and variance calculation of what possible range of values was likely to occur for each statistic. The places where the empirical networks are statistically different are characteristics of structure that are unlikely to have occurred by accident, but instead developed over a century of corporate involvement.

The second test, given multiple networks varying in time, is to see if some change, consistent with that predicted by the Constructal Law, is observed in the differences between the dataset drawn from 1990 and that from 2000. Over the decade, we should see an increase in comparative Krackhardt efficiency, a maximizing of the flow of information or contact through a few established channels, thus cutting down on the time board members need to spend in meetings, and the number of board members that must be paid and consulted. Thus, not only will this work provide a comparison to what interlock networks might be generated normally, but also a test of the prediction power of the constructal law: given that certain factors seem to be considered efficient, is the network of interlocking directorates in fact evolving over the decade from 1990 to 2000 in such a way as to increase these relative measures? A comparison of the same characteristics examined with respect to the simulated networks will provide an answer.

13.6. Analysis

Immediately from the comparison of the degree distribution, the empirical interlock networks are drastically different from a randomly simulated network. The size of the board is naturally limited by the amount of money a company can afford to pay its board, and possibly the size of a board room; however, the maximum number of interlocks present in the empirical data is by far larger than the extreme values generated randomly. For each board member, there exists a limit to the number of boards each can sit on. An additional board position, while increasing prestige and pay, diminishes the amount of time that can be spent for each company. The incorporation of additional board members contains considerations on the part of both the board and the member. From the skewed distribution of number of interlocks shown in Figs. 13.2 and 13.3, some organizations have many interlocks, but most have far fewer.

The first observation is that the distribution of interlocks is highly skewed with a handful of corporations possessing large numbers of interlock connections. Other work on different interlock data suggests that, predictably, larger and more centralized firms tend to have more interlocks (Mintz and Schwartz 1981). The slopes from the 2 years are directly compared in the log-log plot shown in Fig. 13.4. Note (i) that the slope does not approach a power law and (ii) that the comparison of the 1990 and 2000 data reveals that the slopes are similar, albeit with slightly different heads. The second observation from this data is that the system does not change, with regard to the interlock distribution, over the decade. This implies, given the continuity over time, that the way the firms are organized is maintained.

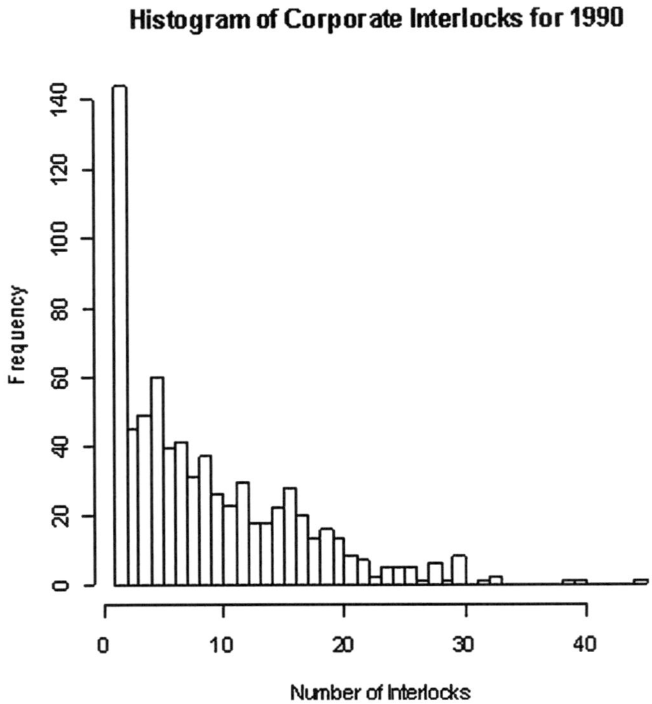

FIGURE 13.2. Histogram of Corporate Interlocks: 1990

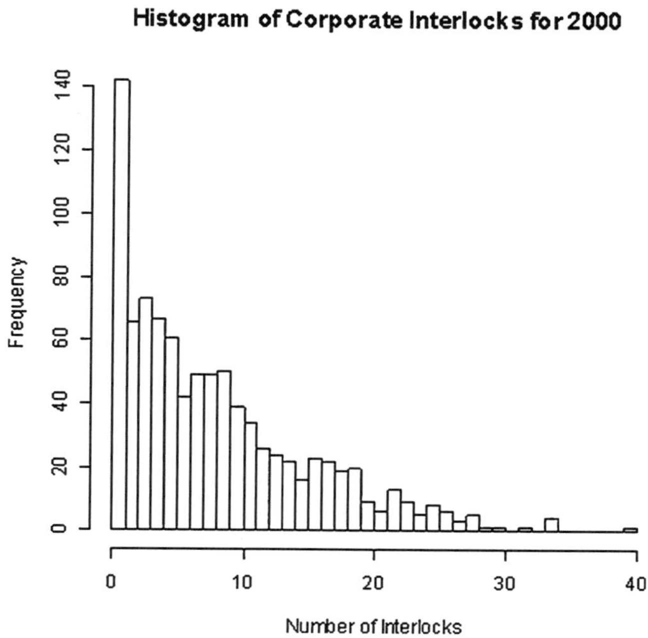

FIGURE 13.3. Histogram of Corporate Interlocks: 2000

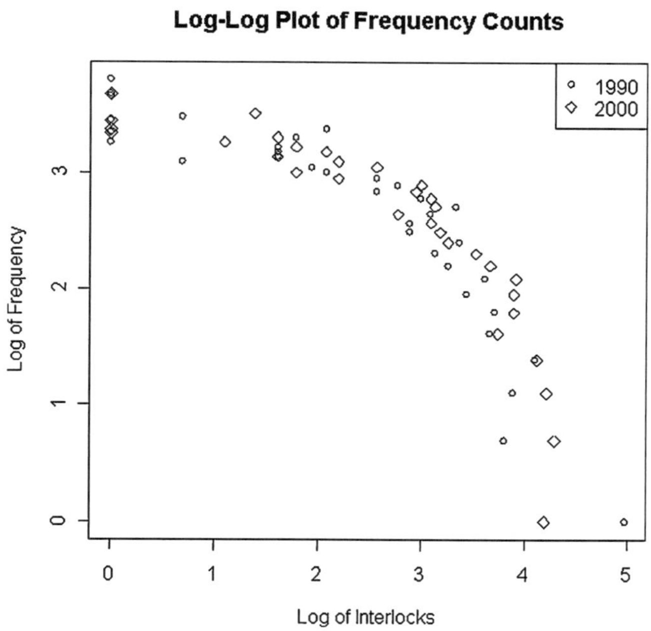

FIGURE 13.4. Log-Log Plot of Interlock Frequency Counts

These results dispute any argument that the interlock networks are evolving to a condition where an ideal number of interlocks is found for every organization. Instead, the equilibrium utility point, stable over the decade in question, is found empirically for the network itself, rather than from the point of view of individual corporations. Although companies' interlocks change over this decade, and many of the corporations in the Fortune 1000 are different at each time point, this structure of degree distribution, highly different than that found in randomized networks, is maintained.

In comparison to the simulated networks, whose degree distributions are shown in Figs. 13.5 and 13.6, clearly there is an organizing principle at work where a handful of organizations operate with many interlocks, but the costs of additional links for all the nodes, along the lines of a normal distribution that the simulated networks display, is either too costly or not optimal for the system to support. Of course, further investigation into the actual networks is necessary, but for the sake of comparison to other similar studies, I present these initial findings to demonstrate the abstract relation being investigated. For a more detailed discussion of the 1990 dataset, see Davis 1991 and 2003.

These distributions approach a normal curve with means of 18.26 and 16.59, variances of 37.39 and 34.13, and maximum values of 42 and 40 respectively. Clearly, these fall well short of the tail of the empirical distribution. Of course, all theories of corporate interlocks account for the benefits some companies receive from these large number of connections, thus this comparison to a random graph

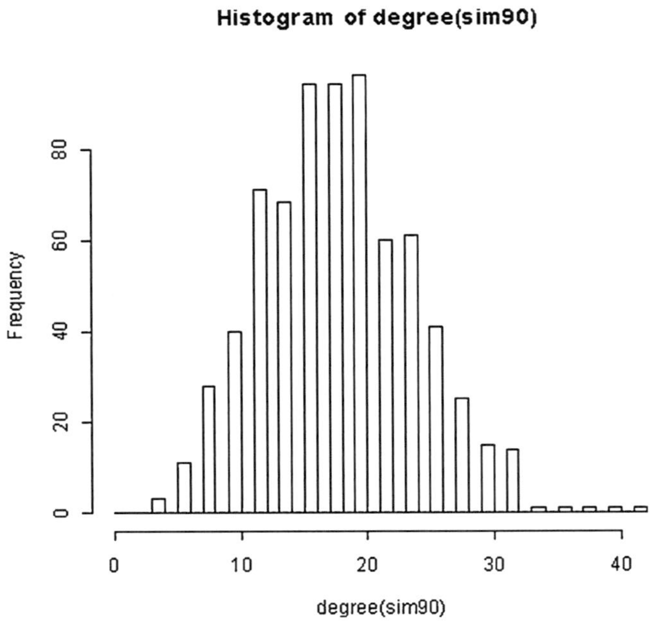

FIGURE 13.5. Histogram of Degree for 1990 Simulated Networks

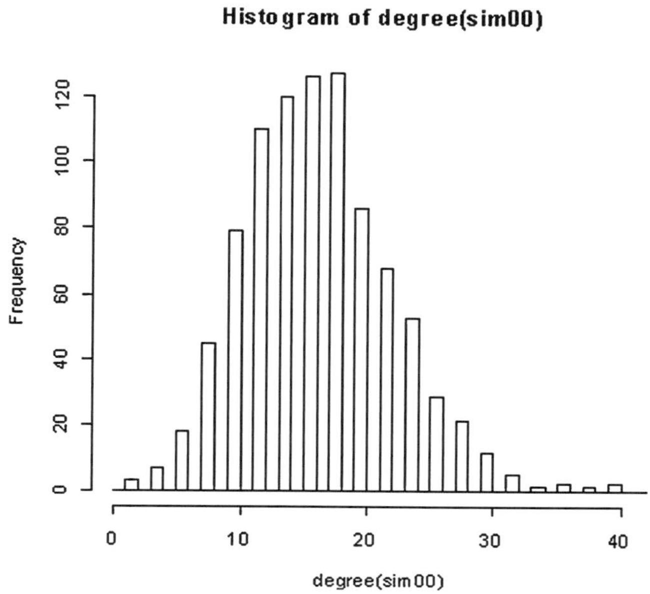

FIGURE 13.6. Histogram of Degree for 2000 Simulated Networks

was expected and unsurprising. The comparison of degree distributions, however, does provide a platform for understanding some of the basic differences between the simulated and empirical networks, and aids in the interpretation of the later, more complicated, network measures (Figs. 13.7 and 13.8). The stability of the skewed degree distribution, completely different from the normal distributions seen in randomly generated networks, classifies the empirical networks, and shows that some structural relation between the corporations is functioning to maintain these results.

Given the skewed quality of the degree distribution, and the large number of companies with relatively few interlocks, one might expect the empirical data to have a very different path length distribution than the randomly simulated networks—in other words that the corporate interlock network would be much more sparse, with distances between companies with few ties much greater than the maximum distances created in the simulated networks.

Both of the empirical networks have a maximum path length of 8 links and, as expected, the empirical datasets have slightly longer tails than the simulated networks. This difference indicates that the random networks are more closely tied together, and the empirical networks have some longer chains than found in the simulations. Still, in the simulations and empirical networks, most pairs of nodes are (on average) at a distance of 3 or 4 interlocks from each other. Take, for example, the efficiency of the interlock network depending on one company being able to spread information to all the other members of a network through some kind of diffusion process. Under that condition, only a very specific

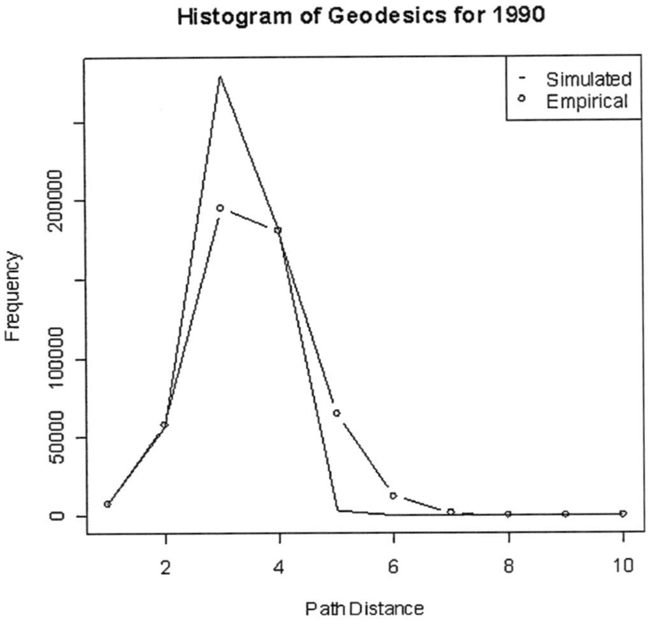

FIGURE 13.7. Geodesics for 1990

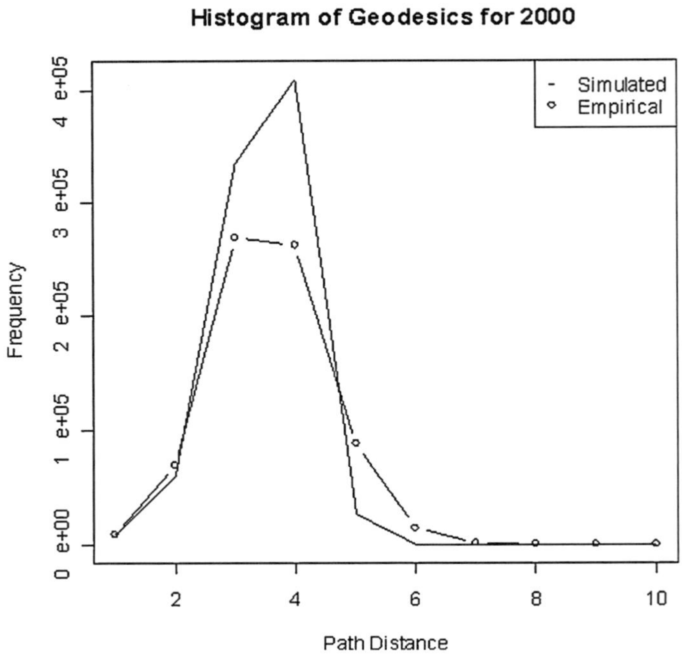

FIGURE 13.8. Geodesics for 2000

mechanism would prefer the empirical network over the randomly linked one, since the shorter average path length indicates that, under general diffusion conditions, such information would spread faster in the random networks rather than those existing in practice. The conclusion is, either that particular scenario is not what the system is used for or there is some impediment that prevents the system from reaching a configuration more efficient than the random network.

The third form of analysis is a triad census. Here, all combinations of three companies are considered, and a count is taken. Null triads contain no links between the three companies, one tie means one company is connected to a second, but the third is isolated, and so on. In the 1990 interlock (Table 13.1), we can determine from the z-score comparisons to the simulated networks, what forms of relationships are more likely to occur than in the random graphs, and thus emphasized by the interlock system. Null triads exist in the empirical dataset with a slightly higher probability than in the simulated networks, and triads with only one tie are suppressed, as seen by the negative z-score. Triads with two ties occur in the data more often than simulated, and the most extreme difference is seen in the presence of completed triads, which occur much more frequently in the empirical data than in the random graphs, as shown by a positive z-score of over 500. Although these links account for only .004% of the triads in the network, these 2803 triads make the empirical network significantly different from the range of networks generated randomly. Given three nodes in a random network, or three nodes from the actual interlock network, those in the interlock network have a higher likelihood of all being connected. It is an interesting pattern: having only one tie is suppressed in the empirical data, whereas null ties or two or more are emphasized significantly over their levels in the simulated network.

Turning to the 2000 interlock data (Table 13.2), we observe the same pattern with the triads containing only one link suppressed, and a significant increase in the number of complete triads, with a z-score of 296.69: less than in 1990, but still more than three times the difference between empirical and simulated data for any other type of relation.

In comparing the two years of data, although similar triad censuses are produced, the emphasis on the complete triad has declined.

The Krackhardt efficiency measures how often a network deviates from an exact hierarchical model. It is calculated as $\left[1 - \frac{\sum_i^I V_i}{\sum_i^I MaxV_i}\right]$ where V_i is the number

TABLE 13.1. Triad Census for 1990

1990 Interlock	Null	One tie	Two ties	Complete
Triad fraction	96.290%	3.645%	0.060%	0.004%
Z-score	70.67	−64.04	43.75	557.28
Triad count	61156625	2315293	38179	2803
Null mean	61142925	2340014	29838.2	123.3
Null variance	37577.73	149003.1	36339.29	23.12222

TABLE 13.2. Triad Census for 2000

2000 Interlock	Null	One tie	Two ties	Complete
Triad fraction	97.318%	2.648%	0.033%	0.002%
Z-score	90.52	−82.28	57.69	296.69
Triad count	124251836	3380286	41614	2924
Null mean	124235694	3409743	31127.3	96.2
Null variance	31800.01	128165.3	33037.79	90.84444

of observed links in excess of $N_i - 1$ for each ith component in the network, i, of size N_i, and $MaxV_i$ is the maximum number of possible edges for that i[th] component. In Fig. 13.9, the one on the left is a perfectly organized hierarchy. The network on the right has additional lines (indicated by the dotted arrows) that interfere with an organized linear hierarchy. In resemblance to constructal theory, Krackhardt efficiency measures exactly how often a network deviates from a tree structure, so often cited as evidence for an efficiently organized system. Given the higher proportion of complete triads, as found in the census above, one would expect these networks to not show as much efficiency as the randomly generated networks.

The hypothesis about lower efficiency is proven in the data shown in Table 13.3, with negative z-scores demonstrating that the empirical datasets are less efficiently organized, in reference to the tree hierarchy structure, than the range of randomly generated networks. Additionally, the difference between the interlock and the simulation network grows over the decade, from a z-score of −13.49073 to −257.1521. Thus, according to this measure, even if some constraint forced down the efficiency of the network, according to this measure, the empirical data is not growing more efficient over time, in fact the reverse.

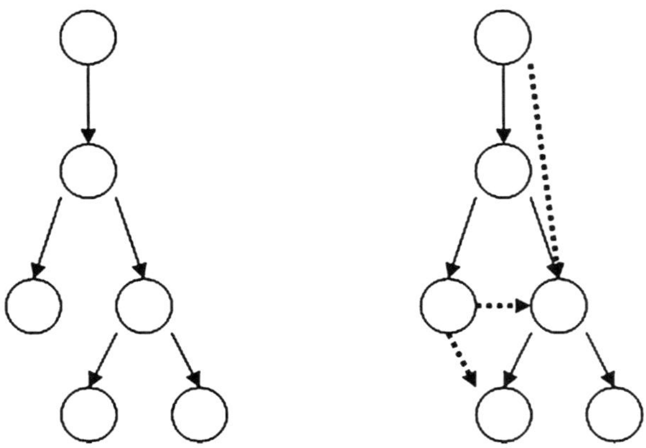

FIGURE 13.9. Krackhardt Efficiency

TABLE 13.3. Krackhardt Efficiency Results

	1990 Interlock	2000 Interlock
Empirical efficiency	0.9885717	0.9904974
Null mean	0.9887592	0.9920158
Null variance	1.931663e-10	3.48652e-11
Z-score	−13.49073	−257.1521

13.7. Conclusion

This data contradicts the result (found in the demonstration of the Constructal premise) that systems evolve over time to a more efficient system for reaching their stated goals. In this paper I have made multiple assumptions: first, that the goals of the corporate interlock network are the prompt and thorough diffusion of information, and second that these concepts translate into measurement by the Krackhardt efficiency statistic. The conclusion from the analysis is that not only are the empirical networks less efficient when compared to randomly generated graphs, but what was considered a measure of efficiency in fact decreased over time. Although the hypothesis behind Constructal theory is convincing in many of the cases cited elsewhere in this book, in many social networks perhaps the structure that is most "efficient" is a redundancy effect, seen in the triad census as a tendency toward two or three ties and a suppression of one tie-triads. These connections provide the network with other attributes rather than a linear hierarchy, and perhaps that is what the system is evolving toward. Redundant ties, the triads with two or three links, are an important figure in social networks. Many theories revolve around systems needing secondary links to reinforce social norms or aid in the transmission of information or disease, but this concept has not yet been applied to corporate interlocks. This would be a logical step in the research; distinguishing between those interlocks with redundant links, and those only linked through one edge, might reveal trends in the structure of the network heretofore unrecognized. The final point is that, while the premise of constructal theory works in many cases, some systems may be organized differently, or have conditions such that the constructal formulation of efficiency does not operate as expected.

References

Allen, M. (1974) The structure of interorganizational elite cooptation: interlocking corporate directorates. *American Sociological Review* **39**, 393–406.

Bejan, A. (2000) *Shape and Structure: from Engineering to Nature.* Cambridge: Cambridge University press.

Bejan, A. (2005) The constructal law of organization in nature: tree-shaped flows and body size. *J. Exp. Biol.* **208**, 1677–1686.

Butts, C. (2006) The SNA Package for R at http://erzuli.ss.uci.edu/R.stuff/, accessed 24 January 2006.

Davis, G. (1991) Agents without Principles? The Spread of the Poison Pill through the Intercorporate Network. *Administrative Science Quarterly* **36**, 583–613.

Davis, G., Yoo, M. and Baker, W. B. (2003) The small world of the American corporate elite, 1982–2001, *Strategic Organization* **1**(3), 301–326.

Faust, K. (2006) Comparing Social Networks: Size, Density, and Local Structure. *Metodološki zvezki* **3**(2), 185–216.

Fennema, M. and Schijf, H. (1978/79) Analysing Interlocking Directorates: Theory and Methods. *Social Networks* **1**, 297–332.

Krackhardt, D. (1994) Graph Theoretical Dimensions of Informal Organizations. In: Carley CM, Prietula MJ (eds) *Computational Organization Theory*. Lawrence Erlbaum Associates, pp. 89–110.

P.L.M. de Maupertuis, (1744) Accord de différentes lois de la nature qui avaient jusqu'ici paru incompatibles. Mém. As. Sc. Paris

Mills, C. W. (1956) The Power Elite. Oxford University Press.

Mintz, B. and Schwartz, M. (1981) Interlocking Directorates and Interest Group Formation. *American Sociological Review* **46**, 851–869.

Mizruchi, M. (1982) The American Corporate Network: 1904–1974. Sage Library of Social Research.

Mizruchi, M. and Schwartz M. eds. (1987) Intercorporate Relations: the Structural Analysis of Business. Cambridge: Cambridge University Press.

Mizruchi, M. (1990) "Cohesion, Structural Equivalence, And Similarity of Behavior: An Approach to the Study of Corporate Political Power." *Sociological Theory* **8**, 16–32.

Mizruchi, M. and Galaskiewicz J. (1993) Networks of Interorganizational Relations. *Sociological Methods and Research* **22**(1), 46–70.

Neustadtl, A. Scott D. and Clawson D. (1991) "Class Struggle in Campaign Finance?

Political Action Committee Contributions in the 1984 Elections." *Sociological Forum* **6**(2), 219–238.

Powell W, Koput K, Smith-Doerr, L (1996) Interorganizational Collaboration and the Locus of Innovation: Networks of Learning in Biotechnology. Administrative Science Quarterly

Zipf G. (1949) *Human Behavior and the Principle of Least Effort: an Introduction to Human Ecology*. New York: Harvard University Press.

Chapter 14
Constructal Approach to Company Sustainability

Franca Morroni

14.1. Introduction

The preservation of our terrestrial environment implies great modifications of the individual and social human behavior, especially in industrial and developed societies (USA, Europe, etc.), at a worldwide scale. This dynamic social phenomenon directly concerns the future of the overall human societies and is so of really great importance. The individual constitutes the smallest building block of a sustainable world and its behavior is the origin of the global resistivity against the required change of mentality at a greater scale such as companies or nations.

The sustainable development is, by definition, a development meeting our current needs without compromising the capacity of the future human societies to meet their own needs. The sustainable company is an enterprise taking into account this issue in its strategy of development in order to guarantee its economical, social, and financial viability.

A first problem is that no sustainable model of company exists and it is difficult enough for an enterprise committed in this process to understand if it is engaged on the good or on the bad way.

A second problem is that it is even more difficult to know if these choices will produce a positive impact on its viability in short term and in particular in the middle and long term. Extra-financial analysts rating the companies face the same kind of problem: the lack of objective criteria to judge the company performances in many of the various domains of their analyses. Created at the end of the 1990s, extra-financial rating agencies evaluate and rate the policy of social and environmental responsibility as well as corporate governance for the companies, most of the time for investors.

Each one has its own rating methodology, what does not simplify the communication between analysts and the rated company or the comparison between the various notations. This chapter addresses the relevance of the promising union of Thermoeconomics and Constructal theory for the construction of a sustainable model of companies, based on a “stakeholder approach,” this last emphasizing the social dynamic aspect of the methodology thought the involvement of the waits of different social actors involved.

From this point of view, companies are nothing but complex bundles of economical, human (social), liquid, gas, material, and various other flows—many of them being non-renewable resources—and following the constructal configuration of maximum flow access, they should have an optimal structure, depending in particular on their sector (energy, transport, service, agronomy, etc.), and allowing them "to persist in time"—i.e., to achieve their sustainability goal.

14.2. Sustainability and Its Evaluation

The sustainable development is a development meeting our current needs without compromising the capacity of the future generations or human societies to meet their own needs. Two concepts are inherent to this notion: first the concept of need, and particularly the essential needs of the more impoverished and stripped, to whom it is appropriate to give the greatest priority, and second the concept of limitation that the current state of development of our techniques and social organization are imposing on the capacity of the environment to answer to the actual and future needs (World Commission on Environment and Development 1987).

The sustainable company is an enterprise taking into account this issue in its strategy of development in order to guarantee its economical, social, and financial viability. A first problem is that no sustainable model of company exists and it is difficult enough for an enterprise committed in this process to understand if it is engaged on the good or on the bad way. A second problem is that it is even more difficult to know if these choices will produce a positive impact on its viability in short term and in particular in the middle and long term. This way of training toward durability—restrictive in its finality—forces each company to raise certain fundamental questions:

- Relative to what is the company engaging in? What is/are the comparison parameter(s)?
- Relative to whom is the company engaging in? Relative to itself or relative to its stakeholders?
- What is the company expecting from its engagement? An investment return concerning its reputation and goodwill? Or also an advantage in terms of financial and extra-financial performance?

In this context of the search of a model of sustainable companies, we propose an approach based especially on two theories: thermoeconomics and constructal theory.

Particularly, we propose to define sector-specific company models, based on a stakeholder approach, relative to a particular economical sector of human activity, that will reproduce an abstract model of such a company in terms of size, number of employees, etc., using multiple-scale thermoeconomics metrics aggregated to produce comparable indexes.

In order to offer new concepts, the approach that we propose is to look at sustainable company and processes as flows environmental, social, goods that can be optimally distributed on the basis of the Constructal Law of generation of configurations for maximum flow access in freely morphing structures (Bejan 1997, 2000). Such a thermo-economical and constructal model will be especially useful in the domain of extra-financial rating.

Effectively, extra-financial analysis evaluates the engagement, policy, and performances of a company in the social, environmental, and corporate governance domains, relatively to its activities. The work of the extra-financial analysts is based on the analysis of the public documents published by the companies, specifics questionnaires, and interviews with companies' responsible or other stakeholders (NGOs, trade union, mass medias, etc.). Public documents can be, for instance, the yearly social and environmental reports—these last being mandatory in France for every company quoted on a stock market (e.g., all the members of the CAC40 financial index) since the 2001 NRE Law (New Economical Regulations) (Journal Officiel 2001, 2002). The FTSE4Good and DJSI (Dow Jones Sustainability Indexes) are two examples of extra-financial index (Baue 2002, 2003). Extra-financial analysts—also known as Socially Responsible Investment (SRI) analyst or Corporate Social Responsibility (CSR)—can work, e.g., for banks, pension funds, and brokers, but work especially for extra-financial rating agencies, these last being the non-financial counterpart of the well-known financial agencies providing financial performance indexes like Standard & Poor's, Moody's, or Fitch. Created at the end of the 1990s, extra-financial agencies are today more or less 30 different actors in the world, mainly localized in Europe, North America, and Asia.

Investors use these ratings or the information delivered (and their variations in time) by the agencies or their internal SRI department in Order to select the companies for their SRI investment portfolios. Several approaches to extra-financial analysis were developed depending on the expectation and willingness of the investors:

- The "Ethic oriented" analysis approach is based on exclusion criteria, defined by the agency itself or by its clients. These exclusion criteria can concern controversial activities such as tobacco industry, weaponry, alcohol, pornography, nuclear, etc., and also activities regarded as non-responsible such as child labor, animal testing, pesticides, etc. (all being especially related to the Anglo-Saxon world and culture; cf. Baue 2003).
- The "Performance oriented" analysis approach is done on the basis of positive selection criteria, established by the agency or its clients. In this case, the goal of this analysis is to identify the financial over-performance sources at medium and long term, of the studied values.
- The "Risk and opportunities" societal approach has for goal to provide to investment funds managers a complete view of the extra-financial risks and

opportunities of the companies which they own investment shares. This societal analysis tries to translate the environmental, social, and governance impacts on the financial performance of the company at short, middle, and long term.

The analysis is usually conducted following rating grids, very variable between agencies. These grids comprise a set of extra-financial criteria selected for their relevance, and weighted depending on their degree of importance.

The result of this scoring is a global rating. This rating enables to define the positioning of the company on a scoring scale, usually sector-specific.

There are several problems arising from these rating methods:

1. The diversity of the methods, frequently not public and so poorly transparent, and the great subjectivity of the work of the analyst. All this does not allow to compare the analysis between different methodologies, and the analysis can even be contradictory! This effect is also known as the Rashomon effect—named after a Kurosawa film—and concerns the different perception of two persons, each of them from its own point of view. For instance, Toyota was rated as second of 16 enterprises in its automotive category by the Innovest agency, whereas the rating company Oekom Research ranked Toyota as 16th (over 20 automotive companies) (Baue 2004). This subjectivity is a clear difference with the financial notation, which reached a threshold of objectivity far beyond than the current state of extra-financial analysis.
2. The fact that the rating is absolute and relating to the company itself, and not relative to the whole sector. In order to produce a truly relative rating, and despite the commercial discourse of the rating agencies, the scoring process should be done relatively to a sector-specific sustainable model of the company, which does not exist, and the company should use the same indicators or the same extra-financial tools. This last is not possible actually because no standard exist neither at the national nor at the international scale.
3. The lack of reliability of the information coming from the companies: the verification of extra-financial data is a universe to be explored, with all the inherent and possible drifts. The Enron and Parmalat scandals, in the financial domain, are representative of the potential dangers of such drifts, extended in the extra-financial domain.

So, the question is, what is really rated? How can we deduce if the company behavior is truly sustainable, and even more how can we predict that it will produce a good financial performance at short, middle, or long term? Several studies were pursued during the last decade in order to demonstrate the reality of the linkage between extra-financial and financial performance.

Results are nothing but ambivalent: investing in sustainable companies does not provide lesser results, neither greater. So, it is not bad, and in addition the money is invested in ethical and sustainable companies—or at least is supposed to be.

But is it the truth? In fact, the more a company spent in extra-financial communication (environmental and social reporting, etc.) for financial and cultural reasons, the best the rating will be, despite the fact that it maybe did not do really much more than the other companies in the sector. The fact is that today the only parameter truly differentiating the companies is the transparency, the quantity of information prevailing over the quality, with all the possible drifts.

Another problem to be added on to the establishment of the financial-to-extra-financial link is the time frame of these two different kinds of analyses: in the first case, extra-financial, middle to long term prevails, whereas short term dominates the financial analysis.

14.3. The Constructal Law of Maximum Flow Access

After this brief review of the current situation in the domain of SRI rating and analysis, the objective of this chapter is to propose a new theoretical approach for the construction of a model of sustainable companies especially dedicated to the monitoring and rating of extra-financial impacts and performance.

The theory is based on the recent formulation of the Constructal principle of the maximization of flow access and the generation of multi-scale architectures in all flow system. All the systems are destined to remain imperfect, but the Constructal approach of Professor Bejan allows the creation of efficient designs by using a new principle that generates their form: the optimal distribution of imperfection.

From this point of view, companies are nothing but complex bundles of economical, human (social), liquid, gas, and material flows—many of them being non-renewable resources—and following the Constructal of maximum flow configuration, they should have an optimal structure, depending, in particular, on their sector (energy, transport, service, agronomy, etc.), and allowing them "to persist in time"—i.e., to achieve their sustainability goal.

Because the Constructal theory is a theoretical framework, it allows making verifiable predictions. In particular, it allows predicting an optimal flight speed for any flying body (insect, bird, and engineered machine-like aircraft) (Bejan 2000; Bejan and Marden 2006). This optimal speed, let us call it the "constructal speed," is the speed maximizing the overflow distance by minimizing the power consumption embarked "on board" of the living organism (food) or machine (fuel).

From our point of view, this constructal speed is the "sustainable" speed of the flying body, and in a certain manner, we are all embarked in the same spatial ship, the Earth, and have to manage the finite, non-renewable resources of our planet in a sustainable—or constructal—way to push ahead the limits of the journey of humankind in the universe.

The constructal theory showed in particular that hierarchical scaling tree structures are patterns arising from this optimal distribution of imperfections in a flow system exhibiting two or more dissimilar regimes at different scales of the system. The Constructal theory is relevant for a wide range of flow system, at many different scales, and can be potentially applied to the following domains of application specific to the sustainable modeling of companies:

- Company design and optimization
- Optimization of flows
- Optimization of relation of different flows
- Optimization of the link between the short and middle/long permanence of the company.

14.3.1. Application to Complex Structures: Design of Platform of Customizable Products

The design of platforms for customizable products can be formulated as a problem of optimization of access in a geometric demand-space based on Constructal Theory (Allen et al. 2004). This approach allows us to develop systematically hierarchic products and processes.

This approach to product platform design relies on the work of H. A. Simon, who received the The Bank of Sweden Prize in Economic Sciences in Memory of Alfred Nobel in 1978. H. A. Simon argues that complex structures grow and evolve more efficiently when organized hierarchically. It is not assembly of components by itself, but hierarchic structure produced either by assembly or specialization that enables complex systems to adapt and respond to changes in the environment. This leads to first two principles:

1. Principle 1. Potential for rapid adaptation and/or response is higher in complex systems when they are organized hierarchically.
2. Principle 2. In hierarchically organized systems, the high frequency (short run) responses tend to be associated with the lowest levels of the hierarchy and the low-frequency (long run) ones with the interactions of these subsystems, i.e., the higher levels of the hierarchic organization.
   Hierarchical structure, morphing in time, different flow regimes at different scales of the system: the links between H. A. Simon ideas and with Constructal theory are clear. A third principle is then necessary, based on Constructal theory (Bejan 2000; Bejan and Marden 2006).
3. Principle 3. Systems complexity results from a natural process of providing paths of easier access.

These three principles compose the theoretical foundation for developing an effective approach to manage complexity which can be used in designing product architectures for mass customization, which is customizing products to satisfy individual customer specifications while maintaining costs and speeds close to those of mass production (Allen et al. 2004).

This, in turn, is one aspect of managing an enterprise system. From a designer's perspective, the advantages of constructal theory–based product/process platform development method include:

- Cost-Effectiveness. The method offers a rigorous approach for balancing effectively the tradeoffs between the various costs involved in product/process customization and linking market and design capability forecasts to design decisions and plans for product portfolios.
- Suitable for small or large variety in the product specifications. The method can be applied to a small number of product/process variants by formulating the space of customization as a discrete one or to a large number of variants—as is typically the case in mass customization—by formulating the space of customization as a continuous one.
- Adaptability. It is possible to alter product/process designs for portions of the space of customization without affecting other products of the family that are not in the same branch of the hierarchic construct.
- The starting point to manage the complexity created by mass customization is, based on the basic principle of constructal theory, to determine the intersection of the demand and technically feasible designs and then to formulate a problem of optimization of access.

14.4. The Structural Theory of Thermoeconomics

During the three decades from 1972 to 2002 various thermoeconomics methodologies have been developed. All of them have in common cost calculated on a rational basis, which is the second law of thermodynamics (Ukidwe and Bakshi 2003, 2004; Yi et al. 2004). This cost is a very useful tool for solving problem in complex energy system, such as rational prize assessment of the products of a plant based on physical criteria, local optimization, or operation diagnosis. These problems are difficult to solve using conventional energy analysis techniques based on the first law of thermodynamics.

There are two groups of thermodynamics methods: (1) cost accounting methods, which use average costs as a basis for a rational price assessment and (2) optimization methods, which employ marginal costs in order to minimize the costs of the products of a system or a component.

It is important to keep in mind that thermoeconomics connects thermodynamics with economics, that is, by sorting the thermodynamic properties of the physical mass and energy flow-streams of a plant, which in turn provide the energy conversion efficiency of each subsystem, thermoeconomics analyzes the degradation process of energy quality through an installation (see Fig. 14.3).

Depending on the scope of the analysis, a subsystem can be identified as a separate piece of equipment, a part of device, several process units, or even the whole plant. Sometimes the objective consists of analyzing a plant in great detail. In this case it is advisable, if possible, to identify each subsystem with a separate physical process in order to locate and quantify, separately if possible, each thermal, mechanical, and chemical irreversible process occurring in the plant.

The objective consists of analyzing a macrosystem composed of several plants, in this case the more convenient approach would probably be to consider each separate plant a subsystem. Thus, thermoeconomics always performs a systemic analysis, no matter how complex the system, oriented toward locating and quantifying the energy conversion efficiency and the process of energy quality degradation. It is not within the scope of thermoeconomics to model the behavior of the process units, which is done by the mathematical equations of the physical model.

The use of thermodynamics permits representation of all kinds of inputs and outputs in consistent units, facilitating the definition of aggregate metrics. In their research of thermodynamics approach in an ecosystem, Ukidwe and Bakshi employ a thermodynamics approach for including contribution of ecological products and services to economic sectors via input–output analysis (Ukidwe and Bakshi 2003, 2004). A thermodynamic approach provides a common currency or a way to deal with a diverse set of units, as any system, economic or ecological, can be considered as a network of energy flows (see Fig. 14.2).

The proposed thermodynamics approach is not meant to replace, but to complement an economic approach. The economy consists of a large number of industry sectors defined according to their standard industrial Classification codes. Materials enter the system in the form of import and exit in the form of exports. Consideration of imports and exports is, however, beyond the scope of the study. Solid lines in Fig. 14.1 represent tangible interactions that include raw materials from and emissions to ecosystems and human resources. The algorithm for thermodynamic input-output analysis focuses on the economic system and its interactions with ecosystems and human resources shown in Fig. 14.1. It consists of three tasks:

1. To identify and quantify ecological and human resource inputs to the economic system.
2. To calculate Ecological Cumulative Exergy Consumption (ECEC) of ecological inputs using transformities from systems ecology (see definition below).
3. To allocate direct ecological and human resource inputs to economic sectors.

The contribution of ecosystems is represented via the concept of ECEC, which is related to exergy and emergy analysis but avoids the latter's controversial assumptions and claims. The definition of emergy and transformities are the following (Odum 1995):

- Emergy is all the available energy that was used in the work for making a product and expressed in unit of one type of energy.
- Transformity is the emergy of one type required to make a unit of energy of another type.

The Emergy Systems school promoted the concept of emergy in order to establish the metric for a rigorous and quantitative sustainability index (Brown and Ulgiati 1997).

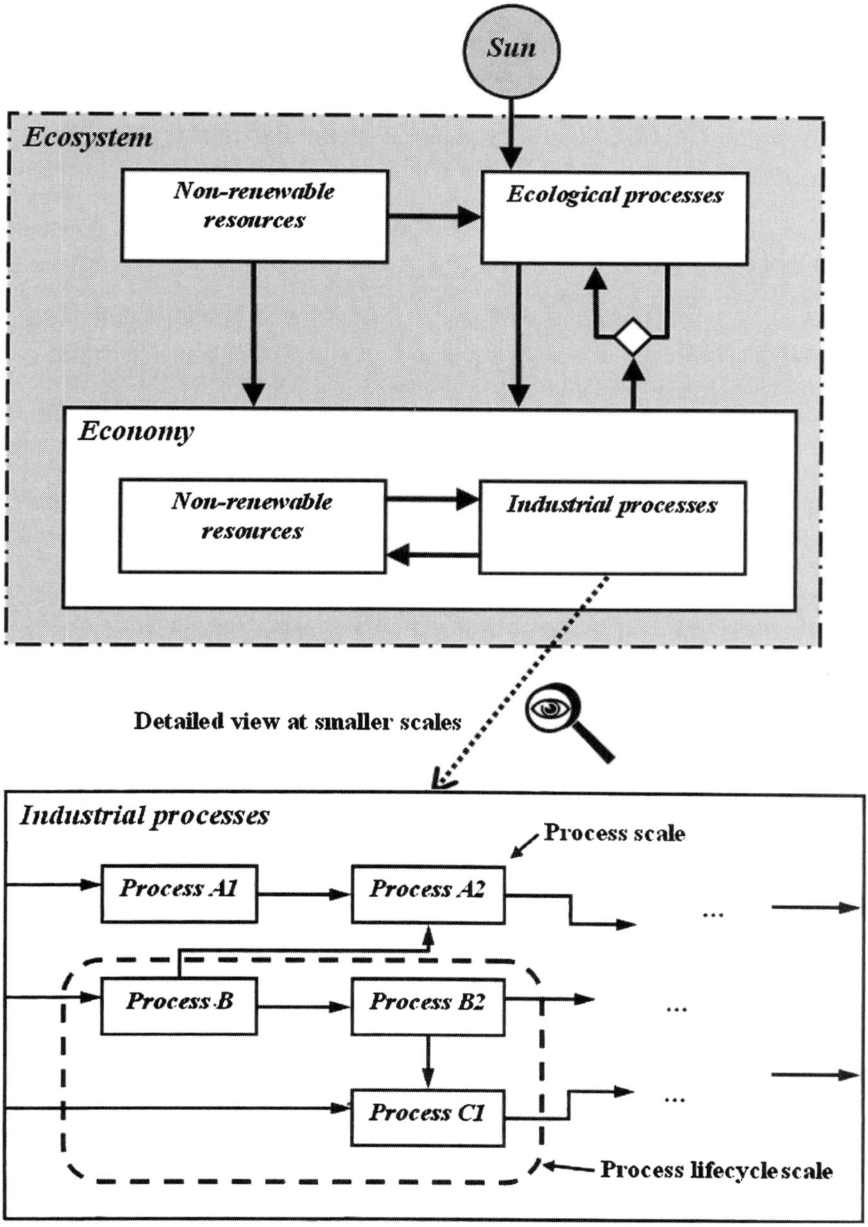

FIGURE 14.1. Energy flow diagram at economy and ecosystem scales, and the more detailed scales of process and process lifecycle (adapted from Yi et al. 2004)

Total ECEC requirement indicates the extent to which each economic sector relies directly and indirectly on ecological inputs. The ECEC/money ratio indicates the relative monetary versus ecological throughputs in each sector, and indicates the relationship between the thermodynamic work needed to produce a product or service and the corresponding economic activity.

In their article, Ukidwe and Bakshi (2004) show ECEC/money ratio of each of the 91 industry sectors on a semi-log plot. It also shows ECC/money ratio for renewable resources, non-renewable resources, human resources, and human health impact of emission separately.

The ECEC/money ratio does not support or debunk any theory of value, but is rather meant to provide insight into the magnitude of discrepancy between thermodynamic work needed to produce a product or service and people's willingness to pay for it. Some observations about this ratio are as follows:

- The ECEC/money ratio for the sectors of non-metallic minerals mining and metallic ores mining are the highest. Sectors with the smallest ECEC/money ratios are owner/occupied dwellings and advertising. Sector of radio and TV broadcastings also has a high ECEC/money ratio due to the human resource inputs.
- Specialized sectors such as tobacco products, drugs, and computer and office equipment have smaller ECEC/money ratios than basic sectors such as petroleum refining and primary iron and steel manufacturing

For example, the average ECEC/money ratio for mining sector is 22 times the average ECEC/money ratio for service industry sectors.

The wide variation in the ECEC/money ratio indicates the discord between natural capital and corresponding economic capital. Thus, sectors with larger ratios seem not to appreciate or value ecosystem products and services as much as those with smaller ratios. This not only corroborates the lack of integration of the "eco-services" sector with the rest of the economy but also quantifies the magnitude of this discrepancy.

As we saw in this section, many of the thermodynamics ideas rejoin the Constructal principles of analysis: multiple scales, hierarchical structures, and is originated comes from the same domain—thermodynamics.

14.5. Application to Company Sustainability

14.5.1. The Stakeholder Approach

As a result, business and policy decisions in a company are usually made with a flawed accounting system that ignores the basic life support system for all activities. The focus of such an approach tends to be on short-term gain, while longer-term sustainability issues get ignored. By now, enduring success for a company is only possible when it pursues growth in harmony with the interests of all its stakeholders.

Therefore, assessing stakeholder interests form the core of our proposed approach to evaluate relevance. Furthermore, this approach emphasizes the social dynamic aspect of the methodology thought the involvement of the waits of different social actors involved.

Sustainability issues can indeed act as early signals of factors which will influence corporate profitability, which have an impact on a company's long-term value creation.

Acknowledging that sustainability issues and stakeholder interests can be very different from sector to sector, we have developed a flexible analytical framework, which can in any given sector distinguish between the more relevant and less important issues. An example of sector can be "Transport," including all different kind of means of transport, and especially Air Transport (see Chapter 6 in this book).

From this point of view, companies are nothing but complex bundles of economical, social (human), environment, and material flows—many of them being non-renewable resources—and following the Constructal of maximum flow configuration, they should have an optimal structure, depending, in particular, on their sector (energy, transport, service, etc.), and allowing them "to persist in time"—i.e., to achieve their sustainability goal. As already stated above, one problem is that actually no sustainable or "optimal" model of company exists. But the previous chapter provided us a set of tools through the thermodynamic, thermoeconomics, and Constructal theory in particular.

A next step would be to define a sector-specific Constructal model of companies, optimally distributing the system imperfection, using a multiple-scale analysis and a hierarchical aggregation of thermoeconomics metrics, in order to provide a theoretical reference allowing to compare a given company performance to its optimal model (absolute reference) and to the other companies of its sector (relative reference).

A first possible approach is to define a tree-structure comprising all internal and external flows categorized in stakeholders' domains. The optimization of all stakeholders' relationships is important for the intangible value creation and long-term growth potential of a company. The stakeholders approach was chosen as it reflects best the reality of a company, creating value through interactions with all its stakeholders, working in a dynamic world.

A company can act in either a proactive or a passive way, i.e., it can actively strive to optimize the relations with stakeholders or it can merely undergo these relations passively. We think that a company's ability to create value and sustain competitive advantage depends on it acting actively, anticipating the requests and needs of stakeholders and utilizing opportunities they present.

This approach will allow the company to truly understand and manage the dynamics of a changing world in order to guarantee its own continuing social and financial viability. The stakeholder's domains of a company are identified as

1. Environment
2. Society

3. Customers
4. Suppliers
5. Shareholder and Bondholders
6. Employees.

14.5.2. The Analytical Tree

The stakeholders approach allows to identify the relevant issues in the company's value creation. In order to capture all issues in a structured way, the analytical tree defines then for each stakeholder domain the potentially relevant sub-domains and themes (i.e., flows).

The starting point is a common analytical tree, in which certain themes can be deactivated or additional themes can be activated for each sector. The six stakeholders' domains taken into account are defined here above.

Following our philosophy, the analytical tree is generic, which can be fully or partially used depending on a sector's issues. Where relevant, there is scope for additional themes to be added, based on the particular responsibilities of the company towards its stakeholder in a specific sector.

For example, the impacts of the "Environment" domain can be classified in three categories like process related impacts, products or service related impacts, and supply chain impacts. Each of these categories will be itself detailed at a sublevel. For instance, products or service related impacts will address the analysis of the product lifecycle (LCA), the impact of the product use. This last sub-category includes especially the climate change impact due to green-house effect gas like CO_2 and the impact on biodiversity. The detailed view of this branch is display on the excerpt of the analytical tree on Fig. 14.4.

14.5.3. The Objectives of Research

This chapter is not presenting an operational solution, but has for objectives to propose some axes of research. Many open questions exist at this point, but we can nevertheless propose some meeting points for such future project:

- Achievement of per-sector studies of Constructal and thermoeconomical model of sustainable company. About 30 sectors are defined (automotive, energy, food, etc.), and so about 30 models have to be specified.
- A thermoeconomics approach (e.g., ECEC) will allow to classify, to set in order, and to weight objectively the different flow of these sector-based systems.
- A rating system based on the scientific thermoeconomics metrics should be fixed, allowing then to rate and to monitor the sustainable performance and the deviation from its theoretical Constructal optimum, and relatively to the other companies of its sector.

The definition of a scientific Sustainable and Constructal model of company is then expected to provide a definitive tool in order to answer to the big

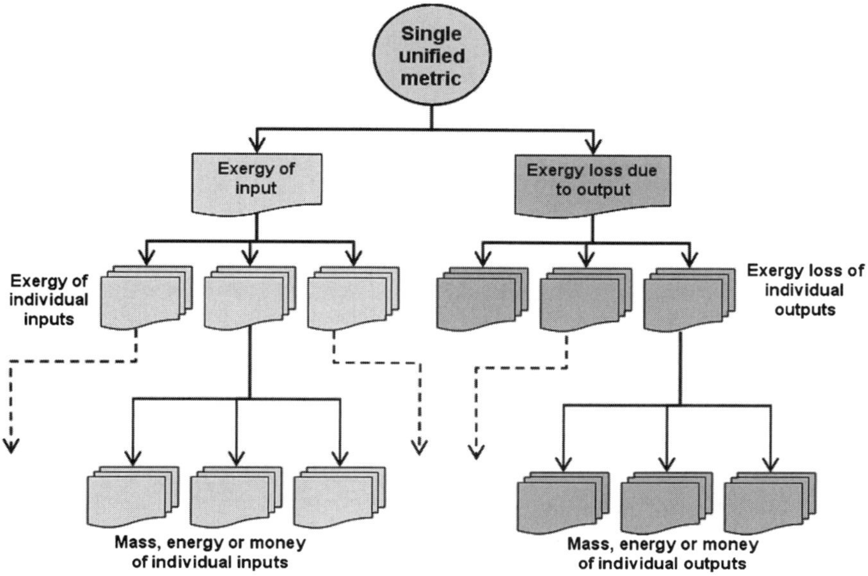

FIGURE 14.2. Hierarchial structure of sustainability metrics for a selected system (adapted from Yi et al. 2004)

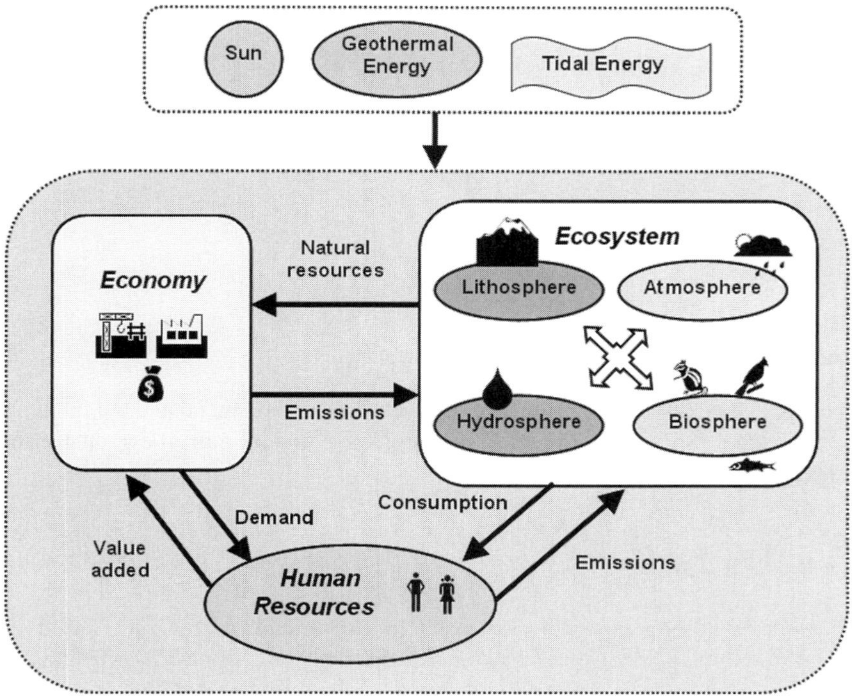

FIGURE 14.3. The integrated economic-ecological-human resource system (simplified view, adapted from Ukidwe and Bakshi, 2003)

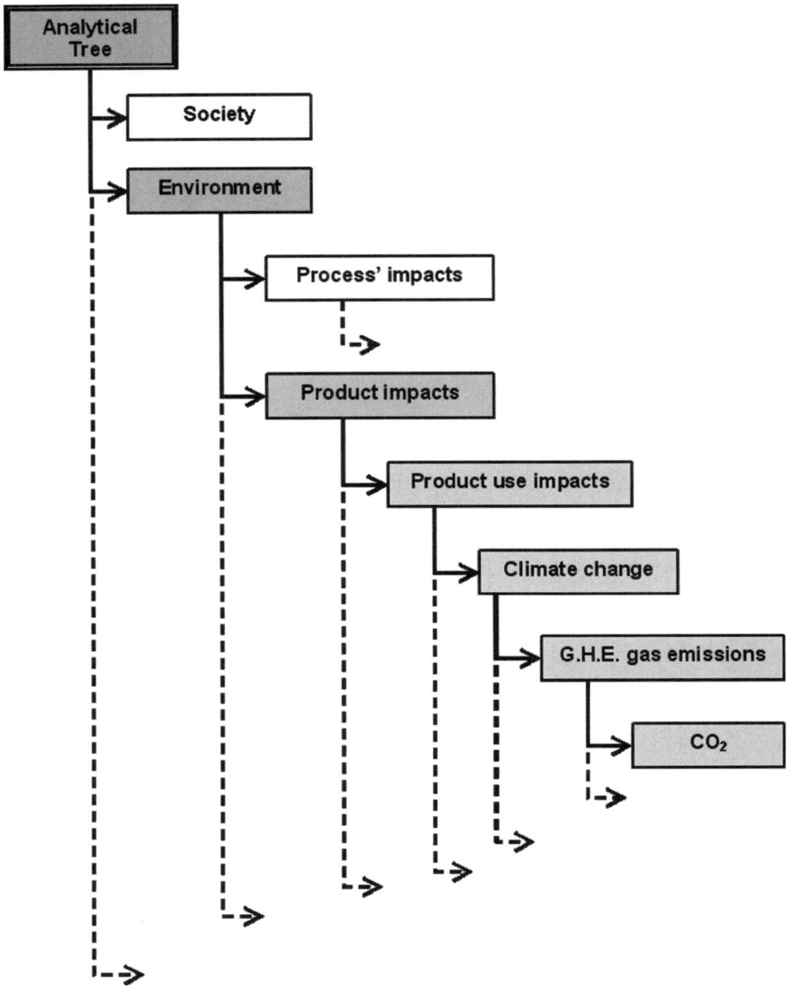

FIGURE 14.4. Detail of a specific branch of the analytical tree

question, i.e., demonstrating the link between the extra-financial and the financial performance. Only such kind of truly objective model can allow answering scientifically to this open problem.

14.6. Conclusions

This chapter proposes a new approach to extra-financial analysis, based on thermoeconomics and Constructal ideas, and concepts, and to briefly look at the promising union of these two theoretical frameworks to define a sustainable model of company.

The previous paragraphs show us the interest of both Constructal theory and thermoeconomics in the domain of company sustainability, in particular through a multiple-scale flow analysis, and a thermodynamics metrics aggregation.

From this point of view, companies are nothing but complex bundles of economical, social, liquid, gas, material, and various other flows—many of them being non-renewable resources—and following the Constructal of maximum flow configuration, they should have an optimal structure, depending, in particular, on their sector (energy, transport, service, agronomy, etc.), and allowing them "to persist in time"—i.e., to achieve their sustainability goal.

Our contribution here is to propose to extend such kind of analysis to the domain of extra-financial rating in order to provide a theoretical model and scientific metrics to compare the companies performance and to shift from a discipline actually glued in the analyst subjectivity to a totally objective and comparable rating.

The resulting Constructal model of sustainable company is expected to help to answer to the open problem of the link between extra-financial and financial performance.

References

Allen, J., Rosen, D. and Mistree F (2004) An Approach to Designing Sustainable Enterprise Systems, Engineering Systems Symposium , MIT Engineering System Division, Cambridge, MA, March 31.

Baue, W. (2002) DJSI Adds "Sin" Stocks, SocialFunds.com, December 06. http://www.socialfunds.com/news/article.cgi/article983.html

Baue, W. (2003) FTSE4Good Indexes Add Bank of America, Delete Molson and ConocoPhillips, SocialFunds.com, September 19. http://www.socialfunds.com/news/article.cgi/article1225.html

Baue, W. (2004) The Rashomon Effect: Why Do Innovest and Oekom Rate Toyota's Environmental Performance So Differently?, SocialFunds.com, January 20. http://www.socialfunds.com/news/article.cgi/article1318.html

Bejan, A. (1997), *Advanced Engineering Thermodynamics*, 2nd ed., Wiley, New York, ch. 13.

Bejan, A. (2000) *Shape and Structure, from Engineering to Nature*, Cambridge University Press, Cambridge, UK.

Bejan, A. and Marden, J. H. (2006) Unifying constructal theory for scale effects in running, swimming and flying, *J. Exp. Biol.* **209**, 238–248.

Brown, M. T. and Ulgiati, S. (1997) Emergy-based indices and ratios to evaluate sustainability: monitoring economies and technology toward environmentally sound innovation, *Ecological Engineering* **9**, 51–69.

Journal Officiel (2001) Loi no 2001-420 du 15 mai 2001 relative aux nouvelles régulations économiques (1), J.O n° 113 du 16 mai 2001 page 7776, Article 116.

Journal Officiel (2002) Décret n° 2002-221 du 20 février 2002 pris pour l'application de l'article L. 225-102-1 du code de commerce et modifiant le décret n° 67-236 du 23 mars 1967 sur les sociétés commerciales, J.O. Numéro 44 du 21 Février 2002 page 3360.

Odum, H. T. (1995) *Environmental Accounting: Emergy and Environmental Decision Making*, Wiley, New York.

Ukidwe, N.U. and Bakshi, B. R. (2003) Accounting for Ecosystem Contribution to Economic Sectors by Thermodynamic Input-Output Analysis. Approach, Technical Report, Department of Chemical and Biomolecular Engineering, The Ohio State University.

Ukidwe, N.U. and Bakshi, B. R. (2004) Thermodynamic Accounting of Ecosystem Contribution to Economic Sectors with Application to 1992 US Economy, *Environmental Science and Technology* **38** (18), 4810–4827.

World Commission on Environment and Development (1987) *Our Common Future*, Oxford University Press, New York, Chapter 2, p. 19.

Yi, H.-S., Hau, J. L., Ukidwe, N.U. and Bakshi, B. R. (2004) Hierarchical Thermodynamic Metrics for Evaluating the Sustainability of Industrial Processes, Environmental Progress, *Sustainable Engineering* **23** (4), 302–314.

Chapter 15
The Inequality Process Is an Evolutionary Process

John Angle

15.1. Summary

In the Inequality Process (IP) randomly paired particles compete for each other's wealth with an equal chance to win. The loser gives up a fraction of its wealth to the winner. That fraction is its parameter. By hypothesis and empirical inference, that fraction scales inversely with the particle's productivity of wealth. Long term wealth flows to particles that lose less when they lose, robust losers, nourishing their further production of wealth. Given a survival function that is an increasing function of wealth, the more robust loser is more losses away from death (Gambler's Ruin) at any given amount of wealth than others. As the level of productivity rises among particles, holding the global mean of wealth constant, expected particle wealth in each productivity equivalence class decreases and the variance of wealth in each equivalence class decreases, i.e., the transfer of wealth from the less to the more productive occurs more efficiently making wealth a better indicator of productivity. As productivity in the IP's population of particles increases, the IP's penalty for low productivity increases, further incentivizing an increase in productivity. The IP operates with no information about how wealth is produced and consequently adapts fluidly to higher productivity, change in constraints on wealth production, and variation in global mean wealth. The IP is a dynamic attractor for a population, maximizing wealth production, minimizing extinction risk. It thus is an evolutionary process in an interdependent population, a colony of organisms, in which each organism depends on the product of the whole population.

15.2. Introduction: Competition for Energy, Fuel, Food, and Wealth

"If living systems can be viewed as engines in competition for better thermodynamic performance, then physical systems too can be viewed as living entities in competition for survival. .... For a finite-size open system to persist in time (to live), it must evolve in such a way that it provides easier access to the imposed (global) currents that flow through it" (Bejan 1997, pp. 806–807).

A population of particles (eddies, molecules, organisms, workers, hereafter "particles") that act as heat engines that work and maintain themselves by converting ambient energy to fuel will likely persist longer through time, perhaps indefinitely, if particles compete for fuel and other necessities according to the Inequality Process (IP) (Angle1983, 1986a,b, 1990, 1992, 1993a,b, 1996, 1997, 1998, 1999a,b, 2000, 2001, 2002a–c, 2003a–c, 2005, 2006; Kleiber and Kotz 2003; Lux 2005). The IP is a stochastic interacting particle system in which particles compete for a positive quantity, e.g., energy, fuel, food, or wealth, in which particles that are more efficient in producing that positive quantity lose less of it when they lose a competitive encounter with another particle. The IP is relevant to explaining the dynamics of wealth and income distributions (macro-level) and the dynamics of the wealth and income of individuals if, empirically, robust losers are more productive. They may be, empirically, if due to their greater productivity they (a) rebound faster from a loss, (b) are treated more gently, (c) have more bargaining leverage, or (d) control wealth, like skills (human capital) that cannot be readily alienated, as for example, a crop can be taken from a farmer. Educating or training a worker is an investment of wealth into that person.

In the long run the IP transfers a positive quantity, be it energy, fuel, food, or wealth—hereafter referred to in this paper as wealth—from particles that are less efficient at producing it to particles that are more efficient. The IP performs this task via zero sum competition by transferring wealth, in the long term, from less robust to more robust losers. The IP thus nourishes the production of wealth by more efficient particles. The IP makes all particles safer from ruin—starvation—by maximizing the production of wealth. In the IP increases in the unconditional mean of wealth are universally shared in proportion to a particle's productivity with the more productive receiving a larger share of the increase. The IP insures that the more efficient are more competitive losses away from ruin at every amount of wealth. So the IP minimizes the extinction risk of the whole population and, in particular, the extinction risk of the more productive.

The IP originated as a mathematical model of a speculative theory by a sociologist, Gerhard Lenski (1966), about the course of inequality of wealth over techno-cultural evolution. This speculation asserts that (a) more human capital (skills) is invested in each worker over the course of techno-cultural evolution, (b) human capital is an inalienable form of wealth controlled by each worker, (c) human capital becomes a larger share of a society's total wealth over the course of techno-cultural evolution, resulting in (d) less concentration of wealth in the hands of a very few rich people. It is certain that most of the wealth of a contemporary industrial population is in the form of human capital, a fact ascertained by "capitalizing" the aggregate labor income stream, that is, estimating the amount of tangible capital required to generate a stream of income of the same size.

15.2.1. The Inequality Process (IP) as an Evolutionary Optimizer

Despite the origin of the IP as a particular social science theory of a particular statistical pattern in inequality of wealth, the IP is a general evolutionary process. Think of a population of particles, each a solution to the problem of how to generate wealth. Assume that nothing is known about each particle except its current wealth and its recent history of wealth. Since, on average, each particle has as many losses as wins, it only takes a short history of wealth to estimate the proportion of wealth lost in a loss for each particle. That proportion, ω, is the particle's parameter. $(1 - \omega)$ measures the particle's productivity. The more robust loser, the particle with smaller ω, is more losses away from death (Gambler's Ruin) at any given amount of wealth than others.

The IP solves the problem of how to allocate wealth to individual particles to maximize aggregate wealth production by the population, while minimizing extinction risk for the population, and consequently, for the IP itself. The IP solves this problem despite changing conditions. The IP moves fluidly from one particular solution to another with changing conditions, using very little information, just knowledge of current particle wealth and particle wealth in the recent past. The IP does not need to measure individual particle productivity directly. In industrial economies there is a great diversity and complexity of ways of producing wealth. It is evident that this diversity and complexity has grown with techno-cultural evolution and is increasing. The IP does not have to know about the diversity and complexity of producing wealth. In the IP, competition between particles for wealth measures particle productivity using the inference rule that the more productive particle loses a smaller proportion of its wealth when it loses. Thus the IP can operate homogeneously up and down the scale of techno-cultural evolution inferring wealth productivity from zero-sum competition for wealth alone without (a) direct knowledge of individual particle productivity or (b) modeling the wealth production process.

The IP is a competition process in a population of particles that transfers wealth (or energy, fuel, or food, or any positive quantity useful for its own production) via randomly decided competitions between randomly paired particles. While, short term, wealth goes to winners of these encounters since the chance of winning is 50%, in the long term, wealth flows to the robust losers. We will see that if the level of productivity rises in the IP's population of particles, holding the global mean of wealth constant, expected particle wealth falls in every productivity equivalence class of particles. In the IP the penalty for being a productivity laggard grows as productivity rises in the population incentivizing particles to become more productive. Further, as the level of productivity rises in the IP's population of particles, the variance of wealth in each equivalence class decreases, i.e., the transfer of wealth from the less to the more productive occurs more efficiently making wealth a better indicator of productivity. So, as the productivity of a population rises, the IP works more efficiently to transfer wealth to the more productive. That is why the IP is a functional attractor. Once a process of competition in a population of particles chances into the IP, the IP increases wealth, reduces extinction risk for the whole population, particularly the more productive particles, and creates incentives for further productivity gains. The IP adapts fluidly to productivity increase or decrease in the population of particles, to change in constraints on wealth production, and to change in global mean wealth. While incentivizing productivity increase, the IP requires no schedule of productivity increase be met.

Thus, once a competition process within a population chances into the form of the IP, the IP tends to become self-perpetuating with increasing stability as the population of particles prospers and grows (assuming greater wealth and smaller risk of individual loss of wealth contribute to population growth). The IP moves the population in which it has become established and itself into the future with increasing assurance. Thus the IP is an evolutionary optimizing process in an interdependent population, a colony of organisms, in which each organism depends on the product of the whole population. As an evolutionary process, the IP models selection rather than search, the groping process by which each particle explores how to become more productive.

15.2.2. Mathematical Description of the IP

In the IP, wealth is distributed to particles via zero-sum transfers. These do not change the amount of wealth summed over all particles. Particles are randomly paired; a winner is chosen via a discrete 0,1 uniform random variable; the loser gives up a fixed proportion of its wealth to the winner. In words, the process is as follows:

Randomly pair particles. One pair is particle i and particle j. A fair coin is tossed and called. If i wins, it receives an ω_θ share of j's wealth. If j wins, it receives an ω_ψ share of i's wealth. Repeat.

The transition equations of the transfer of wealth between two IP particles are

$$x_{it} = x_{i(t-1)} + \omega_\theta d_{it} x_{j(t-1)} - \omega_\psi (1 - d_{it})\, x_{i(t-1)} \tag{15.1a}$$

$$x_{jt} = x_{j(t-1)} - \omega_\theta d_{it} x_{j(t-1)} + \omega_\psi (1 - d_{it})\, x_{i(t-1)} \tag{15.1b}$$

where $x_{i(t-1)}$ is particle i's wealth at time $t-1$ and:

$\omega_\psi =$ proportion of wealth lost by particle i when it loses

$\omega_\theta =$ proportion of wealth lost by particle j when it loses.

and

$$d_{it} = \begin{bmatrix} 1 \text{ with probability 0.5 at time t} \\ 0 \text{ otherwise} \end{bmatrix}$$

The IP has an asymmetry of gain and loss, which is apparent in Fig. 15.1, the graph of forward differences, $x_{it} - x_{i(t-1)}$, against wealth, $x_{i(t-1)}$, in the IP (15.1a,b).

When particle i in the ω_ψ equivalence class loses, its loss in absolute value is

$$\omega_\psi x_{i(t-1)} \tag{15.2}$$

Losses of particles in the ω_ψ equivalence class fall on the line $y = -\omega_\psi x_{i(t-1)}$. When particle i, whose parameter is ω_ψ, wins an encounter with particle j, whose parameter is ω_θ, its gain is

$$\omega_\theta x_{j(t-1)} \tag{15.3}$$

The expected gain of all particles in the IP (15.1a,b) is

$$\overline{\omega\mu} \equiv \sum_{\psi=1}^{\Psi} w_\psi \omega_\psi \mu_\psi \tag{15.4}$$

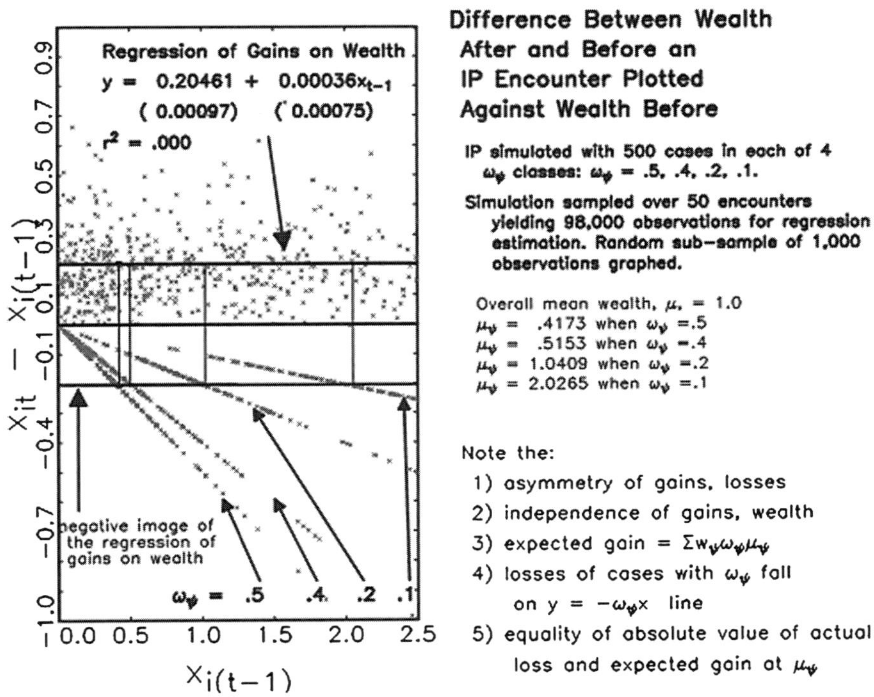

FIGURE 15.1. The scattergram of wealth changes in the population of particles from time t − 1 to t plotted against wealth at time t − 1

where there are Ψ distinct ω_ψ equivalence classes, and w_ψ is the proportion of the population of particles in the ω_ψ equivalence class

$$\begin{aligned} & w_i > 0 \\ & w_i + \ldots + w_\psi + \ldots + w_\Psi = 1.0 \qquad (15.5) \\ & w_\psi = \frac{n_\psi}{n} \\ & n = n_1 + n_2 + \ldots + n_\psi + \ldots + n_\Psi \qquad (15.6) \end{aligned}$$

The expectation of gain of particle i is independent of the amount of its wealth, $x_{i(t-1)}$, resulting in a regression line with near-zero slope fitted to all gains, regardless of ω_ψ equivalence class, for particles with a gain in Fig. 15.1.

15.3. The Gamma PDF Approximation to the IP's Stationary Distribution in the ω_ψ Equivalence Class

15.3.1. The Exact Solution

The long-term properties of the IP (15.1a,b) as a dynamic process are given by its solution. Equation (15.1a,b) is solved by backward substitution:

$$x_{it} = \begin{pmatrix} \omega_{\theta j} x_{j(t-1)} d_{it} \\ +\omega_{\phi k} x_{k(t-2)} d_{i(t-1)} [1 - \omega_\psi (1 - d_{it})] \\ +\omega_{\xi l} x_{l(t-3)} d_{i(t-2)} [1 - \omega_\psi (1 - d_{it})][1 - \omega_\psi (1 - d_{i(t-1)})] \\ + \ldots\ldots\ldots \end{pmatrix} \qquad (15.7)$$

Particle i's wealth is the sum of its gains from competitors, each gain weighted by $(1 - \omega_\psi)$ raised to the power of the number of later losses. The RHS of (15.7), after the realization of d_{it}'s as 0's or 1's, equals

$$x_{it} = \begin{pmatrix} \omega_{\theta j} x_{j(t-1)} d_{it} \\ +\omega_{\phi k} x_{k(t-2)} d_{i(t-1)} (1 - \omega_\psi)^{1-d_{it}} \\ +\omega_{\xi l} x_{l(t-3)} d_{i(t-2)} (1 - \omega_\psi)^{2-d_{it}-d_{i(t-1)}} \\ + \ldots\ldots\ldots \end{pmatrix} \qquad (15.8)$$

Equation (15.8) is the sum of "bites" taken out of competitors multiplied by $(1 - \omega_\psi)$ raised to the power of the number of later losses, i.e., particle i's current wealth, x_{it}, is what it has won from competitors and did not lose at a later time. When $(1 - \omega_\psi)$ is small, x_{it} is determined by the length of a consecutive run of wins backward in time. Where $(1 - \omega_\psi)$ is large, losing is less catastrophic and x_{it} can be considered a run of wins backward in time tolerating some intervening losses.

15.3.2. An Approximation to the Exact Solution

When the ω parameters, ω_θ, ω_ϕ, $\omega_\zeta, \ldots,$ ω_Ψ are sufficiently small, and particle i's parameter, ω_ψ, is also sufficiently small, the "bites" taken out of competitor particles become smaller and there are more of them. In this situation the arithmetic mean of these bites, $\overline{\omega\mu}$, is a better approximation to each bite. It can be shown numerically that the RHS (right hand side) of (15.9) approximates (15.8):

$$x_{it} \approx \overline{\omega\mu} \begin{pmatrix} d_{it} \\ +d_{i(t-1)}(1-\omega_\psi)^{1-d_{it}} \\ +d_{i(t-2)}(1-\omega_\psi)^{2-d_{it}-d_{i(t-1)}} \\ +d_{i(t-3)}(1-\omega_\psi)^{3-\sum_{\tau=0}^{2} d_{i(\tau-1)}} \\ + \ldots\ldots\ldots\ldots \end{pmatrix} \tag{15.9}$$

The infinite series of weighted Bernoulli variables in the brackets on the RHS of (15.9) can be approximated by summing a finite sequence of unweighted Bernoulli variables running from the present, t, back to $t-[k+N_\psi]$ in the past where $[k+N_\psi]$ is the number of previous competitive encounters that make a difference in particle i's wealth at time t, the time horizon of the process in the past for particle i:

$$d_t + d_{t-1} + d_{(t-2)} + \ldots\ldots + d_{(t-[k+N\psi])} \tag{15.10}$$

The number of wins of particle i, k, must be the same in (15.10) as (15.9). However, the number of random losses, N_ψ, $N_\psi = 1, 2, \ldots,$ that terminates (15.10)'s run back into the past varies with $(1-\omega_\psi)$. The larger $(1-\omega_\psi)$, the larger the N_ψ must be so that $(1-\omega_\psi)^{N_\psi}$ is negligibly different from zero. Angle (2006) shows that (15.9) requires $N_\psi + 1$ losses to approximately erase wealth from past wins. So there are $k + N_\psi + 1$ summands in (15.10), and

$$\begin{aligned} N_\psi + 1 &\approx \sum_{\tau}^{T} (1-\omega_\psi)^\tau \\ N_\psi &\approx \left(\frac{1-\omega_\psi}{\omega_\psi}\right) \end{aligned} \tag{15.11}$$

N_ψ, like ω_ψ, is a parameter.

The random variable, k wins before n losses, is distributed as a negative binomial probability function, NB(n,p), where $p = 1/2$:

$$\begin{aligned} P(X=k) &= \binom{k+n-1}{n-1} p^n (1-p)^k \\ E[k] &= n \end{aligned} \tag{15.12}$$

So the expectation of k successes, the number of successes before the number of random losses that ends the run, in (15.9), is N_ψ.

The gamma pdf that approximates NB(n,p) has a shape parameter, α_ψ, equal to N_ψ:

$$\alpha_\psi \approx \frac{1-\omega_\psi}{\omega_\psi} \tag{15.13}$$

where the gamma pdf is defined by

$$f(x) \equiv \frac{\lambda_\psi^{\alpha_\psi}}{\Gamma(\alpha_\psi)} x^{\alpha_\psi - 1} e^{-\lambda_\psi x} \tag{15.14}$$

and
$x > 0$
$\alpha_\psi > 0$
$\lambda_\psi > 0$
$\alpha_\psi =$ the shape parameter, ω_ψ equivalence class
$\lambda_\psi =$ the scale parameter, ω_ψ equivalence class

If the ω_θ, ω_ϕ, ω_ζ, ..., 's are all equal to a single value, ω, then the expression in brackets on the RHS of (15.9) has an expectation equal to $1/\omega$. Thus the expectation of the series in brackets on the RHS of (15.9) is equal to $1/\omega_\psi$. So taking the expectation of both sides of (15.9) gives an expression for the expectation of wealth in the ω_ψ equivalence class, μ_ψ:

$$\mu_\psi \approx \frac{\overline{\omega\mu}}{\omega_\psi} \tag{15.15}$$

Given (15.15) and the fact that the mean of the approximating gamma pdf, μ_ψ, is α_ψ/λ_ψ:

$$\mu_\psi \approx \frac{\overline{\omega\mu}}{\omega_\psi} \approx \left(\frac{1-\omega_\psi}{\omega_\psi}\right)\frac{1}{\lambda_\psi} \tag{15.16}$$

which implies

$$\lambda_\psi \approx \frac{1-\omega_\psi}{\overline{\omega\mu}} \tag{15.17}$$

Equation (15.4) defines $\overline{\omega\mu}$ in terms of w_ψ's and ω_ψ's which are known and μ_ψ's which are not. λ_ψ can be solved for in terms of knowns, ω_ψ, w_ψ, and the grand mean, μ, also known, in the following way:

$$\mu = w_1\mu_1 + w_2\mu_2 + \ldots + w_\Psi\mu_\Psi \tag{15.18}$$

and from (15.15):

$$\mu \approx \left(\frac{w_1}{\omega_1}\right)\overline{\omega\mu} + \left(\frac{w_2}{\omega_2}\right)\overline{\omega\mu} + \ldots + \left(\frac{w_\Psi}{\omega_\Psi}\right)\overline{\omega\mu} \tag{15.19}$$

which implies that

$$\overline{\omega\mu} \approx \frac{\mu}{\frac{w_1}{\omega_1} + \frac{w_2}{\omega_2} + \ldots + \frac{w_\Psi}{\omega_\Psi}} \tag{15.20}$$

so the RHS of (15.17) can be expressed in terms of known quantities:

$$\begin{aligned} \mu_\psi &\approx \frac{1-\omega_\psi}{\overline{\omega\mu}} \\ &\approx \frac{(1-\omega_\psi)\left(\frac{w_1}{\omega_1} + \frac{w_2}{\omega_2} + \ldots + \frac{w_\Psi}{\omega_\Psi}\right)}{\mu} = \frac{1-\omega_\psi}{\tilde{\omega}\mu} \end{aligned} \tag{15.21}$$

where $\tilde{\omega}$ is the harmonic mean of the ω_ψ's. Given (15.15):

$$\overline{\omega\mu} \approx \omega_\psi \mu_\psi$$

a loss occurring to an ω_ψ class particle at the conditional mean, μ_ψ, approximately equals expected gain, $\overline{\omega\mu}$. See the vertical lines in Fig. 15.1 at the conditional means, μ_ψ. The length of the vertical line segment above the x-axis, $\overline{\omega\mu}$, approximately equals the length of the vertical line segment below the x-axis, $\omega_\psi\mu_\psi$. $\overline{\omega\mu}$ can be estimated over any range of income sizes, in particular close to the income size that the definitions and collection practices of large-scale household surveys are optimized for: the median. Estimating $\overline{\omega\mu}$ from gains does not require the identification of the ω_ψ of particles. $\overline{\omega\mu}$ can be estimated either as the intercept of the linear regression of gains on wealth or as the mean gain of particles with a gain. If the ω_ψ's are known, $\overline{\omega\mu}$ can be estimated as the mean loss (in absolute value) of cases in the ω_ψ equivalence class or as the actual loss at the conditional mean, μ_ψ . Given ω_ψ, an estimate of $\overline{\omega\mu}$ can be used to estimate μ_ψ, or given μ_ψ, $\overline{\omega\mu}$ can be used to estimate ω_ψ.

15.4. The IP, an Evolutionary Process

The expectation of particle x_{it}'s wealth in the ω_ψ equivalence class when the unconditional mean of wealth, μ_t, can change, as well as the proportion of the population of particles in any given ω_ψ equivalence class, is

$$\begin{aligned} E[x_{it}] &\approx \frac{\alpha_\psi}{\lambda_{\psi t}} = \mu_{\psi t} \\ &\approx \frac{1-\omega_\psi}{\omega_\psi} \cdot \frac{\tilde{\omega}_t\mu_t}{1-\omega_\psi} = \frac{\tilde{\omega}_t\mu_t}{\omega_\psi} \end{aligned} \tag{15.22}$$

i.e., the ratio of the current harmonic mean of particle productivity in the population to particle productivity in the ω_ψ equivalence class multiplied by the current unconditional mean of wealth. Note that smaller $\tilde{\omega}_t$ and ω_ψ indicates greater productivity. So the smaller $\tilde{\omega}_t$, the greater, μ_t. While this relationship is not explicitly modeled, the relationship can be inferred from data statistically. Is the failure to model the relationship between $\tilde{\omega}_t$ and μ_t a drawback? To model this relationship requires modeling the very large number of ad hoc ways that wealth is created. It is an evolutionary advantage if the mechanism which evaluates how productive individual particles are and then transfers wealth to the more productive does not require information about how wealth is created. Wealth creation is a vast and complex subject that continually ramifies as new technology and industrial organization appear. Evaluation of worker productivity is a complex, difficult issue. If the IP is the way wealth is transferred to the more productive of wealth, the IP does not need to model wealth production or directly evaluate worker productivity. Instead the IP bypasses these difficult issues and evaluates worker productivity indirectly. All the IP needs to transfer wealth to the more productive, if indeed the more productive lose less proportionally when they lose wealth, is to notice who loses less proportionally when they lose and then to transfer wealth, long term, to those workers. Specifically, what (15.22) asserts is that an individual particle's (to revert to the more abstract language of statistical mechanics) expected wealth is approximately the ratio of the product of the harmonic mean of all proportional particle losses at time t and the unconditional arithmetic mean of all particle wealth at time t, $(\tilde{\omega}_t\mu_t)$, to the proportion of wealth that the individual particle loses when it loses, ω_ψ. What might at first glance appear to be an incomplete model is an example of nature's extreme economy in the use of information to maximize the production of wealth.

Assuming that wealth is an input into its own production, the IP differentially nourishes the production of wealth by the more productive, maximizing the aggregate production of wealth. Everyone in the population benefits since increasing μ_t decreases $\lambda_{\psi t}$ and decreasing $\lambda_{\psi t}$ stretches the distribution to the right over larger wealth amounts. See by comparing Fig. 15.3 to Fig. 15.2 how a smaller $\lambda_{\psi t}$ stretches a gamma pdf to the right over larger x values. In Fig. 15.2, the gamma scale parameter is $\lambda = 2.0$, but in Fig. 15.3, in which the mass of the distribution is stretched to the right over larger x values, $\lambda = 0.5$. Figure 15.3 shows how the smaller λ stretches gamma pdfs to the right, putting more of their probability mass over larger x's (wealth amounts).

So the whole population shares in the increase in aggregate wealth due to the IP's transferring resources to the more productive, and given survival as an increasing function of wealth, the consequent decrease in extinction risk.

Given fixed ω_ψ's, $\mu_{\psi t}$ becomes a smaller fraction of μ_t in every ω_ψ equivalence class as $\tilde{\omega}_t$ falls. With fixed ω_ψ's, $\tilde{\omega}_t$ falls as the proportion of particles in equivalence class ω_θ grows at the expense of the proportion of particles in equivalence class ω_ψ where $\omega_\psi > \omega_\theta$. Following (15.21), $\lambda_{\psi t}$, the gamma scale parameter of wealth in the ω_ψ equivalence class where $\tilde{\omega}_t$ and μ_t can change is

$$\lambda_{\psi t} \approx \frac{1 - \omega_\psi}{\tilde{\omega}_t \mu_t} \tag{15.23}$$

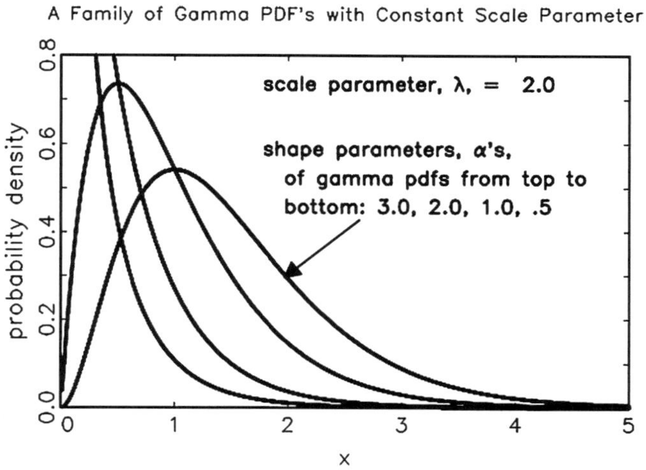

FIGURE 15.2. Gamma pdfs with common scale parameter, $\lambda = 2.0$, and different shape parameters

If the proportional decrease in $\tilde{\omega}_t$ is greater than the proportional increase in μ_t, the product $(\tilde{\omega}_t\mu_t)$ decreases and the distribution of wealth in all ω_ψ equivalence classes is compressed left over smaller wealth amounts, as in the comparison of Fig. 15.3 to Fig. 15.2 since $\lambda_{\psi t}$ is larger. The variance of wealth in the ω_ψ equivalence class decreases rapidly as $(\tilde{\omega}_t\mu_t)$ decreases since the variance of the gamma pdf approximation to the stationary distribution of wealth in the ω_ψ

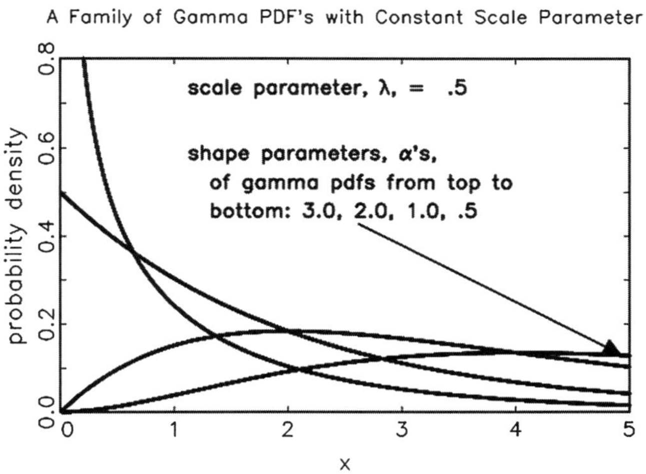

FIGURE 15.3. Gamma pdfs with common scale parameter, $\lambda = 0.5$, and different shape parameters

equivalence class is

$$\begin{aligned}\operatorname{var}(x_{\psi t}) &= \frac{\alpha_\psi}{\lambda_{\psi t}^2} \\ &\approx \left(\frac{1-\omega_\psi}{\omega_\psi}\right)\cdot\frac{(\tilde{\omega}_t\mu_t)^2}{(1-\omega_\psi)^2} = \frac{(\tilde{\omega}_t\mu_t)^2}{\omega_\psi(1-\omega_\psi)}\end{aligned} \tag{15.24}$$

So luck in winning IP encounters advances the wealth of the lucky less than when $(\tilde{\omega}_t\mu_t)$ was larger. If particle i were able to decrease ω_ψ (increase its productivity) in order to maintain its expected wealth, it would, in a finite population of particles, lower $\tilde{\omega}_t$, further increasing particle i's need to lower ω_ψ to maintain its expected wealth. Smaller $\tilde{\omega}_t$ means that the wealth of particle i is more closely tied to ω_ψ at any given level of μ_t because $\operatorname{var}(x_{\psi t})$ decreases in proportion to the square of decreasing $\tilde{\omega}_t$. Thus, not only does the IP cause μ_t to drift upward (something the IP, however, does not explicitly model), the IP creates an incentive for particle i to lower ω_ψ (to become more productive) to maintain its wealth whenever μ_t decreases or fails to increase proportionally as fast or faster than any decrease in $\tilde{\omega}_t$. Thus, the IP acts as a dynamic attractor for a population by decreasing its extinction risk via wealth maximization by the transfer of wealth to the more productive and incentivizing increasing productivity throughout the population. An evolutionary process is a dynamic attractor through time, a way of creating more time for a population.

15.5. The Empirical Evidence That Robust Losers Are the More Productive Particles

Clearly, the IP empirical relevance turns on the issue of whether the more robust loser, the particle with smaller ω, is in fact the more productive particle. The easiest way to substantiate this point is to empirically test the IP's implications against data. The IP's empirical explanandum includes the following:

- The decrease of the Gini concentration ratio and the increasing dispersion of wealth and income over the course of techno-cultural evolution (Angle 1983, 1986a, 2006); a decrease in the Gini concentration ratio of the stationary distribution of the IP with smaller ω_ψ is a logical requirement of the model [given its derivation from Lenski (1966)]; see (15.25).

$$G_\psi = \frac{\Gamma\left(\alpha_\psi + \frac{1}{2}\right)}{\sqrt{\pi}\,\Gamma\left(\alpha_\psi + 1\right)} \tag{15.25}$$

  G_ψ is the Gini concentration ratio of a gamma pdf (McDonald and Jensen 1979). Given (15.13), G_ψ is an increasing function of ω_ψ as stipulated in the social theory from which the IP was abstracted.

- The sequence of shapes of the distribution of wage incomes of workers by level of education and why this sequence of shapes changes little over decades (Angle 1996, 2006); see Figs. 15.4–15.6:
  Note that sequence of shapes of wage income with education level in Figs. 15.4 and 15.5 is as expected under the theory from which the IP was derived, i.e., smaller ω_{ψ} is associated with a higher level of education since that distribution is fitted by a gamma pdf with a larger shape parameter α_{ψ}. See (15.13).
- The dynamics of the distribution of wage income conditioned on education as a function of the unconditional mean of wage income and the distribution of education in the labor force (Angle 2001–2003a, 2003c, 2006).
- Why a gamma pdf is a useful model of the left tails and central masses of wage income distributions as ω becomes smaller and why their far right tails are approximately Pareto pdfs (Angle 2006).
- Why the IP's parameters estimated from certain statistics of the wage incomes of individual workers in longitudinal data on annual wage incomes are ordered as predicted by the IP's meta-theory (Angle 2002a) and approximate estimates of the same parameters from fitting the gamma pdf model of the IP's stationary distribution to the distribution of wage income conditioned on education.
- The difference in shape between the distribution of income from labor and the distribution of income from tangible assets (Angle 1997).
- The sequence of shapes of the distribution of personal wealth and income over the course of techno-cultural evolution (Angle 1983, 1986a).

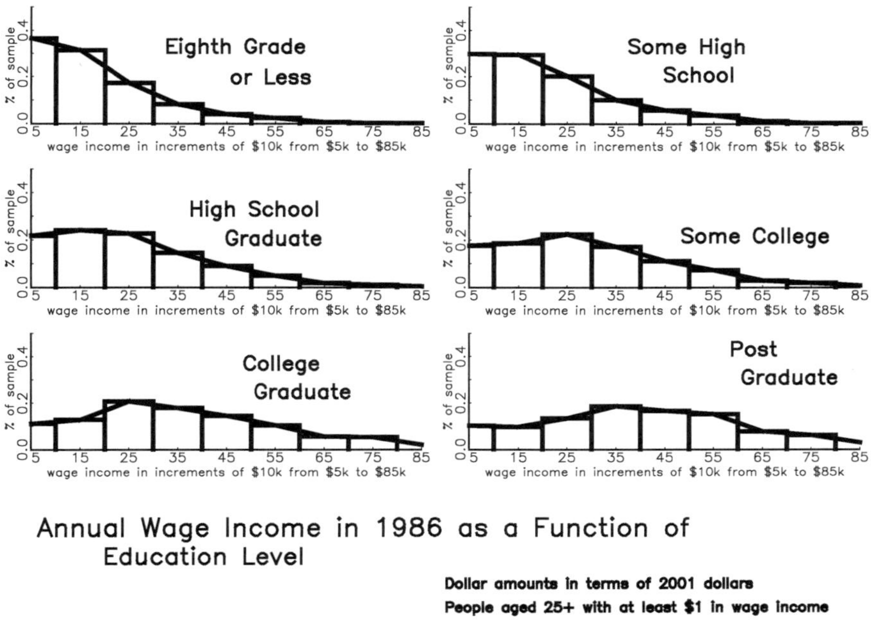

FIGURE 15.4. Source: author's estimates from the March CPS

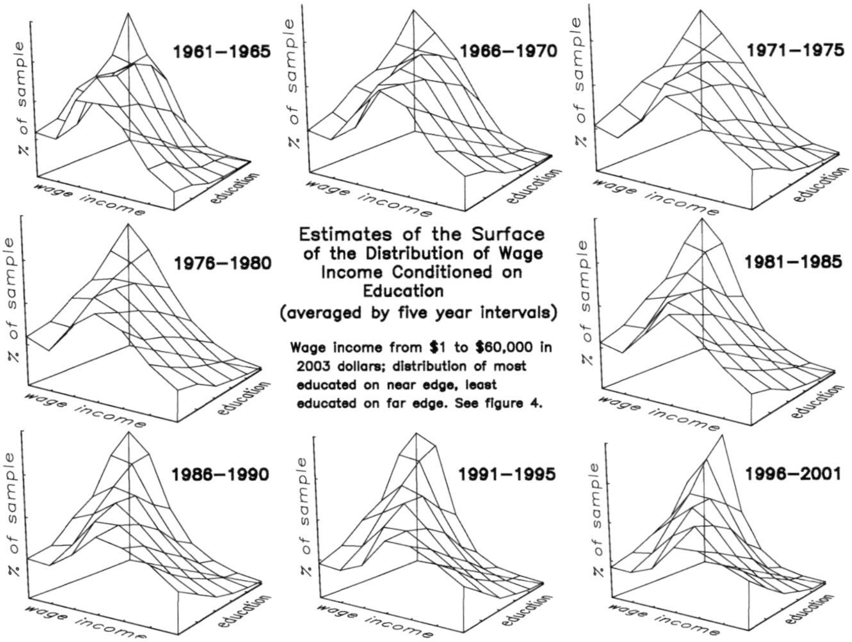

FIGURE 15.5. Source: author's estimates from the March CPS

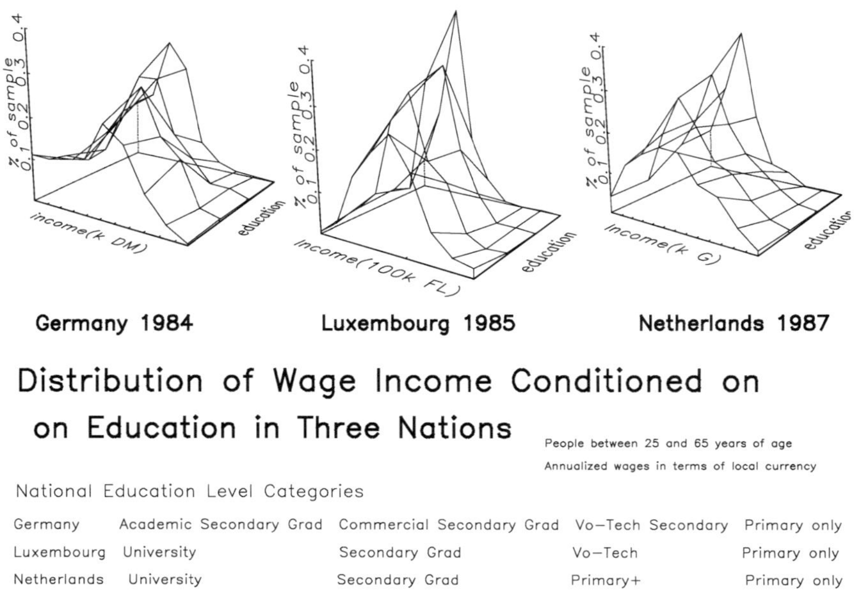

FIGURE 15.6. Source: author's estimates from surveys of the Luxembourg Income Study

- The universality of the transformation of hunter/gatherer society into the chiefdom, society of the god-king, with the appearance of storable food surpluses (Angle 1983, 1986a).

If one allows a particle in a coalition of particles to have a greater probability than 50% of winning a competitive encounter with a particle that is not a member of that coalition, then the IP so modified reproduces features of the joint distribution of income to African-Americans and other Americans such as:

- the % minority effect on discrimination (the larger the minority, the more severe discrimination on a per capita basis) (Angle 1986b, 1992);
- the relationships among variables as specific as (a) % of a US state's population that is non-white; (b) median white male earnings in a US state; (c) the Gini concentration of white male earnings in a US state; and (d) the ratio of black male to white male median earnings in a US state (Angle 1986b, 1992).

15.6. Conclusions

The mechanism that the IP uses to transfer wealth to the more productive is the asymmetry of gain and loss. If the more productive lose less when they lose—because they rebound faster, or are treated more gently, or have more leverage, or have more wealth directly under their control in an inalienable form, such as human capital—then an IP-like empirical process of competition will transfer wealth to the more productive, nourishing their more efficient production, causing upward drift in aggregate wealth production, lowering the extinction risk in the population, particularly of the more productive, and creating incentives for all particles in the population to become more productive in an increasingly precise way. The IP implies faster upward drift in aggregate wealth production, the higher the level of productivity in a labor force, the longer it operates, since its mechanism of wealth transfer to the more productive works more efficiently in a more productive labor force. Thus the IP becomes an "attractor" for any competition process in a population over time (Angle 2006), i.e., an evolutionary process, a process by which an interdependent population of organisms that produce wealth, what they need for survival, maintains itself and expands over time by more efficiently producing what it needs and differentially preserving the more productive organisms.

Zero-sum competition is the perturbing mechanism that transfers wealth from the less productive to the more productive in the IP via the asymmetry of loss and gain of wealth. Zero-sum competition is how the IP learns about particle productivity, dispensing with the need to measure individual wealth productivity in all its great variety and complexity. The only information the IP requires to work is knowledge of a particle's current wealth and past wealth back a few encounters in time, just long enough to estimate its parameter, ω, the proportion of wealth it loses when it loses a competitive encounter. Thus the IP can operate homogeneously up and down the scale of techno-cultural evolution: in today's industrial economies with tens of millions of workers and a complex division of labor as in a hunter-gatherer society with several dozen workers and a simple division of labor. As an empirical process, the IP operates in parallel in a population, without central direction. If nature chose extreme parsimony for its algorithm to maximize wealth production, it may have chosen zero-sum competition as in the IP.

The IP was derived from a speculative theory of the course of inequality of wealth over techno-cultural evolution, an extension to industrial society of a theory from economic anthropology about the origin of substantial inequality of wealth. The IP is social science. However, the discovery of the IP in social science may simply be a function of the fact that it was more noticeable there than in, for example, the fields of comparative animal behavior and population biology. The IP's statistical signature is all over US data on wage income at the micro- and macro-levels. The work of empirically confirming the universality of the operation of the IP in human society is just beginning, but there is nothing in the IP's mathematics to suggest that it can only operate in the US or only an industrial society or, for that matter, only among humans. Examination of the IP's mathematics shows that the IP is designed to operate independently of the details of wealth production in any population. The IP's mathematics are entirely general. One should expect the IP to emerge out of almost any process of competition for a positive quantity within a population of particles although for populations of particles with initially large ω values the likelihood of the process' taking hold at any particular time may be small. So one might speculate that the IP is the human expression of a process of intra-species (intra-specific) competition at work in all species. One might speculate even more boldly, following Adrian Bejan's (1997) suggestion that competition for energy can characterize non-living systems, that the IP can operate pre-bioticly among organic molecules competing for energy via chemical reaction, i.e., that the IP is the mathematical expression of evolution driven by dissipating ambient energy, the formation of an eddy in this dissipative flow. The empirical confirmations of the IP cited in this chapter are, however, exclusively of a process of wealth and income distribution among people.

Acknowledgement The author is indebted to Profs. Kenneth Land, Adrian Bejan, and the other participants of the Conference on the Constructal Theory of Social Dynamics, Duke University, April 2006, for stimulating discussions about the Inequality Process.

References

Angle, J. (1983) The surplus theory of social stratification and the size distribution of personal wealth. *1983 Proceedings of the American Statistical Association, Social Statistics Section*. pp. 395–400. American Statistical Association, Alexandria, VA.

Angle, J. (1986a) The surplus theory of social stratification and the size distribution of personal wealth. *Social Forces* **65**, 293–326.

Angle, J. (1986b) Coalitions in a stochastic process of wealth distribution. *1986 Proceedings of the American Statistical Association, Social Statistics Section*. pp. 259–263. American Statistical Association, Alexandria, VA.

Angle, J. (1990) A stochastic interacting particle system model of the size distribution of wealth and income. *1990 Proceedings of the American Statistical Association, Social Statistics Section*. pp. 279–284. American Statistical Association, Alexandria, VA.

Angle, J. (1992) The Inequality Process and the distribution of income to blacks and whites. *J. Mathematical Sociology* **17**, 77–98.

Angle, J. (1993a) Deriving the size distribution of personal wealth from 'the rich get richer, the poor get poorer'. *J. Mathematical Sociology* **18**, 27–46.

Angle, J. (1993b) An apparent invariance of the size distribution of personal income conditioned on education. *1993 Proceedings of the American Statistical Association, Social Statistics Section*. pp. 197–202. American Statistical Association, Alexandria, VA.

Angle, J. (1996) How the gamma law of income distribution appears invariant under aggregation. *J. Mathematical Sociology* **21**, 325–358.

Angle, J. (1997) A theory of income distribution. *1997 Proceedings of the American Statistical Association, Social Statistics Section*. pp. 388–393. American Statistical Association, Alexandria, VA.

Angle, J. (1998) Contingent forecasting of the size of the small income population in a recession. *1998 Proceedings of the American Statistical Association, Social Statistics Section*. pp. 138–143. American Statistical Association, Alexandria, VA.

Angle, J. (1999a) Contingent forecasting of the size of a vulnerable nonmetro population. *Proceedings of the 1999 Federal Forecasters' Conference*. pp. 161–169. U.S. Government Printing Office, Washington, DC.

Angle, J. (1999b) Evidence of pervasive competition: the dynamics of income distributions and individual incomes. *1999 Proceedings of the American Statistical Association, Social Statistics Section*. pp. 331–336. American Statistical Association, Alexandria, VA.

Angle, J. (2000) The binary interacting particle system (bips) underlying the maxentropic derivation of the gamma law of income distribution. *2000 Proceedings of the American Statistical Association, Social Statistics Section*. pp. 270–275. American Statistical Association, Alexandria, VA.

Angle, J. (2001) Modeling the right tail of the nonmetro distribution of wage and salary income. *2001 Proceedings of the American Statistical Association, Social Statistics Section*. [CD-ROM], American Statistical Association, Alexandria, VA.

Angle, J. (2002a) The statistical signature of pervasive competition on wages and salaries. *J. Mathematical Sociology* **26**, 217–270.

Angle, J. (2002b) Contingent forecasting of bulges in the left and right tails of the nonmetro wage and salary income distribution. *Proceedings of the 2002 Federal Forecasters' Conference*. U.S. Government Printing Office, Washington, DC.

Angle, J. (2002c) Modeling the dynamics of the nonmetro distribution of wage and salary income as a function of its mean. *2002 Proceedings of the American Statistical Association, Business and Economic Statistics Section*. [CD-ROM], American Statistical Association, Alexandria, VA.

Angle, J. (2003a) The dynamics of the distribution of wage and salary income in the nonmetropolitan U.S. *Estadistica* **55**, 59–93.

Angle, J. (2003b) Inequality Process, The in T. Liao, et al. (eds.), *The Encyclopedia of Social Science Research Methods* **2**, 488–490. Sage, Thousand Oaks, CA.

Angle, J. (2003c) Imitating the salamander: a model of the right tail of the wage distribution truncated by topcoding. *Proceedings of the Conference of the Federal Committee on Statistical Methodology*, (November). [http://www.fcsm.gov/events/papers2003.html].

Angle, J. (2005) Speculation: The Inequality Process is the Competition Process Driving Human Evolution. Paper presented to the First General Scholarly Meeting, Society for Anthropological Science. February, 2005, Santa Fe, New Mexico.

Angle, J. (2006) The Inequality Process as a wealth maximizing process. *Physica A: Statistical Mechanics and Its Applications* **367**, 388–414 (DOI information: 10.1016/ j.physa.2005.11.017). A draft version available at http://www.lisproject.org/publications.

Bejan, A. (1997) *Advanced Engineering Thermodynamics*. Second Edition. Wiley, New York.

Kleiber, C. and Kotz, S. (2003) *Statistical Size Distributions in Economics and Actuarial Sciences*. Wiley, Hoboken, NJ.

Lenski, G. (1966) *Power and Privilege*. McGraw-Hill, New York.

Lux, T. (2005) Emergent statistical wealth distributions in simple monetary exchange models: a critical review. pp. 51–60 in A. Chatterjee S. Yarlagadda, and B.K. Chakrabarti (eds.), *Econophysics of Wealth Distributions*. New Economic Windows Series. Springer Verlag, Milan.

McDonald, J. and Jensen B. (1979) An analysis of some properties of alternative measures of income inequality based on the gamma distribution function. *J. American Statistical Association* **74**, 856–860.

Chapter 16
Constructal Theory of Written Language

Cyrus Amoozegar

16.1. Introduction

The topic of this chapter began as a way to merge two of my interests: the study of foreign language and the study of engineering. I was introduced to constructal theory initially though Prof. Bejan in his thermodynamics course and then subsequently through his course on constructal theory and design. Because of the nature of my study, I was invited to attend the First International Workshop on Constructal Theory of Social Dynamics.

Language serves as a fundamental aspect of social dynamics. It acts as the medium for nearly all communication and interaction. Language can both bring together and separate people. As it is such an essential element for society, it is important to understand its evolution and the movements language takes. This chapter is my attempt to explain this evolution through constructal theory.

The elements of this evolution, as they are presented here, should not be thought of as unique to the development of written language. One thing made clear by this book and the history of constructal theory in general is that where there is flow, there is configuration, and thus configuration can be predicted and explained.

16.2. Written Language

16.2.1. What Is a Written Language?

In order to see how constructal theory applies to written language, it is crucial to understand what language is. Language is simply a tool (a construct) one uses to express one's ideas about the world. It allows for individuals to formulate ideas as well as for these individuals to communicate and share their ideas. Society is able to function because individuals are able to communicate. The need for communication within a society is the driving force behind the development of language.

Language can take both spoken and written forms and throughout time these forms change, they evolve. This evolution of written language is *natural*. Written languages were not developed by groups of linguists who merely wanted to develop a language. Their creation was natural, driven by the need of the common person to communicate with large numbers of persons. Languages are constructs that occur naturally in our minds. Writing is a manmade construct along which language can be used to convey, store, and express ideas.

16.2.2. How Does Constructal Theory Apply?

Because languages are used to describe the world, there is a flow of ideas from the world to the mind of the user. This flow can be described as a flow from a volume (the world) to a point (the mind of the user). Constructal theory follows the idea that "For a finite-size open system to persist in time (to live), it must evolve in such a way that it provides easier access to the imposed (global) currents that flow through it" (Bejan 1997).

Constructal theory can be applied to written languages because they are, in fact, "finite-size open systems." There are only so many elements that make up a written language. The number of elements is generally extremely large, but always finite. The "imposed currents" are the ideas that language is used to express. At this point it becomes necessary to define "easier access."

There are certain criteria that a written language needs and that are universal to all written languages. Firstly, the language needs to be able to accurately describe the ideas the user desires to present. A written language is useless if it cannot be used to portray what the user wishes. Secondly, the language needs to be learnable. In modern society, written languages are used by a broad spectrum of people and therefore they must allow for everyone in this spectrum to learn the language. From there different cultures have different customs that will be reflected within their written language.

Languages are used for description. There are an infinite number of concepts in the world and ideally written language would be able to describe each of these concepts. However, in order for a written language to fully describe the infinite set of ideas in the world, the written language would itself also need to be infinite. Written languages cannot be infinite because they would not be learnable and thus be unusable. Therefore a written language can, at best, only approximate all of the ideas in the world. The difference between reality and what written language is able to communicate is a resistance to the flow of ideas. It also follows that generally the greater the number of elements within the language the more accurately the language describes reality. In terms of trying to describe reality, the greater the number of elements within a language, the easier it is for the flow of ideas to progress.

A written language is useless if no one is able to learn it. In modern society, mastery of written language is almost essential for survival. The greater number of people within a society who know the language, the more communication that can take place, thereby allowing the society to function more efficiently.

In order for a large portion of society to be able to master the language, the language has to appeal to a great number of people. To accomplish this, written language should have fewer elements. The number of elements a language has affects how easily one can master the language. That is, it is easier for a large group of people to remember ten different elements than it is for a large group to remember twenty different elements. Thus in terms of making the language appeal to a large number of people, the fewer the number of elements within a language, the easier it is for the ideas to flow.

These two requirements are in direct contrast with one another. More elements of a written language allow the flow of ideas to progress easier in terms of being able to describe reality, but fewer elements allow the flow to progress easier in terms of allowing more people to master the language. Through time these language elements, or constructs, evolve in both form and number to maximize "access" to the language mainly by balancing these two requirements. However, these are not the only two factors that influence written language.

The complexity of the language constructs also affects the flow of ideas. On one side, there is a need to make the constructs simpler. If a construct is complicated it will likely take too long to write and would be more difficult to remember. The global resistance would be increased and thus the flow of ideas would be slowed. On the other side, the constructs cannot be too simple. If a construct were too simple, there is a higher probability that the users of the written language would confuse constructs and the meanings of writings would be misconstrued. Too simple a set of constructs would also cause the global resistance to increase. The natural evolution of written language must head for a balance between the complicated and the simple.

16.2.3. Origins of Written Language

The origins of writing come from a combination of writing and art: pictographs. Pictographs are written constructs whose meaning lies directly in what is written. Examples of pictographs are the cave paintings from the prehistoric period found throughout the world. These paintings depict various images of animals and humans. The meaning of each image lies directly in what is painted. That is, the pictograph symbol for an ox looks like an ox. These pictographs are more art than writing, but a form of art that would continue through to the development of writing. The practice of directly depicting an idea through picture continue through time and can be found as the predecessor to many of the world's first writing systems.

The use of pictographs to express ideas can be represented by a flow from a single point to a volume without pairing levels. Figure 16.1 represents this flow. The point in the center is the mind of the user of the writing and the outer circle is world. The use of pictographs causes the thoughts of the user to travel directly from the center to the circumference. That is, there are no pairing levels. Each line from the center to the circumference expresses the use of a pictograph or a single thought. If the user wishes to express a large number of thoughts, the path from the center to the circumference would contain a considerable number of pictographs.

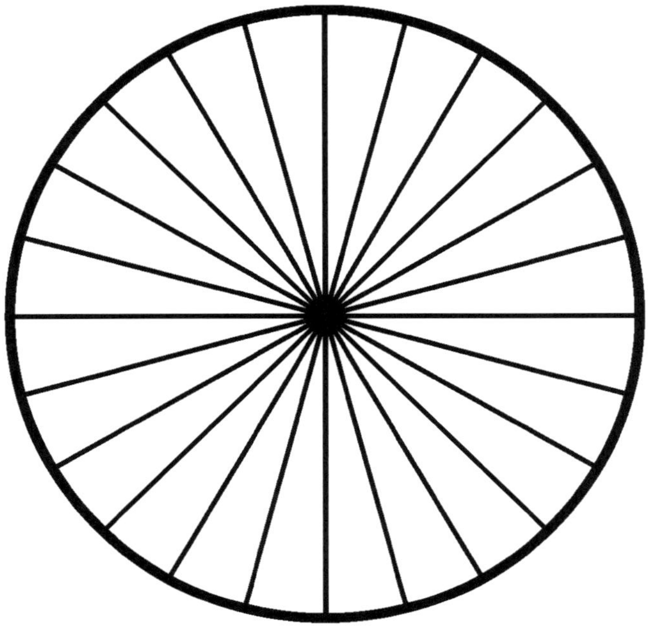

FIGURE 16.1. Representation of the flow of thoughts from the user of written language to the outside world through the use of pictographs

Such a method of conveying thoughts is extremely inefficient. There arises a need for an incredible amount of pictographs. The users of the written language must learn all of these symbols. However, this task is not necessarily as difficult as it seems because the symbols are essentially drawings of what the user sees. The use of such a large number of pictographs leads to other issues. Each of the pictographs must be different, so a fair amount of detail must be applied, making the writing process a slow one. Also, the use of the language is limited to only ideas that can be expressed through pictures. How does one easily express conditional or potential ideas through pictures?

No modern language functions solely off of pictographs. Why is that? The answer comes from constructal theory. The strict use of pictographs is far too inefficient to survive the test of time. The global resistance brought by the limitations of pictographs led to evolution to reduce this global resistance. One way this evolution manifested itself in the development of pairing levels that reduce said resistance.

16.3. First Pairing Level

16.3.1. Creation of First Pairing Level

Three of the earliest written languages, Egyptian hieroglyphics, Sumerian Cuneiform, and Chinese, were all derived from pictographs (Wilson 2003;

Glassner 2003; Björkstén 1994). The global resistance resulting from the use of pictographs was extremely high for the reasons outlined earlier. If follows from constructal theory that in order for these languages to persist in time, they had to evolve to lower their global resistances.

The global resistance arises from the number of pictograph constructs, the necessary detail needed in differentiating these constructs, and the natural limits of what can be expressed with these constructs. To lower the global resistance, one or more of these issues must be resolved. The number of constructs cannot be lowered on its own because that would severely compromise the effect of using the language. The natural limit of pictograph constructs can be resolved by moving away from the sole use of pictographs and toward the use of other forms of constructs. The detail needed in differentiating different constructs can be resolved by the development of a common set of elements used in the creation of all constructs, or simply, the creation of the first pairing level. The existence of a first pairing level is a common thread amongst nearly all modern written languages. The origins of the first pairing level lie in the earliest languages.

The use of Egyptian hieroglyphs for writing text began around 2100 BC and the entire writing system was made up of about 700 constructs (Wilson 2003). These 700 constructs represent the first pairing level of the written language. From these constructs, close to 17,000 words could be made (Allen 2000). Thus in order for one to learn to write, he need only learn 700 constructs as opposed to 17,000. These constructs can be used together in combination to make all of the words, thus eliminating the need of excessive detail to distinguish between constructs. It is easier to make a set of 700 unique constructs than it is to make 17,000.

The problem regarding the limits of pictographs was solved by the development of other types of constructs, or graphemes. The constructs of Egyptian hieroglyphs occurred as five different types: unilateral, bilateral, trilateral, ideograms, and determinatives (Wilson 2003). Unilateral, bilateral, and trilateral constructs were phonograms or constructs that represented a sound. Ideograms were constructs that represent ideas and determinatives were constructs that aid in understanding the meaning of words created with phonograms (Wilson 2003).

The development of phonographs represents an important step in the evolution of written language. The significance of such a step will be discussed in Section 16.3.2.

Sumerian cuneiform followed a similar development. It began as pictographs and developed into a writing system. The cuneiform system began around 3400 BC and was originally constructed with 800–900 constructs (Glassner 2003).

Chinese had a somewhat different first pairing level. The earliest known evidence of the Chinese writing system comes from writing on turtle shells and the shoulder blades of oxen created during the Shang Dynasty (1600–1100 BC) (Björkstén 1994). The characters found on these bones were extremely inefficient. Many of them were pictographs and ideographs and they were independent of one another. Also, there was no discernable set of constructs from which the characters were made, so each character was written differently (Zhou 2003). Up until around 200 BC the flow of ideas in written writing had no pairing levels, as in Fig. 16.1.

Around 200 BC, the Clerkly style was introduced. It was the first style to introduce the concept of strokes and stroke order (Zhou 2003). The idea is simply that there are eight basic strokes, exemplified by the character 水, and the characters are generally written from the top left to the bottom right. These eight basic strokes are the constructs from which Chinese characters are made. What is different about the first pairing level of Chinese is that this pairing level tells the user nothing about the pronunciation of each character.

Constructal theory explains the development of the written language in all of these cases. The languages simply became too complex to exist as pictographs, in both number of constructs and the complexity needed to distinguish each construct. This complexity was the global resistance hindering the flow of ideas. The only way for these languages to continue to exist was to evolve in such a way to lower this global resistance. The development in the first pairing level in all of these cases allowed the language to become more efficient and the problems associated with global resistance were all somewhat alleviated.

Figure 16.2, in comparison to Figure 16.1, represents this transition. The center is still the user of the written language. The paths from the center to the first circle are the constructs that make up the first pairing level. The second set of paths, from the inner to the outer circle, is the flow of ideas to the outside

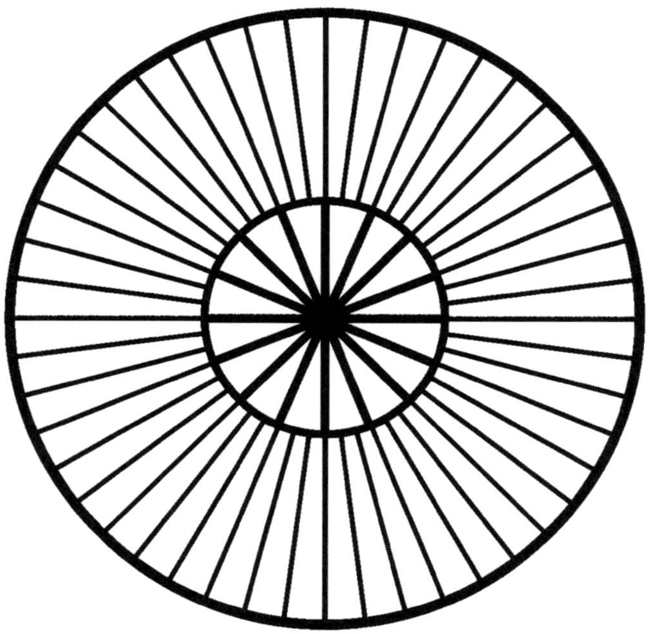

FIGURE 16.2. Representation of the flow of thoughts from the user of written language to the outside world with implementation of one pairing level

world. The ideas in this second set are represented through combinations of the constructs of the first pairing level. The increased efficiency from the creation of the first pairing level allows the number of paths in the second set to be greater than the paths in Figure 16.1. The idea here is simply that the presence of a set of constructs from which ideas are formed permits the written language to be more efficient and leads to the ability to express a wide variety of ideas.

16.3.2. Evolution of First Pairing Level

The creation of the first pairing level, while important, is not necessarily the most interesting step. The original first pairing levels were far from perfect. Though more efficient than the pictograph system, the early writing systems that used the early first pairing levels fell prey to the many same problems. In the case of Egyptian, 700 constructs was still a lot to learn and the constructs were fairly elaborate. The first pairing level of a written language, like the entire written language itself, must evolve in such a way that lowers the global resistance to the flow of ideas.

The source of global resistance must once again be identified in order to determine how to best lower it. One source, which has already been dealt with once before, is the number of constructs. Having too many constructs increases the global resistance because in order to properly use the language one has to learn, remember, and be able to replicate more constructs. On the other hand, having too few also increases global resistance because the language must be able to fully describe any thought. The fewer constructs in a language, the less concise and accurate this description can be.

The development of phonograms provides a mechanism by which the number of constructs of written language can be lowered. Generally the existences of written language and spoken language are intertwined. That is, they tend to be used in conjunction with one another. Both are used to express essentially the same ideas, so it would be inefficient for both to exist simultaneously while being completely independent of each other. Phonograms exist as a link between the two systems.

In first-language acquisition, one tends to learn to speak a language before he learns to write the language. It would then follow that one of the best way to learn to write the language would be to relate it to the already known spoken language. Furthermore, this serves to lower the number of constructs from the number used in pictograph systems because spoken languages tend to use the same sets of sounds to express more than one idea. Due to needing fewer constructs, the global resistance to the flow of ideas is lowered. Sometimes phonograms will contain more than one construct in the first pairing level. In these situations, the sound the phonogram represents is usually similar to the sounds for which the individual constructs are symbols. An example of this is "sh" in English. The sound is a cross between the sound associated with "s" and that associated with "h."

Another source of resistance is the complexity of each construct. The more complex a construct is the longer it takes to replicate and the greater the difficulty associated with remembering it. The less complex the constructs, the greater the difficulty in distinguishing between constructs.

In both of the cases it seems that balance must be reached. If both having too many constructs and having too few increase global resistance and if both having constructs that are too complex and having constructs that are too simple increase global resistance, then there must be a number and complexity of constructs at which the global resistance is lowest. The history of language shows that the first pairing level in writing has followed this movement toward a decrease in global resistance.

The use of phonograms achieves this balance. Generally, the number of phonograms needed to describe a language is low. Furthermore, the combination of sounds works in expressing ideas verbally, so having just enough symbols to express those sounds should work in expressing those ideas through writing.

Nearly all of the modern written languages in the world have evolved to having a first pairing level consisting of phonograms. English has the Roman alphabet, Farsi has the Persian alphabet, and Japanese has katakana and hiragana. These phonograms express different levels of sounds. The first pairing level of English consists of the letters in the Roman alphabet. Each one of these represents a sound and the letters are used together to form syllables. The constructs of Japanese hiragana and katakana, on the other hand, represent entire syllables.

One notable exception to this is the Chinese written language. The Chinese written language begins to loosely implement phonemes in the second pairing level. Beyond the phonemes in the second pairing level, Chinese is being taught in schools by using a phonetic counterpart to the language. In the People's Republic of China, Hanyu pinyin is currently used to teach the language. Hanyu pinyin is essentially the Romanization of the language. The letters of the Roman alphabet are used to represent the sounds of the language. In Taiwan, the Zhuyin Fuhao system is used to represent the sounds of the language. The Zhuyin Fuhao system is derived from an originally Mainland Chinese system for representing the sounds of the language.

A detailed explanation of how the Egyptian language followed this first pairing level evolution can be found below.

16.3.2.1. Egyptian

The writing system of the Egyptian language evolved from hieroglyphs to cursive hieroglyphs and hieratic, then to Demotic, and then finally to Coptic (Allen 2000). Through each one of these steps the constructs of the first pairing level became progressively simpler in both design and number. The move from pictographs to the hieroglyphic constructs introduced phoneme symbols, but the phonograms existed in categories of one, two, or three syllables and were not the only constructs in the system (Allen 2000). The large number of phonograms and existence of other constructs led there to be a large total number of constructs.

The cursive hieroglyph system had the same number of the constructs as the original hieroglyph system. The change between the two systems was that the constructs of the cursive system were of a simpler design (Allen 2000). Figure 16.3 shows a sample of cursive hieroglyphs and original hieroglyphs side by side. The cursive hieroglyphs tend to be made up of fewer strokes and the pictographs contain less detail (Allen 2000). For instance, the figure shows two constructs in which the original construct was a detailed bird, while the cursive construct was only the outline of the bird. While the cursive hieroglyph constructs do resemble their original hieroglyph counterparts, the removal of detail shows the movement away from use of pictographs (Allen 2000).

Like the cursive hieroglyph system, the hieratic system had the same number of constructs as the original hieroglyph system. The design of the constructs continued to become simpler (Allen 2000). Figure 16.4 shows a comparison between hieratic and original hieroglyphs. The constructs contain even less strokes and are much farther removed from their pictograph origins. While the cursive hieroglyph constructs still resembled the original hieroglyphs, the hieratic constructs are almost completely separate (Allen 2000).

The original hieroglyph system began to be used around 2100 BC and shortly after the cursive hieroglyph and the hieratic systems were developed and put into use (Allen 2000). The original hieroglyph system had too large of an internal resistance, so through time the system was used less, while the cursive and hieratic systems were used more. By 1300 BC the majority of writing was done using the hieratic system (Allen 2000). Because the design of the constructs was

FIGURE 16.3. A sample of cursive hieroglyphs and their original hieroglyph counterparts. The original hieroglyphs are on top and the cursive hieroglyphs are on the bottom

FIGURE 16.4. A sample of hieratic constructs and their original hieroglyph construct counterparts. The original hieroglyphs are on top, and the hieratic constructs are on the bottom

simpler, the internal resistance was lower. Because the internal resistance was lower, the system was able to survive the longest.

The hieratic scripts remained dominant until around 650 BC when the Demotic texts were introduced (Allen 2000). Figure 16.5 shows a sample of Demotic text. The Demotic constructs continued the trend and were of an even simpler design than the hieratic constructs (Allen 2000). The Demotic constructs tended to contain fewer strokes and the strokes were usually shorter. Fewer and shorter stokes generally indicate that it would take less time to convey the same message. By lowering the internal resistance, the language is able to more effectively serve its purpose.

The final step in the evolution of written Egyptian, taking place around the first century AD, was the development and use of the Coptic system (Allen 2000).

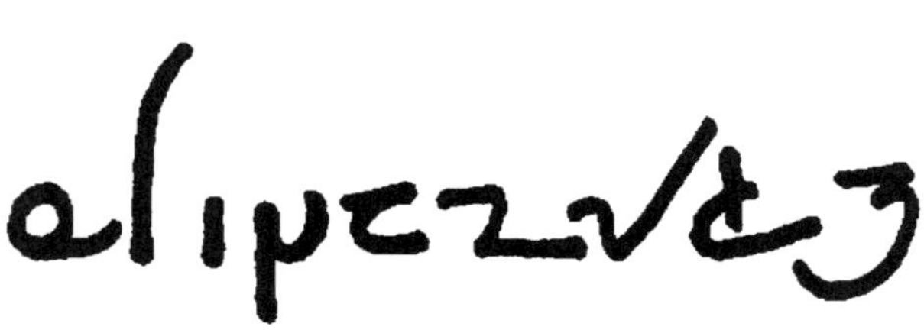

FIGURE 16.5. A sample of Demotic text. The Demotic constructs were simpler in design than the hieratic constructs

FIGURE 16.6. A sample of Coptic text

Figure 16.6 shows several of the Coptic constructs. The constructs of the Coptic system were the simplest of all the systems. All of the constructs consisted of three strokes or fewer and were small variations of common geometric shapes. In total there were only 32 constructs in the first pairing level, 24 of which were taken from the Greek alphabet. Each construct represented a sound of the Egyptian language (Wilson 2003).

This is a large change from the original 700 constructs in the first pairing level. The decrease in number and complexity of constructs shows the elimination of the factors that cause global resistance to the flow of ideas. The natural evolution of the language made it such that written Egyptian was easier to use on a large scale. Due to the complexity and number of original hieroglyphic constructs, their use would make writing a much lengthier and harder process compared to the use of the Coptic constructs.

16.4. Second Pairing Level

16.4.1. Creation of Second Pairing Level

As time progresses, the number of ideas that a written language needs to be able to represent increases. For instance, the number of words associated with the Internet has skyrocketed over the past 30 years. As the number of ideas that need expressing increases, the global resistance to the flow of ideas also rises. In order the further lower the global resistance to flow of ideas, written languages tend to naturally develop a second pairing level.

Once again, the first task to understanding the development of the second pairing level is to understand from where the global resistance to the flow of ideas arises. In order to express ideas, the constructs of the first pairing level need to be used in combination with one another. Going straight to making ideas from the first pairing level, the combinations of constructs have no sense of organization when expressing ideas. This lack of organization prevents the user from having some sort of standard by which to use the language. Every idea appears as a random assortment of constructs from the first pairing level.

The creation of a system to organize the constructs in the first pairing level would lower the global resistance to the flow of ideas because it would allow the user to more easily use the language. This system is the second pairing level. In order to organize the constructs, the second pairing level should implement a method that relates similar ideas. Essentially, the second pairing level is similar to the morphology of the language or the rules by which words are put together.

Throughout their evolutions, languages all over the world have developed a second pairing level of some kind. Like the variety in first pairing levels, there is a large spectrum of second pairing levels, but the underlying idea of a unifying system is the same throughout. The second pairing levels of both modern English and modern Chinese are analyzed below, with Chinese being the more interesting and detailed case.

16.4.1.1. English

In written English, the second pairing level involves the different methods by which words are formed. These methods include processes such as combining prefixes, suffixes, and roots or combining two words to create a new idea. Each of these consists of a combination of constructs from the first pairing level, in this case the Roman alphabet. Each prefix, suffix, and root indicates some meaning. When used in a word, the meaning of the word does not exactly reflect the meaning of the prefix, suffix, or root, but the two meanings are related.

For example, the prefix trans- indicates a meaning of "across." It is used in the word transport, which means to move across a distance; transparent, which is the property of allowing rays of light through; and translate, which is the action of converting an idea from being expressed in one language to being expressed in another. Neither transport, transparent, nor translate indicate a meaning of "across," but all of them have something to do with the concept of "across." Transport is across a distance; transparent is the property of allowing light across a material; and translate is across languages. The prefix trans- is able to relate words which have meanings dealing with the concept of "across."

The same idea applies to all of the other techniques for making words in the English language. Instead of the user of the language having to memorize every word in the language, he can apply the concepts in the second pairing level to determine the meanings of words. That is, in the case of the words transport, transparent, and translate, the user would not have to memorize each of these words independently of one another. He could understand that trans- indicates dealing with the concept of across and from there relate the three words. This relation increases the ease by which memorization occurs and thus lowers the global resistance to the flow of ideas.

16.4.1.2. Chinese

The second pairing level of written Chinese came into existence when the number of constructs in the first pairing level began to become too great for people to learn. New characters consisting of two or more basic characters began to form (Zhou 2003). These combinations added a new level of complexity to the written language, and through time these combinations evolved to the most efficient form. This movement allowed for the introduction of two new elements into written Chinese that make up the second pairing level: the concept of the radical and the concept of the phoneme.

Radicals aid in indicating the meaning of a character. Like prefixes, suffixes, and roots in English, the meaning of a Chinese radical and the meaning of the Chinese character that contains that radical are not the same, but are related. Radicals usually express simple concepts. Parallel to the trans- example is the example of the character 人. 人 is a character that, by itself, means person. When used as a radical, it appears in characters such as 众, which means crowd, and 他, which is the pronoun "he." Neither of these characters have the exact meaning of person, but both deal with the concept of person.

Figure 16.7 shows a small sample of the radicals that can be formed from the eight basic strokes of the first pairing level. Each radical has a general meaning to which other characters can relate. Figure 16.8 shows these relations. Thirteen

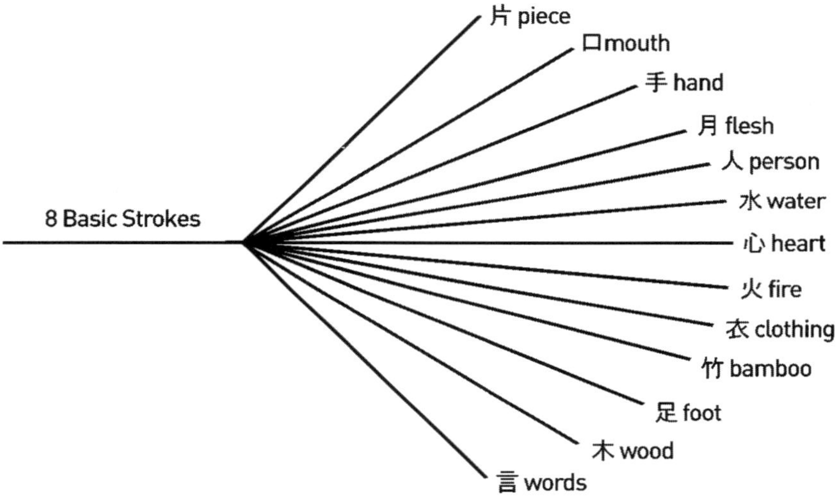

FIGURE 16.7. A sample of radicals and their respective meanings

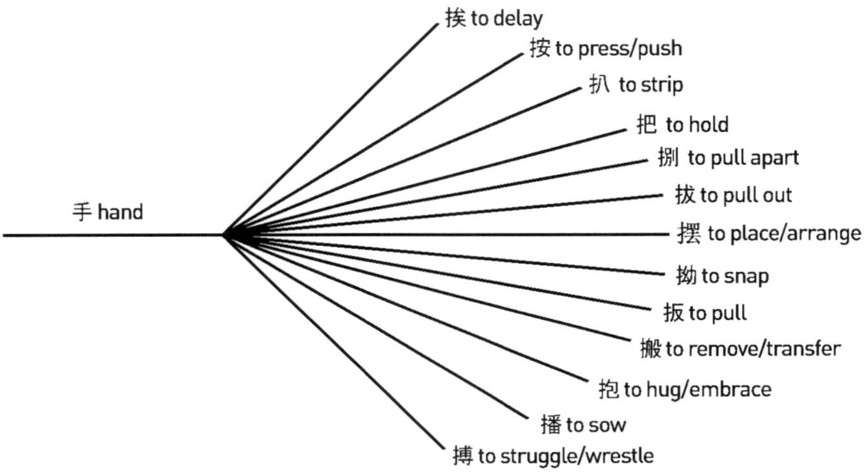

FIGURE 16.8. A sample of the characters that can be formed with the radical for hand and their respective meanings

characters that contain the radical for hand are shown. These characters have meanings such as to delay, to press/push, to strip, and to hold. None of these mean hand directly, but all are associated with the idea. One can use a hand to be delayed, such as not doing work that should have been done. Usually a hand is used to press or push an object. A hand is used to strip something and objects are held in hands.

In the 3500 most common characters, 188 radicals are used (Zhu et al. 2003). These radicals add a sense of organization and make it easier for one to learn to read and write. They allow the user of the language to understand the nature of unknown character by making connections to known, related characters. They are a basic unit applicable to all characters dealing with a particular subject. This allows the user to have to memorize less and draw more connections, thus lowering the global resistance.

One notable feature that appears at this level is the intertwining of language and culture. In any given society, it is likely that one who is exposed to the language is simultaneously exposed to the culture of the society. By the same logic that led to the connection between spoken and written language, a connection exists between written language and culture.

In the case of Chinese, this relation takes the form of how the radicals are used in particular characters. One of the best illustrations of this idea comes from the character 好, which means "good." This character is made up of two radicals: the 女 radical, which means "woman," and the 子 radical, which means "child." In ancient Chinese culture, the more children a couple had, the more fortunate that could was considered. From there the association was made that in society a woman with a child is considered "good" (The Composition of Common Chinese Characters 1997). It follows that since in society a woman with a child is good, the character for woman and the character for child together should also indicate the idea of "good." The global resistance to the flow of ideas is lowered because one can view the connections made in relation to society instead of having to memorize each character independently.

As described earlier, the presence of phonemes is important in lowering the global resistance to the flow of ideas. Also as mentioned earlier, the phonemes in Chinese are introduced in the second pairing level. Disregarding tones, there are 397 different syllables in the Chinese language. Out of these 397 syllables, 334 syllables have characters that share a non-radical grapheme with another character with the same pronunciation (Shanghai 1992). These shared non-radical graphemes are sound-indicators. In first-language acquisition, one generally is able to speak before they are able to write. The sound-indicators allow for connections to be made between spoken and written Chinese. While radicals indicate a strong sense of meaning, sound-indicators only weakly indicate the pronunciation. For instance, the sound-indicator 包 can be used to indicate the pronunciation "bao" or the similar pronunciation "pao". Also, a single pronunciation can have more than one sound-indicator associated with it.

It is understandable that radicals have a stronger presence in characters given the state of language in China. Almost every area in China has its own dialect with a unique pronunciation. However, the characters used in writing are more or less the same. Since there is more uniformity between written styles, it naturally follows that connections with the spoken language are loose.

The same pathway exists from the eight basic strokes to the phonemes as exists from the eight basic strokes to the radicals. Figure 16.9 shows the same branching scheme from the eight basic strokes to a small sample of the phoneme. Figure 16.10 follows through to the next step to show the branching scheme from the phoneme for "ba" to a number of characters with the pronunciation "ba".

The presence of two different sets of constructs—one from the radicals, the other from the phonemes—is what makes the second pairing level of the Chinese language so interesting. These two systems work together to make a complete second pairing level. The presence of both systems indicates that the Chinese written language has evolved over time to allow for greater access.

A simple model can be used to view this evolution. Without the radicals and phonemes, 100 different ideas would require 100 different independent characters, even if the ideas were related. However, with pairings between sound-indicators and radicals, one would only need to know 10 radicals and 10 sound-indicators. Each radical can be combined with a phoneme to generate a character.

Reality is not this simple as such a model makes a number of assumptions, including that the original set of 100 ideas divides into 10 groups of 10 related ideas and the idea that any phoneme can be combined with a radical to make a new character. In modern Chinese, there are many combinations that do not exist. So while this model does not hold exactly true, the general idea remains. The presence of the second pairing level decreases the amount of effort needed to memorize and use a written language to synthesize ideas.

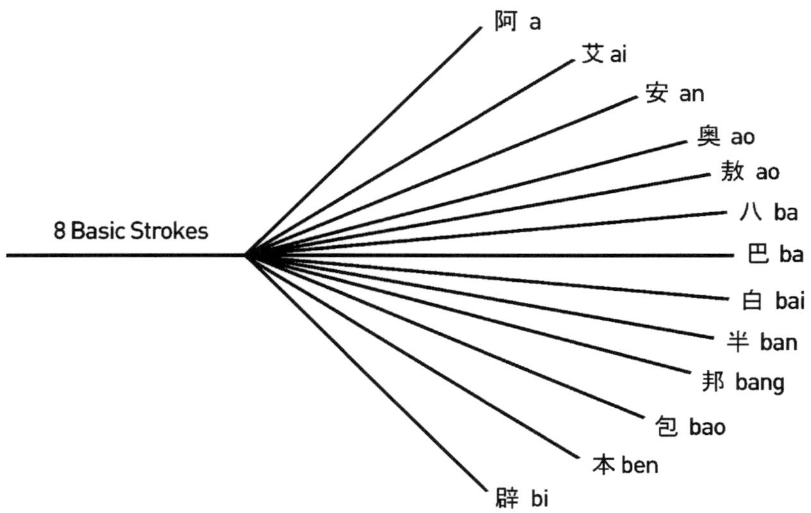

FIGURE 16.9. A sample of phonemes and their respective sounds formed from the basic strokes of the first pairing level

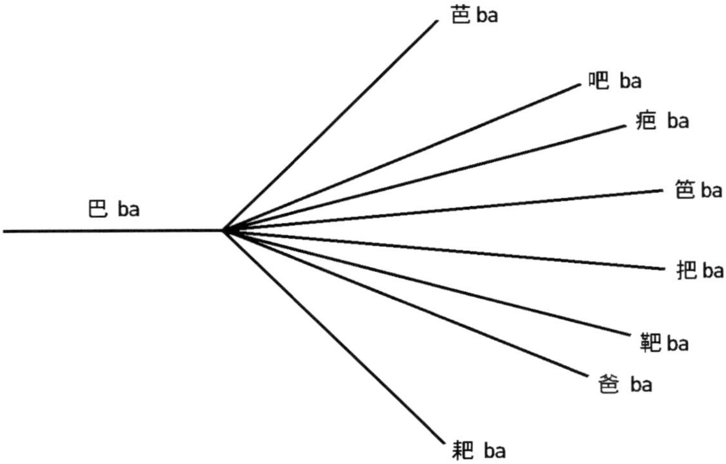

FIGURE 16.10. A sample of the characters that can be formed with the phoneme "ba" and their respective sounds

16.4.2. Evolution of Second Pairing Level

Like the first pairing level, the second pairing level also evolves with time and in much the same way. The sources of resistance are still the complexity and the number of constructs. According to constructal theory, for a language system to persist in time, the language would have to evolve in order to reduce the global resistance to the flow of ideas. Constructal theory predicts the evolution of the constructs in the second pairing level. Languages show that an evolution occurs that reduces the complexity and the number of constructs, while still allowing for the constructs to remain unique and the language to fully express ideas. An example of how the second pairing level of the Chinese language evolved through time is given below.

16.4.2.1. Chinese

Numerous times throughout the history of the written language, Chinese characters have undergone massive simplification. The most recent large- scale simplification of the language took place during the 20th century. The People's Republic of China moved from using what are termed "traditional" characters to "simplified" characters. In the 1950s the government of the People's Republic of China implemented the "General List of Simplified Characters," which put into effect a switch of 2,235 commonly used characters to a more simplified form (Zhou 2003).

Figure 16.11 shows an example of the change a character underwent while changing from traditional to simplified. The radical changed from consisting of eight strokes to consisting of only five. Such an evolutionary process took place

針 ⟶ 针

FIGURE 16.11. The evolution of the character for needle from traditional to simplified

numerous times in the history of Chinese writing, but the results were always similar.

As languages evolve and new constructs form, it should be expected that similar constructs come into existence. Often, multiple constructs with the same meaning come into existence. Multiple times throughout history, characters of similar meaning have been eliminated. In the 1950's, over 1000 characters were eliminated on the basis that more frequently used characters existed that shared the same meaning (Zhou 2003).

Other simplifications also took place, such as the elimination of single characters with disyllabic pronunciations (Zhou 2003). These simplifications allow the user of the language to have an easier time in learning and using the language thus reducing the global resistance to the flow of ideas.

16.5. Conclusions

Languages exist as finite-size, open systems and these systems have changed through time. Constructal theory can be used to explain this evolution that languages take. The movement toward greater access drives the creation of first and second pairing levels and even the evolution of those pairing levels to more efficient forms. Such predictions are general and apply to a wide range of languages.

This analysis stopped at the word level. The future of constructal theory explaining the evolution of language would involve looking beyond the word level. Such analysis would involve looking at grammar and how it changes through time and seeing if the path it takes allows the language to be learned by more people. Also, some of the claims presented can be tested empirically so further study would involve collecting data to support or negate these claims.

The workshop on constructal theory of social dynamics ended with looking at the theory itself. As a theory, it is quite general, which allows it to be applied in many applications, but this generality also makes it somewhat unclear how to apply it in certain situations. As alluded to earlier, its applications in engineering are quite clear, but it is making its movement toward non-engineering disciplines. It is during this movement that the problem associated with how one would use constructal theory appears. This issue causes resistance to the theory and if the theory follows itself, in the future, constructal theory will have to evolve to minimize such resistance.

References

Allen, J. P. (2000) *Middle Egyptian: An Introduction to the Language and Culture of Hieroglyphics*. Cambridge University Press, Cambridge, UK.

Bejan, A. (1997) *Advanced Engineering Thermodynamics*. Second ed., Wiley, New York.

Björkstén, J. (1994) *Learn to Write Chinese Characters*. Yale University Press, New Haven, CT.

The Composition of Common Chinese Characters: An Illustrated Account. (1997) 谢光辉. Beijing: 北京大学出版社.

Glassner, J. J. (2003) *The Invention of Cuneiform: Writing in Sumer*. Johns Hopkins University Press, Baltimore.

Shanghai (1992) 范, 可育. 常用汉字实用手册, 华东师范大学出版社: 新华书店上海发行所经销, 1992.

Wilson, P. (2003) *Sacred Signs: Hieroglyphics in Ancient Egypt*. Oxford University Press, New York.

Zhou, Y. (2003) *The Historical Evolution of Chinese Languages and Scripts*. Ohio State University, Columbus, OH.

Zhu, Y., Wang, L., and Ren, Y. *Pocket Oxford Chinese Dictionary*. Oxford University Press, Oxford, UK.

Chapter 17
Life and Cognition

Jean-Christophe Denaës

17.1. What is Life?

If one wishes to model and then simulate cognitive capacities, one has to wonder *what cognition is* and *what computation is*. Finally, question by question, one comes to ask *what is the Nature*, "Being" for some philosophers, "Void" for some others. To have an almost safe journey, we shall follow the same road. Starting from Psyche and its hylemorphic cognition, we will head toward Nature and natural cognition to return to our point of departure, which, because of our journey, will be seen under quite a different perspective. It is indeed a journey across the lands *of cognition* that we are about to embark on, and not a journey *toward cognition* as if it were an objective.

Journeys are the mark of confidence of those who embrace life. This one is an affair of History and, therefore, can only be told as a story. Therefore, it would be vain to explain it mathematically, reducing it to an infinite number of points in an infinitesimal space which is always too narrow. History does not lent itself to explanations. It only lets itself be told in a never-ending story. Likewise, and for the very same reason, Life does not allow itself to be reduced to a serial explanation made of inferences, but comes to light through our intuitions and, sometimes, for the most sensitive and subtle of us, in an act of serendipity[1] (Van Andel 1994) like Bejan's description of his own experience (Bejan 2000, p. 3): "I *became interested* in this topic purely *by accident*, not because I was trying to solve the puzzle" (emphasis added).

Intuition and serendipity are both unknown and unfashionable cognitive capacities that relate not only to the acquisition of knowledge but also to the unveiling of a reality that is always slipping away. Widely ignored by the contemporary scientific field of cognitive science to which they belong, they urge us to question its foundations and ambitions: Do the Cognitive Sciences aim at fundamental knowledge, or do they already partake in a more instrumental practice?

---

[1] "I define serendipity as the art of making an unsought finding" (Van Andel 1994). Serendipity is the way to make discoveries, by accident but also by sagacity, of things one is not in quest of. Based on experience, knowledge, it is the creative exploitation of the unforeseen.

From Nature to Psyche, from natural "cognition" to "higher" cognition, it is possible that there is not even a single step, but a continuous flow. Not recognizing this would probably preclude us from hearing a simple and elegant explanation of higher cognition compared to what we shall call natural cognition,[2] the rising form or *in-formation*,[3] which, by lasting, transforms and is transformed, making the History of Nature and its historicity, that is the present state of its duration. From this point of view, the brain itself would not be of another nature.

Given the tortuous history of the vitalism, to venture in its realm can augur a difficult journey. However, in order to allow us to tell this story without implying others, we shall immediately state the context of this one: our story is about a geometric[4] form of vitalism, the constructal theory, which we would like to make more alive by aiming it toward a *Vis formandi*.[5] It is worth saying that in our journey, fundamental research is the field we are concerned with. Simply said, this is the story of *geometric vitalism* and its metaphysics, a *materialist vitalism*.

As in any story that respects itself, the promiscuity created by the co-presence of two terms that are apparently antithetic, vitalism and materialism, is not a coincidence but indeed a providence, an event that comes at the right moment to save a "desperate" situation, a dramatic turn of events.

17.2. Psyche, the "Higher" Cognition

Certainly the most striking thing in human perception is the tension that exists between those two perceptive qualities: the inclination to perceive reality in a discrete way, therefore to conceive it in the same way, and the feeling of continuity, of duration.

It seems obvious that both are expressions of the same reality. In this way, from one to the other and reciprocally, there is only a difference in the degree of

[2] Cognition is the *faculty to know*, and this faculty is *perception*. A matter which can take form is necessarily perceptive. Consequently, the lowest expression of the cognition has to be found in Nature.

[3] We will write "in-formation" to mention the principle of the rising form, which is a continuous process in a continuous reality, as we will see. Our aim is to avoid the misunderstandings brought about by the discrete concept of "information" and, by the same token, the operational theories to which it belongs (Communication, Information, and Computation theories). More generally, we will use "in-" to say "from inside", i.e., by counteraction, and not from "outside", i.e., by composition, which is the accepted view linked to the term "information."

[4] "In geometry, there is no heuristics, you need to resume everything from zero according to the problem, contrary in what takes place in algebra" René Thom (Nimier 1989).

[5] "*Vis formandi est vis ultima ex ultimis*," "The formative force is the ultimate force from ultimates" (Whitehead 1911). *Vis formandi* is also used by Cornelius Castoriadis to mean the same thing, where *Vis* means "force," "power," and *formo* "to make rising/take form," "to form," "to construct." In this chapter, one has to understand that the notion of *force* comes from Physics. We do not talk about "a spiritual vital force" but "a physical force" which forms and animates beings.

perception. This differential degree, this tension, led to two natural metaphysics known today under the names of atomism[6] and vitalism.[7] Both certainly started with the birth of Psyche a long time ago. One of the oldest examples of a materialist vitalism can be found in ancient Egypt with the notion of Ka,[8] a flow which animates all living beings, humans, and gods.

17.2.1. From Aristotle's Hylemorphism to the Rationalization of Probabilities

Hylemorphism implies a necessary composition. It is the imprint of primordial matter by substantial form. Gilbert Simondon made an important critique of hylemorphism (Simondon 1964) that appears to be characteristic of the *natural* way of thinking. It can be summarized as follows: Forms are not inside substance. They are not given but generated (produced) by preindividual matter. Even a construction, made by man or animal, is constrained by the nature of the material.

Simondon's philosophy tends to embrace natural beings and machines in their genesis (Simondon 1989), demonstrating that they are not so different in the way they are built with regards to their own nature, which comes from Nature. "Built" means "not totally independent" from Nature. Simondon says that they are in *individuation*. Individuation brings concretization, but natural beings are a lot more concrete than machines since they are mostly built from the *inside*. They come from a centrifugal mechanism. Machine are the inverse, they are mostly built from the *outside*, from a centripetal mechanism (Lima-de-Faria 1988, p. 145). We say "mostly" because matter, even if it does not have any shape in substance, does not allow any kind of shape. You cannot build a statue only with water, nor can Nature make animals only with clay. Therefore, *concretization* implies *natural realization*. The important idea is that even if a machine is handmade or an insect's nest is made by mandible, both are *natural* above all. They are more constrained by the nature of the matter than the work, intentional or not, of the maker. Simondon's point is that beings, natural and artificial, are *in* individuation. They are individuals never totally individualized, never totally concrete, always *in* individuation. Similar conclusions are reached today concerning the natural construction of honeybee combs (Pirk et al. 2004; Karsai and Pézes 2000) and swarms (Bejan 2000, p. 45).

---

[6] Atomism is the metaphysics according to which the universe is constituted by indestructible building blocks which combine in a mechanical way. The foundation of Atomism is the discrete.

[7] Vitalism is the metaphysics according to which nothing exists but a "vital force," an "élan vital," a flow which animates beings. The foundation of the Vitalism is the continuous.

[8] There are many interpretations of what was meant by *Ka* at that time, but none of our contemporary words approximate it. We decided to keep what seems to be its constant characteristic through all Egyptian periods, which is as a term to indicate the sustaining and creative power of Life. Later in this chapter, we will use it to designate what we will call *Kaos*, but only in the sense of sustaining force.

But certainly the most interesting analysis has to be found in the "Origins of Geometrical Thought in Human Labor" by Gerdes. This professor of mathematics, quoting Aleksandrov and Molodschi, writes that human beings "first gave form to their material and only then recognized form as that which is impressed on *material* and can therefore be considered in itself as an abstraction from the material. [...] As human beings made more and more regular shapes and compared them with one another, they learned to perceive form unattached from the qualitative particularity of the compared object" (Gerdes 2001, pp. 396–397). Human beings do not learn geometry by contemplating Nature, since the contemplation of Nature's geometries can only come after the mental construction of the concept. Thus we cannot state that "geometry is *in* Nature" or "*by* nature." Geometry arises and grows with and by the labor of human beings. Gerdes" work echoes Bejan's study of human constructions and flow fossils in general, especially the pyramids (Bejan 2006b) and, even more, his remarks on *science and civilization as flow systems* (Bejan 2006, pp. 815–816). Gerdes shows in his paper how geometrical thinking develops and then follows its own path to the science of geometry. A strong connection has to be made with the work of De Waal (2001) and Janmaat et al. (2006) and their studies of non-human primates.

But, since matter is not hylemorphic, what then is *hylemorphism*? Is it *real* or is it an *illusion*?

Rather than trying to find out who invented the modern notion of probability, Hacking (2002) wonders how the notion of probability became possible and what historical transformations had to occur during the time of the Renaissance so that we can think in terms of modern probabilities. Just as Gerdes, he proposes an *archeology of knowledge* that unveils the discontinuation of the then common sense of *probability*, which was expressed by the *probation*, into two interlaced trends sufficiently individuated to be identified. Those are *logical probabilities* (epistemic hylemorphism) and *statistical probabilities* (frequentism). Because of their historical evolution, they reinforce one another, and contemporary science certainly contributes to the crystallization of this state of affairs, this dynamic. The expression of this intertwining is the *probability of probability*. It is the logical probability (an episteme is based on a degree of certitude) of a statistical one (a measure is based on frequencies) that, in the sciences, determines the degree of "objectivity," in fact a degree of approbation by use of descriptive methods, in positive sciences.

Hylemorphism is what Bergson called the *natural intelligence*, so does only concern for intentional beings. Because there are two complementary processes, statistic (measure) and logic (episteme), we will use two different words to express them.

First, *hylerealities* points out that the possibilities, the thoughts, are hylemorphic marks. They are produced (morphos) by the brain (hyle). Hylerealities are at the same time virtual possibilities at their higher existential degree (i.e., if regarded as entities), and real embodied productions at their lower existential degree. They tend to fall into Reality when an intentional being tries them in It. *Action* is the proof test of hylerealities. If the action succeeds, the hylerealities related to the action tend to appear more real, even to become real, and of course, if the test fails, they tend to be more virtual, even to disappear. Second, we will use Castoriadis" word, *ensidic*, which is the ensemblistic-identitary perception of the intentional being, the perceptive discretization and categorization of the Reality. Whereas ensidic perception belongs to the statistical process, the measure, ensidic logic belongs to the logical process, the episteme.

These two intertwined capacities, the *hylerealism* and the *ensidic*, are goal oriented toward the action. A better action, for intentional beings, is homolog to a better flowing for non-intentional beings. The differential degree is that non-intentional beings are not goal oriented since optimization is a consequence of flows counteraction (the counteraction of the flows), the balance between two flow regimes, that is, "the interplay between satisfying global constraints while meeting the local requirements" (Bejan 2000, p. 7). This is the direct consequence of the constructal law, the principle of the individuation. For intentional beings, this consequence is canalized, hence catalyzed, by the Psyche. This is our *natural logic*, our *natural intelligence*. As Wittgenstein said, "Logic is not a theory but a reflection of the world" (Allott 2003).

Yet, natural logic reinforces itself by transforming itself into a powerful instrument of description (ensidic logic), hence of persuasion (inference, rhetoric). Mathematics, from geometry to arithmetic and algebra, can be seen in this way: a qualitative (perceptive) capacity transforming itself into a quantitative (operative) one (Gerdes 2001; Thom 1977, pp. 4–7), the ensidic perception folding in an ensidic logic that turns and folds over itself (Dhombres and Kremer-Marietti 2006, pp. 52–54). Computation (operative logic), with computers and cellular automata, is the actual ultimate state of this trend (Thom and Noël 1993, pp. 15–17). Again, this is not a surprise. It can be considered as pure determinism if one dares to consider that *natural intelligence is made for action*, and consequently for instrumentation, even of itself.

Most vitalists, especially modern vitalists like Canguilhem, bring attention to the fact that the "symptom" is too easily transformed into the "sign," that the result of a measure cannot express what is measured. This is why ensidic logic, the "higher" cognition, cannot explain Nature or itself by its very own mechanism, which is *action* by means of perceiving, here and there, *discrete entities*, real or hylereal, on which it can act. As Caws (1978) says about Valéry (1919), "We struggle against our enemies whom we cannot but admire. Valéry makes a marvelous enemy for the intellect, and what higher compliment could be paid the writer in struggle against himself?". Thus, for this prime reason, *Atomism* is doomed to its very first purpose: action, hence instrumentation.

Another argument (Cummins et al. 2004) is that, since one of the cornerstones of naturalistic epistemology is "you cannot require someone to do something they cannot or, as it is usually put, *ought implies can*." However, if "ought implies can" is forbidden, the reverse is possible according to the context. Therefore, differences in existential degree imply substantial differences in capacity, and that some of these differences are ineliminable. If an intentional state is the production and use of hylerealities, which is the mastering of *the probabilities of the probabilities*, then Nature is not made of intelligence or probabilities. This is why *in Nature*, the Reality, there are no such things as *random* mutation (except if one remembers what Darwin said about randomness), natural *selection* (a better *chance* to survive), *struggle* for life, natural *information*, natural *computation*, *intelligent* universe, *intelligent* design, *copying errors* in the genetic material, *selfish* gene or *selfish* idea (memetic). They exist only in our mind. They are possibilities we use to act *in* and *on* the Reality. All are hylerealities we, intentional beings, try to make more real in regards to our need of action.

But if the scientific community is looking for a fundamental knowledge of Reality by using his ensidic capacities too intensively, trying to make them more real than they have to be in the hope to unveil Reality, it can only sink into action and drift away from explanation. If one wants to explain primal matter, one must not follow his *natural intelligence*, as Bejan (2000, p. 4) says, "too much discipline is poison to the individual's innate creativity," but use trickery with it. In a way, one must overflow it, putting it in a metastable state where the intuition's becoming can take place, allowing the possible production of an act of serendipity. One has to think, or at least to try, the genesis of the individual, the individuation, which is the continuous in-formation of matter. This conclusion is not new; many authors since the Pre-Socratic period, and even probably before then, have discussed this point and will continue to do so. Just remember Zeno's Paradox[9] on Achilles and the Turtle (Castoriadis 2005, pp. 410–413). It reminds us that if one needs an infinite number of atoms to describe Reality, one also needs an infinite number of gaps that is an abyss. In the end, because Reality cannot be explained by an abyssal gap, one has to agree that this is a virtual hylereality. We intentional beings journey across the native land of our psyche, and its name is *Hylereal*.

17.2.2. The Cognitive Implication

The cognitive implication is twofold.

(1) The hylemorphic stance implies an *intensional* implication: the logical relation to the referents is never logical but implicit. Referents are given. Rationalized logics are shaped expression of the history of our evolution (Gerdes 2001).

(2) The intentional stance implies an *intentional* implication: not all beings are intentional. And before using this term intentionally, *one has to define what "intention" is*. We said that intentions are utterly psychic activities. Hylerealities can only arise in beings that have a memory (a particular persistence of the historicity) and one or several sensory–motor apparatus, all sufficiently

[9] Plurality and movement are connected in the thought of Zeno. They belong to the world of appearances. However, this movement is the one of the bodies. It is an ensidic movement. We will see later that Zeno stance is justified, and that preindividual matter is also preindividual movement (qualitative inhomogeneity).

individualized and echoing each other, allowing the historico-perception, so intention hence intentional action. Hylerealities are the *virtualities* that a being attempts to affix to Reality.

17.2.3. Empirism Probabilis and Vis Formandi

Once Hobbes said, "Nothing universal can be concluded from the experience" (Hacking 2002, p. 84). We agree with this conclusion since this is the inevitable consequence of Empirism probabilis, the probabilities of probabilities as natural intelligence expressions.

The *proof* is the sign, the fact. It allows the conjecture: "what proves too much proves nothing" but "to explain is not to prove." And if proof forbids the explanation, this last one can only *be* in regard to the probation, that is the approbation, the epistemic vote as the epistemic probability of a dice to unveil one of its faces. Probation is episteme, and episteme is rationalized in logical probabilities based on statistical probabilities. But, then, is episteme only rationality? Not really. The episteme is the terminal but dynamic process of reasoning, and *reasoning is rationality in counteraction with intuition*. If not dynamic, the episteme of probation can only turn into belief, which is the compost of ideology.

Ultimately, *if there is an explanation, it will have to be pregnant, that is full and, most of all, creative*. It will have to be done without *ad hoc* adjunctions and be able to explain all beings, realities or hylerealities, in all fields, and, in the end, be able to give birth to utterly new possibilities. This is to say that only a harmonious alliance between *knowledge acquisition* and *intuition* is capable of permitting those acts of serendipity that, with patience, would allow finding such an explanation, a *true* explanation.

Modern Science allows mostly a certain kind of knowledge and bases its probation on it. It is empirical knowledge. But empirical knowledge, observations of realities or hylerealities (as the studies of the dynamic of numbers or functions in mathematics), is always ensidic. It is always descriptive, thus biomimetic in its deployment. Its method veils the foundation of the scientific activity that is intuition. We can thus argue that Science opposes intuition by giving recursive inferential rhetoric, that is, inferential descriptions and validations based on biomimetic proofs, where one wants explanation. Science bases its probations under the cover of inferential explanations. Consequently, those probations turn into beliefs. Again, to prove is to show, and to explain is to embrace all things. Therefore, it is certainly better to *try to explain* than *to want to prove*, if the aim is to have a fundamental understanding and not an instrumental finality. In regard to this—natural—stance of Science, materialist vitalism is a metaphysical proposition of a never-ending journey from Psyche to and from Nature.

Finally, one can think that the constructal theory "proves a lot but does not explain much." If that were the case, we would have to ask again what a proof is and what an explanation is. This theory certainly explains a lot, and perhaps will one day be subject to *ad hoc* adjunctions, but, without doubt, it proves little since it brings only a single proof which, furthermore, is not new: *Man is—naturally—more inclined to be geometer than scholar*. And one, especially if this *one* is a researcher, has a right to ask Science to what degree scientists are geometers today.

In their last book, Dhombres and Kremer-Marietti (2006), the first philosopher interested in the constructal theory and its philosophical implications (ibid., pp. 90–94; Kremer-Marietti 2006), have written an enlightening chapter, "What epistemology to come?". They argue that "Popper's falsifiability principle does not come under a good history of science, and it does not give to think philosophically" and that this is mainly due to the historical process of mathematization. But, they say, "We can and ought to wonder to determine if certain standards of rationality do not hide, in their way, an interested origin, for example a social practice which tries to give itself values of universality." For us, this origin, which can become a social practice, is the natural inclination of the natural intelligence to instrument, even to instrument itself.

We agree, and certainly Thom would have done the same (Nimier 1989, pp. 100–101), with Dhombres and Kremer-Marietti's conclusion that the history of science must become more sociological and anthropological. It "has to begin by knowing why so many persons dedicate themselves to the science presented as a research activity" and to wonder if this could be due to the fact that "the profession of researcher does not need any more to be qualified by the adjective *scientific*. A part of the research became a common part of the service, and not an upstream from (a prior of) a new production. Furthermore, a part of the research is, under the name of a *proved knowledge*, even of an *expertise*, become an alibi" (emphasis added).

17.3. Nature as Matter, Unique-ness and Kaos

17.3.1. The Impossible Emergence of the Emergence

In systemic, when one speaks of emergence, one usually borrows this sentence from the Gestalt theory: *the whole is more than the sum of its parts*. That is the whole is a system, a discrete entity made of discrete entities where gaps are called relations. We argue that this idea has to be dismissed if one wants to make fundamental instead of applicative research.

Another justification for this can be found in Occam's argument for nominalism: Creatures and objects do not have their form in substance, a form that could have an independent existence. They do not live in the mind of the Creator. In fact, a substance constitutes a limitation to the creative liberty of God. God, he says, did not create the world based on his preexistent ideas, but *shaped as it seemed good* to him. This is where Occam's razor helps since one can speak of general properties without the help of God's substantial ideas. In the same way, we wish to use Occam's razor to say that one does not need a whole with parts, since parts are a limitation even with infinite number. This limitation is incompatible with Nature's expression as a whole without parts.

One has to think the whole as a unique uniqueness, a Reality as Uniqueness, at least if research is aimed at *foundations* (quantitative) or *non-foundation* (qualitative) of Reality.

17.3.2. Matter as Unique-ness

Unique-ness, the *metastable preindividual matter* for Simondon (1964, p. 8), is not an emergent whole but a whole in individuation. It has to be noted that both *Uniqueness* and *Kaos* will be used to designate the metastable preindividual matter. But whereas Unique-ness will be used to point the whole without parts, *Kaos* will be used to emphasize its in-determination, that is, its metastability or qualitative inhomogeneity (Castoriadis 2005, p. 416).

17.3.3. Matter as Kaos

Castoriadis remarks that Chaos means all and nothing, any place we look at where we find "disorder," that is, things that are not "simple." Quoting Ruelle's definition of the (determinist) chaotic phenomena which "are the processes of temporal evolution in which there is a sensitive—an appreciable, an important—dependence on the initial conditions, namely, upon what was there at the outset or upon the limit conditions, as is said in mathematics, that is to say, upon what surrounds the phenomenon" (ibid., p. 382), he replies by arguing that if this is what was at issue, then there is not anything new in that idea of chaos since we use it in our everyday life: "if only I had left home a half second earlier or a half second later, I wouldn't have had that accident" (ibid.). This is the consequence of the *perception* of the determinism of the independent causal series of events, or Cournot's notion of "chance" (Cournot 1875).

Cournot conceives implicitly the *chance event* as the unforeseen and deterministic result of the crossing of independent causal series of events for the intentional being involved as participant. One has to remember that an observer is a causal actor too (Morowitz 1987). Of course, a non-intentional being is unable to foresee anything, hence if one gives it an intentional stance for the occasion, one can say that, for it, all events are unforeseen, that is all events are *deterministic chances*. But there is no *a posteriori* or *a priori* in the non-intentional realm. There is only *a praesenti*. It means that non-intentional begins have no abilities to foresee or not (i.e. to miss) chances or events. They are not capable of historico-perception, hence they have no intentional capacities. They only have a perceptive capacity which is that capacity of matter to be in-formed, to be in individuation.

In the end, seeing sensitive dependencies is seeing discrete states as possibilities in a reality where there are none. Therefore, where an intentional being sees sensitive dependencies (sensitivity to initial conditions), Reality is in fact made of extreme dependence which is the continuous variation of initial conditions leading to a discontinuous variations, to a *catastrophe* (Thom 1977). "And if the thing has been well formulated in this way, we may discover that all forms of modern science have until now been living upon the following implicit postulate: the postulate of continuity of physical phenomena and even of extant phenomena in general. Moreover this continuity is not even mathematical continuity in the full sense of the term: it is mere linearity" (Castoriadis 2005, p. 384).

At the same time, Castoriadis remarks that Aristotle's *hyle* has the same meaning as *chaos*. And because *hyle* is primal matter, one can say that chaos is in fact the deterministic, continuous, and formless primal matter. Primal matter is preindividual matter. It is defined by its *in-determination*, which must not be confused with the word indetermination that means randomness, uncertainty, or imprecision. In-determinism is the quality of the Unique-ness and is expressed by the deterministic, or constructal, rise of the forms.

To speak about this in-determinate chaos, Castoriadis uses the term *Void*. But *void* is an intentional word. Bergson justly remarked that the void, in the first place, expresses the feeling of absence. There is a void, *an absence of*, only for an intentional being, a being capable of remembering, so having expectations (Bergson 2001, p. 281). For non-intentional beings, as for preindividual matter, there is only existence, an existence without expectation.

A middle-way can be found in the term of *Kaos*. Created by taking the Egyptian *Ka*, in the materialist vitalistic way in which we have defined it, Kaos is the metastable preindividual matter. It is in-determinate matter, with no shape at all, therefore an in-deterministic force. The in-determination is necessarily a source of metastability since the force can flow in any direction, but at the same time metastability brings turbulence, thus determination. To understand why, one has to look at the force flow as a fluid flow, such as air. Experiments in thermodynamics show that a flow becomes turbulent when natural channels cannot absorb anymore the pressure forces arising from its acceleration. The channels overflow, it is as if they do not exist anymore. Therefore, this situation can be described as an in-determinate formless flow, and formless flow can only give rise to turbulence (Bejan 2000, p. 149). Since there is no shape, no channels, insight can therefore be achieved into "what's going on" in the Kaos. Turbulence is a consequence, the expression of an in-determinate state which is just able to give rise to a deterministic one.

Kaos, or Unique-ness, is a plasma where matter and force are the same. It cannot be described in terms of topology but only in terms of qualitative inhomogeneity.

17.4. Consequences

17.4.1. The Intentional and Non-intentional Beings

Intentional or not, beings are *individuals in individuation*. In this sense the question of knowing if a being is animated or not, since always animated (in-formation is flows counteraction), is a question of referential.

One can argue that all beings are “living” since shapes are the expression of flows counteraction. But this is not a question for us to answer. It is up to biologists to define the limit(s) between the living and non-living. But, as cognitive scientists, we can say that there is a difference of existential degree (Fig. 17.1) between (1) non-intentional beings, non-psychic “individuals” (from particles to cells and plants) and (2) intentional beings that are psychic individuals (insects, animals), i.e., with a mental “life.”

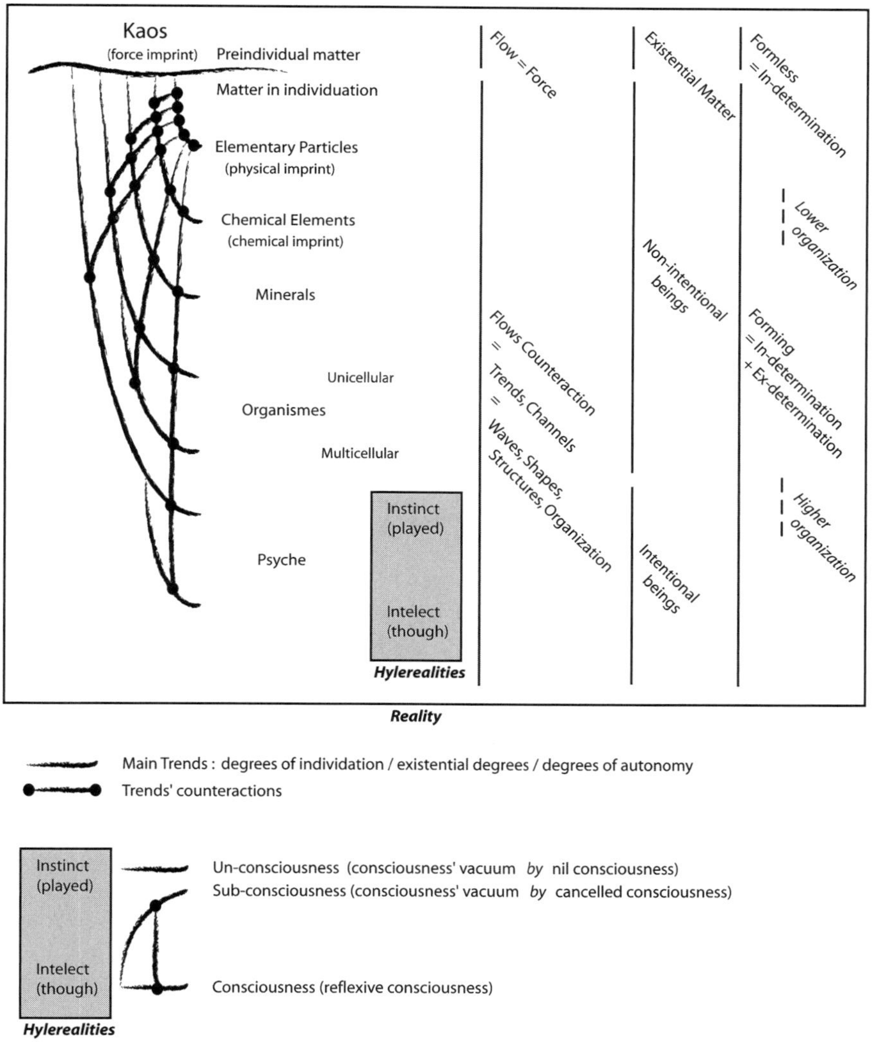

FIGURE 17.1. Kaos” existential degrees: investigation on the evolutionary trends of the individuation

17.4.2. The Descent of Darwin, and Selection in Relation to Ideology

For Darwin, species are made of members and natural selection is more than an eternal mechanism, it is a reversive continuum, a trend as in the metaphor of the Möbius strip (Tort 2002). At least intuitively, Darwin seems to know that (natural and artificial selections) are channeling trends, *canalizations*. In the same way, Simondon (1964) and Lima-de-Faria (1988) have seen discontinous interactions as trends but explicitly asked for a deterministic physical principle with a non-intentional connotation, to explain the *individuation* for the first, the *autoevolution* for the second.

Natural selection designates the indirect and dynamic equilibrium resulting from the interaction of a population and its environment, resources included (Tort 1996, pp. 4173–4175). Because of this interaction, a competition takes place between the individuals, which results in the survival of the fittest. Natural selection is a *persistent differential pressure*, a phenomenon that lasts and so is under the influence of its own mechanism. In those terms, one can say that Darwin's intuition was correct and that the consequence is that the natural selection as a canalizing trend can be applied to any field. Thereupon, it can be said that canalization is not so much natural selection but individuation, which is explained by the continuous optimization of Nature, i.e., by the constructal theory.

Competition is in fact a flow counteraction and "to survive" means "to last." If one wants to continue to use the term "selection," one has to understand that it is an intentional term, relative to an intentional stance, for something that does not have one: the trend or canalization. Canalizations are deterministic consequences, and so is the persistent differential pressure.

Natural selection, in the strict sense of *intentional selection*, is an *intentional process* (this is the artificial selection that inspired Darwin's natural selection) and has to be restricted to this very particular field. In other words, there is a huge gap between artificial selection, which involves natural intelligence, the perception of Cournot's chance based on historico-perception, and natural selection that does not involve anything but a trend among others in Nature. Mechanisms simply cannot be analogs in these terms. Consequently, Neo-Darwinists fail to see Darwin's original problematic, so to keep it with his theory. One has to remember that

(1) Randomness, hence mutation, is a deterministic process, and if randomness exists, it is of the Cournot type.
(2) The *reversive effect of evolution* (Tort 2002) is that "Natural selection, the guiding principle of evolution, that implies the elimination of the least capable in the struggle for existence, selects in humanity a form of social life in which the move towards civilization progressively tends to exclude, through the linked interplay of ethics and institutions, the behaviors that eliminate others. In simpler terms, natural selection selects *civilization, which opposes itself to natural selection*." Therefore, it appears obvious

> that Darwin's theory, not constrained anymore by his thought, has been distorted so as to be instrumentalized. Even so, there is a striking feature in this reversive effect. To those who know Hofstadter's famous book, *Gödel, Escher, Bach: An Eternal Golden Braid*, we would like to note that there is an *atavism* of the Möbius strip, which is the expression of the trends" dynamic, made and unmade continuously, and that this expressive explanation is common among theoreticians and artists who try to figure, construct, and use trends.

An interesting example of the Neo-Darwinian "explication" is the case of cancer. Many papers talk about "the gene of" or "the gene of the mergence of" some cancer or other. Of course, this is not totally false, but most certainly is quite far from the truth, far from the nature of it, since this talk of "the gene of" or "the gene of the mergence of" is deeply ensidic. As we have argued, emergence is ensidic and homolog to the expression "the gene of." We do hope that biological research will take into account the constructal theory to follow and help the works of some of the most intuitive scientists in this field. For example, Sonnenschein and Soto have demonstrated in their experiments that cancer is more a question of cell proliferation and tissue organization than a question of genes (Sonnenschein and Soto 1999). Constructal theory can help here since it shows, instead of taking it for granted, that *auto-organisation* is a consequence of the constructal law. This means that the organizational behavior (or "*mechanism* of global resistance minimization") of growth, assembly, or aggregation is a deterministic one (Bejan 2000, pp. 57–58).

With the understanding of constructal theory and individuation, one has to admit that it is not only genetic theory that has failed in its explanation, but information theory in its entirety (not in their descriptive stance, since they allow an efficient instrumentation). Non-intentional beings are not made of information. They certainly are informational entities for us, but not by and for the environment. Nature only gives rise to in-formation. DNA and genes, like the body or any beings, are Nature's expressions, individuals in individuation, in-formations. The body expresses them which, in turn, express the body. In the end, we have to admit that none of the non-intentional individuals are "selfish," only intentional ones can be so. Non-intentional individuals, such as genes, are just existential beings, individuals individuated by a deterministic individuation, that is, not by themselves or others, but by a whole without parts, a Unique-ness.

Indivi-duals existences, indivisible and dual, are Unique-ness" expression. There is a volume-to-point dependence, *a uniqueness-to-place dependence*. The place is the scene where flows, because of their counteraction, give birth to a singularity that is not a "point," an entity or a particle, but a *problematic*, an imperfection. It is the place of the *expression* of the flows, the forces. When the problematic is solved not by a computational but a continuous process, a solution is produced as a local in-formation, the optimization of an imperfection.

Solutions give birth to new problems that, once again, have to be solved in a never-ending story. Problems of some places are solutions for others and vice versa. Obviously, problems and solutions belong to the same reversive continuum. There is a continuous process, a Möbius strip trend, allowing us to go from problems to solutions, and *inversely*, without jumps or gaps. Consequently, symmetry and asymmetry have to be seen as a complementary aspect of the Nature's dynamic.

17.5. Historicity, Instinct, Intelligence, and Consciousness

17.5.1. History Versus Historicity, Continuous Versus Discreet

As we said, historicity is Kaos" consequence. Where the historicity of Nature is the actual state of its in-formations, History is the continuous historicity of Nature and its in-formations. Where History is historicity across time, fossils and souvenirs are durations of singularities of historicity across time. Since Kaos is in-determinist, historicity is historico-determinate. There is a historico-determination of History.

The genesis of new in-formations is made by the individuation of Nature and its actual individuals (in individuation) that when being transformed become partly or totally inaccessible. Individuation goes ahead and does not authorize the reconstruction by a sort of reversal of time, or reverse engineering, applied to History. Although there is only one History and it is not accessible to intelligence because of, among other causes, the lack of knowledge, the loss of historicity and the isomorphism of the in-formation. Atavism is one aspect of the isomorphism: there is no reappearance of a trait in an individual after several generations of absence, but an utterly new and homolog solution because of the similarity of the environmental condition. Homology does not mean resurgence of the past.

Yet, we do not go back in time by winding it up, but by making our thought jumps to a possible past by means of historic residuals, the knowledge (e.g., fossils, souvenirs), trying then to reconstruct History by approximation. This is to say that we tell *a story of History*, one among other hylerealities, hoping to find the one that would be the most correct, the one with the highest existential degree, the one which tends to fall in Reality, that is, to fit better with it. Finally, one has to understand that History is duration, and this has the consequence that historicity is not memory, but that memory is historicity. Memory is historicity unveiled by the historico-perception of the psyche, that is, historicity as memory belongs to intentional beings. And since historicity evolved by means of the individuation of in-formation, one can see that there is in historicity, just as in memory for intentional beings, maintenance, as well as gain or loss, of in-formation.

The rising form is the lowest degree of the cognitive phenomenon but also the most widely spread. This is what we call *natural cognition*, the growing existence (this is to say that cognition has nothing to do with computation).

But where, by its atomistic stance regarding artificial intelligence (AI), Dennett seems to implicitly ask for an infinite number of homunculi that lack intelligence enough to be considered as non-intentional beings, "as agents so unimpressive that they can be replaced by machines" (Dennett 2005, p. 137). But trying to explain intelligence in AI is trying to explain natural intelligence indirectly. Therefore, we argue that only Kaos is needed. From this plasma, as Dennett for his "horde of demons," we can already say that it is neither stupid nor intelligent since it is in-determinist, in-perceptive.

But Kaos is not a machine. It is only inhomogeneous quality producing forms in quantity and in quality. Flow only flows, just in-formations evolved. Flow is genesis principle and individuals are evolutions. Kaos" consciousness, if it has to have one, is *perception* since Kaos is only able of the lowest degree of perception: in-formation. Therefore, psychic cognition (hylereal perception or information) is intentional where natural cognition (real perception or in-formation) is non-intentional.

17.5.2. The Psyche

We would like to categorize different psyches. But because our stance is a materialist vitalism, categories are obviously hylerealities. One has to see them as intertwined trends, so that the way through states of the psyche is continuous. There is a constant rise of psychic forms and, even if their transformation is continuous, some of those can be individualized enough to be considered as individuals or quasi-systems, as thoughts.

(1) In the case of beings that are only instinctive, the void of consciousness is *nil*, that is, there is no reflective consciousness. As Bergson says, the hylerealities are *played*, that is, the lived experiences are not relived but *lived again* as realities, not virtualities. This consciousness is un-conscious, the thinkable is un-thought. This is what is called *instinct*.
(2) In the case of intellection, the hylerealities are *tough* (evaluated), the thinkable is though. The lived experiences, with which reflective consciousness participates (the first person), stays virtual until (a) decision-making process end in an successful action, which implies the probation of the hylereality that was chosen by testing it in Reality (by applying it to Reality) or until (b) a new stimulus comes in during the process. Therefore, in both cases, the action resumes, canceling consciousness and restoring the un-conscious psyche. But this time it is not a nil consciousness but a *cancelled* one. Here, instinct is not un-conscious but *sub-conscious*. The reflectivity could originate from the cancellation of physical action by its suspension and the concordance of historico-perception with the sensory–motor organs, especially the language apparatus.
(3) "Higher" cognition, or reflective consciousness, is linked to the quality of in-formed matter. This quality can be defined as the dynamic of a flow "system," i.e., the dynamics of the expression of an individual's individuation. The one we take into account is the "echo-chambers" discussed by

Nagarjuna (2005) and Dennett (2005, pp. 164–165, 169–171). Within this framework, consciousness is a dialectic between the sensory–motor apparatus and the brain as memory. An example that is generally given is about the micro-vibrations of vocal cords when a person is thinking, but it may be possible that the language apparatus is not restricted to one's speech. We could expect to observe, as during dream phases, some micro-movements of the visual apparatus, and more generally of the sensory–motor (muscular) apparatus, that is, the body language as a whole.

However, it cannot be denied that the expression (by way of individuation) of the speech apparatus played a major role in the emancipation of reflective consciousness of humans. Yet, language (not only speech) is merely present among the primates and more generally among "higher" vertebrates (Bertalanffy 1964, pp. 35–36; De Waal 2001; Pepperberg and Lynn 2000; Pepperberg and Gordon 2005). Once again, "things" are not as simple as a clear-cut categorization would let us think. Real or hylereal individuals, that is, categories included, are never individuals but individuals in individuation. Therefore, language is not the fact of an individual's interiority, but that of Nature. Its genesis is of both individual interiority and individual exteriority. In a quasi-systemic position, we can say that individuals become more individual through their environment. They belong to it not as parts but as expressions. This is to say that language is a construction as much interior as social (Allott 1985, 1992, 2003; Oudeyer 2003, 2005). Similarly to the origin of geometrical thought, the language is not in Nature but is the work of a laborious History.

17.6. Nature and Cognitive Computer Science

Today, we can see two computational stances: the conventional or classic computation and the unconventional one (Stepney et al. 2005). Both aim to model and simulate the mind, thus Reality. We would like to add our position to these two: unconventional "computation" has to be thought of, not only as computation, but as a constructal individuation. Information processing has to start by the genesis of the in-formation. In fundamental research, a distinction has to be made between computation and individuation, information and in-formation. We do not encourage the use of one or the other but of both as different stances, information for the intentional stance, in-formation for the non-intentional stance, and start from the second to reach the first in terms of explanation. We agree that, in order to make a simulation, we have to use information. But to simulate in-formation is not to simulate operations on it (like waves" interactions as logical gates with cellular automata or Belousov–Zhabotinsky reaction).

There is a common misrepresentation, by way of misunderstanding, between the sign as symbol, or information, and the sign as shape, or in-formation, that is the foundation of the capacity of intentional beings to perceive or "read" information. This point can be crystallized in Kurzweil's quotation, regarding the views of scientists such as Wolfram, Fredkin, and Lloyd or, before them, Ulam: "The information is not embedded as properties of some other substrate (as in the case of conventional computer memory) but rather information is the ultimate reality. What we perceive as matter and energy are simply abstractions, i.e., properties of patterns. [...] Wolfram joins a growing community of voices that believe that patterns of information, rather than matter and energy, represent the more fundamental building blocks of reality".[10]

Obviously, as we have said, information is not Nature's fundamental reality, but an expression of it in intentional beings. This does not mean that cellular automata or chemical computing are not useful, they are indeed powerful computational tools, but that the ultimate reality, the basis of explanation we are searching for, is the Kaos, the preindividual matter or in-determinate plasma. And this has to be tackled in a very different way than information and systems theories. This way has to be found, without a doubt, with the help of intuition and a plain awareness of our knowledge, its limitations, and implications. Thus, perhaps, some will be able by an act of serendipity, by accident and sagacity (to trick his intelligence long enough), to find a *possible* answer.

This is to say that natural computation, which is in fact Nature as in-formation and not Nature as information processing, has nothing to do with our ensidic conception of classical computation that describes more than explains it. Ensidic conception of information and computation belongs to Nature of course, but only *in* intentional beings. Simply said, *Nature is neither a computer nor computational.* Nature does not contain information but only in-formation. It is intentional beings, with their psyche, their hylerealistic capacity based on historico-perception, who give sense to Nature's historico-determinism. They transform in-formations in information, that is, historicity in possibilities, in stories.

17.6.1. Neural Networks Versus Constructal Architectures

GasNet (Philippides et al. 2005) based on chemical (Nitric Oxide) diffusions in the brain, Liquid States Machines (Maass and Markram 2002; Maass et al. 2002; Natschläger et al. 2002) and Echo States Machines (Jaeger 2001a,b) are among the most advanced theories and simulation on the brain activities, and one can understand that the question of the intelligence overflows the "neuronal" answer. Those recent fundamental researches in Neural Networks are going far beyond the point of view of the formal neuron and follow in this way the remark of McCormick (2005), that is, most neuroscientist, and computer scientist, when asked how the brain works, "would respond that the brain is a large, web-like structure in which neurons *gather information* from other neurons, make

[10] In: Kurzweil, R. (2002) *Reflections on Stephen Wolfram's "A New Kind of Science."* http://www.kurzweilai.net/articles/art0464.html

a decision to discharge or not, and then *pass this information* onto other cells; magically, somehow, through this large interaction of neuronal elements, information is extracted, decisions are made, and responses are executed" (emphasis added). The received view is that this activity "is imagined to flow through the neural net with the spatiotemporal path determining the outcome of the particular computation, be it a thought, perception, feeling or action." Although this is certainly true in an ensidic framework, "there are several complicating factors. One of these is the presence of spontaneous, or persistent, activity that can rapidly flip between stable states."

Morowitz (1987) raises in philosophy of mind, and thus in cognitive computer science, a related but deeper problematic: the irreducible problem of the irreducibility of the mind. He shows how the problem of measurement in quantum mechanics is tightly coupled with the Mind–Body problem in cognitive science. He reminds us, based on Brillouin's conclusion in his analysis of the Maxwell's demon paradox, that actual researchers looking at neural networks and the central nervous system, down to the atomic and quantum levels, are modern dualists. And "accepting dualism involves either giving up the second law of thermodynamics or limiting that law to situations not involving mind," since it makes the incommensurable economy of the measurement, that is, the same as taking for granted that mental states can be measured without spending energy, without influencing the physical state, or to have knowledge of the latter without taking any measurement.

In the end, Lima-de-Faria (1988, p. 299) is right when he writes, "It is not biology that must be modified but physics." In fact, not only physics but all sciences which aim fundamental research have to evolve. Many researchers are concerned with the cognitive implication of our hylemorphic capacity, where modern probabilities are the most complete aspect of it. And by unveiling it, we think that it can be of some help for those researchers who are aware, in some ways, of this cognitive implication.

This is the case of the quantum physicist Mugur-Schächter (2005, 2006) who wrote "But why . . . ?" and "On the weaving of the knowledge." Her conclusion is that "it is the cognitive situation that has orchestrated the construction of the quantum mechanics [...]. In turn, the formulated hypothesis suggests a project: *to disregard completely the quantum formalism* and attempt to sketch by oneself only the main lines of an exclusively *qualitative* representation of the structure of the descriptions that we conceive as corresponding to the name "*microsystems" states*" ("*microstates*")" (Mugur-Schächter 1993, 1997, 2002). We have to say that we come to the same conclusion but in a different way since we make a distinction between natural intelligence and intuition, whereas Mugur-Schächter does not. For us, intuition is the metastable state of the natural intelligence, the individual problematic in constant search of a solution, an optimization. Therefore, intuition is not about probabilities, hylerealities, but is certainly based on them since they are the main characteristic of natural intelligence, the psyche. A fruitful discussion certainly lies here, but we do not have the space for it. The idea is to present researchers from different fields that question the cognitive implication.

Beyond the discrete lies the question of the continuity, so the question of the historicity. "Real brain" activity is continuous, persistent,[11] and context-dependent (McCormick 2001, 2005). Even spike trains are canalized waves, which themselves transform the entire brain. For example, there are chemical level waves in neural and glial channels. All cells, all chemicals inside and between cells are flows counteractions. Therefore, to model the brain is not modeling a particular dynamic state of matter but its morphogenetic principle. Brain's historico-capacity is more efficient than those of water or rocks because it has the best of both worlds. *Where Kaos is like plasma, brain is like magma*. Plasma, in Latin, means a fluid, formless substance; magma, in Greek, means "kneaded dough" and *maza* means "to knead." In German, this last term gives *Maßeinheit*, which means "unit (of measure)." Therefore, a unit, for it to be defined, needs a scalable size. This scalable size is given by the information, which becomes, degree by degree, information for the intentional beings (Gerdes 2001). This is to say that statistical probabilities (frequencies), the historico-perceptions in natural intelligence, have to be stable. *Historico-perception is what can define an intentional being*.

In summary, Kaos" own activity is *maza*. Kaos is a plasma that turns itself into magma since this plasma is preindividual, formless, but is also able to produce shapes (waves, rhythms, turbulences are shapes). Kaos is a perceptive preindividual existence, a qualitative inhomogeneity, an in-tension.[12]

"Cognition" basically means "knowledge acquisition," and because information is a solution to a local problematic that is solved in regard to Unique-ness, in-formation is "knowledge acquisition," or "natural cognition," where historicity is "natural memory." Kaos is in-existence, existence before the ensidic one of intentional beings. Where Kaos produce shapes, intentional beings perceive them.

In-formations (shapes) are expressions, the expressiveness of the perceptive quality of Kaos. However, one must not forget that the "human brain" is not unique but singular and there is a need to investigate other brain configurations, such as in the recent and marvelous restructuring of the avian brain nomenclature.[13] History is the individuation of the matter; therefore historicity is about historico-determinism whereas the Mind–Body problem is about historico-perception. Thus, a brain is not a question of "neural networks discretization" but of the individuation of Kaos.

[11] "Perhaps the best-known example of persistent activity in the nervous system is the one that occurs in the prefrontal cortex during the performance of tasks that require information to be stored for brief periods (seconds)" (McCormick 2001).

[12] In-tension is non-intentional, since Kaos" quality is only a *perception* and not *historico-perception*.

[13] http://www.avianbrain.org/new_terminology.html

17.6.2. Cellular Automata and the Belousov–Zhabotinsky Reaction

Cellular automata (CAs) are an ensidic metaphor of the individuation. A cellular automaton is a discrete model that consists in a regular grid of finite dimensions made of "cells." Each cell possesses a finite number of states that are determined by a set of rules based on the values (the states) of the cells in its neighborhood. All cells are identical, so have the same set of rules.

Figure 17.2 shows the possible states of a cell considering its neighborhood in Conway's Game of Life (Gardner 1970). Each cell has two states: deactivated (dead) and activated (alive). Those states are given by three rules that compute the new generation of cells from the previous one and are

(1) "Birth": An inactive cell with exactly three active neighbors comes to life in the next generation.
(2) "Survival": An active cell with two or three active neighbors survives in the next generation.
(3) "Death": An active cell with one or less active neighbor dies from isolation. An active cell with four or more active neighbors dies from overcrowding.

The last example is the "Still Life" configuration. The medium is active since cells are alive, but it misses something to make it move. In fact, it *moves statically*, which is due to the equilibrium of the flows counteractions. This form is autonomous since its organization is maintained. Other forms, such as the "Glider," which are animated forms in movement, can be defined as autopoietic systems since there organizations are maintained *in spite of* the change of their constituents, the cells, but also *by* this change. Most importantly, what has to be remembered is that those forms, in movement or not, are animated. Forms and interactions between forms have to be seen as flows counteractions. Movement is the displacement of the counteraction of flows.

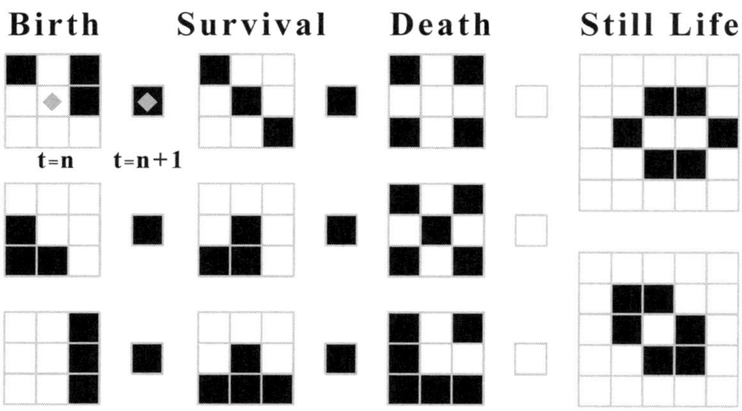

FIGURE 17.2. Cellular Automata rules and "still life" patterns

The capacities exhibited by Cas are used as logical computation by some researchers. Adamatsky (2001, 2004) has implemented a waving computation in reaction-diffusion (RD) and excitable media that behave like Cas. For example, the Belousov–Zhabotinsky (BZ) reaction exhibits interactions of waves and so can be seen as an RD processor. Therefore, the RD medium can be used as "super-computer in a goo." Chemical computing devices "have parallel input of data (spatial distribution of the reactant concentrations), massively parallel information processing (via spreading and subsequent interaction of waves) and parallel output of results (commonly, the results are represented by patterns of reactants or a colored precipitate/product that enables the use of conventional optical reading devices)" (Adamatzky et al. 2003, 2004).

But at this point, we shall say that there is no data, but a flow, so there is no massively parallel information processing but massively parallel in-formation in individuation. Here is the cognitive implication that we would like to overcome. RD media are historicity, even "memory" for a robot, but the medium does not compute, it is in individuation.

17.6.3. RD-Computation or Simulation of the Individuation?

To think what can be a neuron in a constructal architecture, one just needs to take a look at Jupiter's Great Red Spot. It is an autonomous being in individuation. If one imagines that Jupiter is a *body*, then the *brain* is analog to Jupiter's Great Red Spot. Neurons and glial cells have to be seen in the same way. But we can go further by making a homology between the individuation of a neuron and the one of the *Dictyostelium discoideum*, a soil-living amoeba.

Often called a *cellular slime mold*, the fact is that Dictyostelium discoideum neither is a mold, nor is it always slimy. When the organism is individualized, the entities are called *myxamoebae*. When they aggregate into a slug, the organism is called a *pseudoplasmodium*, also named *the grex*. In Greek as in Latin, *grex* means flock or to flock, the "desire" to be with others. And this aggregation into a *unitary grex* may involve tens of thousands of individual amoebae. Accordingly, some propose that a more accurate term would be *social amoeba*. But this phenomenon of aggregation is now explained with the help of constructal theory (Reis et al. 2006). Yet, "social" is still too ensidic (and intentional) since the amoeba like the grex are both individual in individuation. Perhaps an encompassing term should be something like *amoebamedium's individuation*.

Höfer has shown the aggregation of *Dictyostelium discoideum* on an agar plate on which we can see the formation of spiral cAMP waves and, after a certain time, the derived configurations (Höfer et al. 1995, p. 250). A more detailed description can be found in the thesis of Marée (2000). Those deterministic aggregations belong to wave patterns in excitable media like the intracellular calcium waves in the brain (Pasti et al. 1997; Volterra and Meldolesi 2005).

What is important to keep in mind is that cells and waves are individuals in individuation. Shapes, like cells, organelles, and molecules, are flows counteractions, just like Jupiter's great red spot.

Therefore, our aim is to build, not the cells, but the network of cells (Fig. 17.3) from an amorphous activity, the flows counteractions, to simulate and study

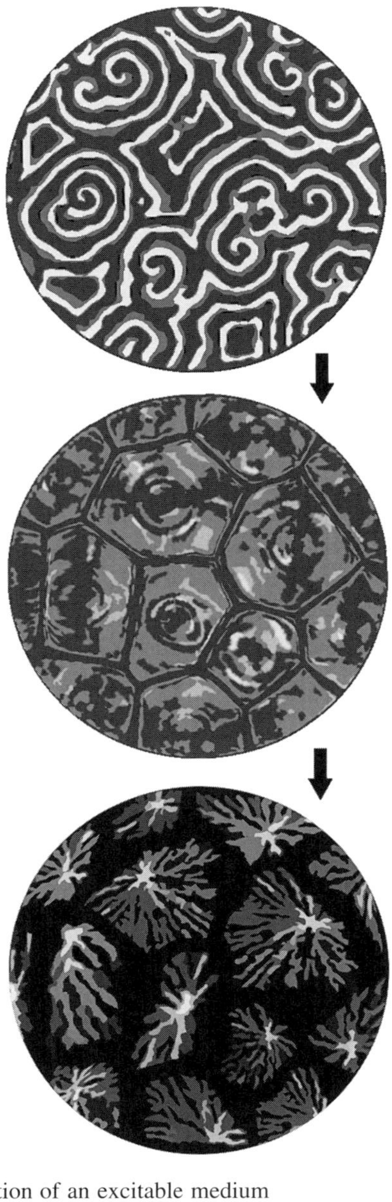

FIGURE 17.3. Individuation of an excitable medium

different qualitative historicity (water, rocks, and brain like), hence memories. We do not aim to simulate an intentional being, but to find ideal "constructal brains," ideal memories. The constructal theory allows the building, from Kaos, so turbulences, to a differentiated network, of such a system, an individual in individuation. It allows us to escape the ensidic problematic, that is, the problematic of composition and emergence, and to tackle the individuation in a continuum.

17.7. Constructal Law, in Depth

17.7.1. The Geometric Vitalism of the Constructal Theory

Today, Constructal Theory is based on a geometrical principle. It encompasses René Thom's catastrophe theory that is, by its qualitative geometry, "the most beautiful aesthetic theory in the world."[14] Where Thom's Theory was the study of the dynamics of shapes, like Mandelbrot's Fractals but in a way that Thom called a *geometric vitalism* (Thom 1977, pp. 158–159), Bejan's Constructal Theory does not need a "given shape." On the contrary it starts by stating that shapes and structures are not given. Instead of Mandelbrot's discrete approach or Simondon discontinuous approach (Thom 1994), Bejan's, like Thom's, is a geometric approach based on the continuous. Constructal theory explains how shapes and structures are produced, and how we can construct them. It follows that their dynamics, like the construction of discontinuities (Bejan and Gobin 2006) are also consequences. But, if one can build a mathematic that will not be ensidic, then perhaps materialistic vitalism will be able to shift from metaphysics to physics, to pass from virtualities (the virtual possibilities of our psyche's hylerealist capacity) to reality (the real product made by our psyche's hylerealist capacity) by increasing its existential degree.

17.7.2. The Constructal Law Definition

For a flow system to persist in time (to survive) its configuration must evolve (morph) in time in such a way that it provides easier flow access.

Before questioning a few specific terms of this law, we would like to briefly recall its meaning and implications.

A flow system and an open system are the same. In thermodynamics, they are called a non-equilibrium system, a system with gradients hence currents (Bejan 2000, p.1). To maximize the access between the volume and the point (to balance or optimize its imperfections), a building block (elemental as construct) needs a finite-size. "Finite-size" means that its volume is finite in size, its external

[14] Dalí, Salvador, *Gala, Velásquez and the Golden Fleece*, 9 May 1979; source: en.wikipedia.org

shape and internal structure can vary. This is why a construct can acquire shape and structure. An elemental system (or volume) is the smallest building block, the smallest length scale characteristic of the flow medium, "so small that it houses only one channel (strip) of fast material" (ibid., p. 9). This intrinsic property of the constructal law is what is missing in fractal geometry. It prevents us from making the shape smaller ad infinitum: under a certain block size, no shape can exist.

A last, quite important, point is that constructal theory establishes that the relation between geometric forms in successive scales (ibid., p. 293) is made by scale covariance and not scale invariance as in fractal geometry (ibid., p. 309).

Now that we have reviewed the fundamental terms of the definition, we would like to present two open questions.

(1) Can we use the expression "to survive"? This is an old question that must be out of fashion. But, as we have seen, we cannot use it without the risk of misrepresenting it since it suggests not only "to last" but also "selection," which has cognitive implications. For Darwin, they were equivalent but he knew, and we have to remember, that in this inversion lie serious consequences. But, since we can explicitly say (a) that selection is a particular case of the individuation, a local trend inside a more general, as Boolean logic is for fuzzy logic; or (b) that natural selection is individuation. Therefore, one has to encompass all trends in a single and more neutral verb, and instead of the expression "to survive," we shall propose "to last."

(2) Can a flow "system" really exist? A small hint before discussing this point is that a flow system is certainly not a shape system since shape, and its displacement if there is one, is flows counteraction. There is no shapes interaction, but only flows counteraction. This being said, is flows counteraction a flow system? Since we show that a systemic approach is ensidic and cannot endorse Unique-ness, we argue that flows counteraction is not a flow system, except if one needs use constructal theory in a systemic framework, as for engineering. In fundamental research, *space* and *system* have to merge. This is where, we think, the geometric vitalsim can find a way out to a materialist vitalism.

17.8. A Never-Ending Story

Like minerals, cells, or plants, theories, ideas, and instruments are non-intentional beings as well (if one wants to call them "memes," then one has to put "non-intentional" before the word) and like all beings are not "parts" of Nature but *expressions of it* (Fig. 17.1).

Beings are not finalities, they do not have to become "fixed trends," for intentional beings, "ideologies."[15] They are *perspectives*. Optimal ideas, in the constructal sense, are ideas that persist in time, but they are neither "good" nor "bad," they do not have intention. They are only optimal. And "optimal ideas" are not necessarily "optimality for human beings." When one learns something new, learns a "new" idea, one is within the crossing of independent causal series. It raises a problematic (an imperfection, so an in-tension) that has to be solved. The solution, a new individuation of the individual, ends in a synthesis or dissolution, partial or total, of old and new ideas. Determining the "good" or "bad" effect of an idea belongs to intentional individuals, within the context of their beliefs and cultures. Yet, for someone who tries to encompass human freedom, there are some clues to ideas that are more dangerous than others, since they force the trend to civilization. When an ideological trend is intentionally canalized under deceit or self-deception that goes in search of a global optimality, it appears that the search is in fact a selfish one. Since human beings are intentional, their choice to canalize the others can only be related to a lack of curiosity, hence a lack of knowledge. Of course, there are many causes for this lack of knowledge, but where analphabetism is certainly the saddest one, beliefs are without a doubt the fiercest.

Ideas, like theories, which are presented without being tightly coupled with the morals of their inventors, can be taken away from them. Once ideas are separated, one can transform them, intentionally or not, into tools. Consequently, ideas become ideological tools by deceit and self-deception.[16]

Like Darwin's Theory, Bejan's Constructal Theory is not an end to our reality, a fixed trend with a fixed finality. It was a mistake to use Darwin's Natural selection, to use it like a tool, by not taking into account its inventor's intention to frame the natural selection in "the reversive effect of evolution." If there is a lesson to learn from Darwin and Neo-Darwinism, it is that a theory must not be separated from its inventor. If one wants to use the theory, one needs to take both the theory and the thought of its inventor together, as a package deal. If one keeps the *thought*, one will keep the *perspective*.

Constructal Theory belongs to a thought that is Bejan's. If we use the theory without his thought, it will turn into a tool that will produce a fixed trend which is the inevitable mark of our *natural self-deception*, our *natural intelligence* which pushes us to live almost only in the action.

If a sentence has to remain from Adrian Bejan *and* his theory, it is certainly the one given by his son William quoting Sophocles (Bejan 2006, pp. 818–819): "For a man, though he be wise, it is no shame to learn—learn many things, and not maintain his views too rigidly."

---

[15] Even those are never "totally" fixed since they are always in individuation. However, they slow down the evolution so much that it can become a dangerous state for intentional living beings. Where is *intention* if there is no more *dynamic*?

[16] See "Discussion with Noam Chomsky and Robert Trivers" http://ww.chomsky.info/debates/20060906.htm

References

Adamatzky, A. (2001) *Computing in Nonlinear Media and Automata Collectives.* Philadelphia Institute of Physics Publishing, Bristol, UK.

Adamatzky, A. (2004) Collision-based computing in Belousov-Zhabotinsky medium. *Chaos, Solitons and Fractals* **21**: 1259–1264.

Adamatzky, A., De Lacy Costello, B., Melhuish, C., Ratcliffe, N. (2003) Liquid Brains for Robots. *AISB Quarterly* **112**: 5.

Adamatzky, A., Arena, P., Basile, A., Carmona-Galan, R., De Lacy Costello, B., Fortuna, L., Frasca, M., Rodriguez-Vazquez, A. (2004) Reaction-diffusion navigation robot control: from chemical to VLSI analogic processors. *IEEE Transactions on Circuits and Systems Part 1:Fundamental Theory and Applications* **51** (5): 926–938.

Allott, R. (1985) The Origin of Language: The General Problem. Presented at 1st Meeting of Language Origins Society, Cracow. In: Wind, J. et al. (1989) *Studies in Language Origins I.* Benjamin, Amsterdam, NL.

Allott, R. (1988) The motor theory of language: Origin and function. In: Wind, J. et al. (1992) *Language Origin: A Multidisciplinary Approach*, Kluwer Academic Publishers, Amsterdam, NL, 105–109.

Allott, R. (2003) Language as a mirror of the world: Reconciling picture theory and language games. *Cogprints: Cognitive sciences eprint archive.* Presented at 30th LACUS (Linguistic Association of Canada and the United States), July 29–August 2, University of Victoria, CA.

Bejan, A. (2000) *Shape and Structure, from Engineering to Nature.* Cambridge University Press, Cambridge, UK.

Bejan, A. (2006) *Advanced Engineering Thermodynamics*, 3rd edn., Wiley, Hoboken, New Jersey.

Bejan, A. and Gobin, D. (2006) Constructal theory of droplet impact geometry, *International Journal of Heat and Mass Transfer*, 2005, **49**: 2412–2419.

Bejan, A. (2007) The constructal law in nature and society, Chapter 1 in this book.

Bergson, H. (2001) *L'évolution créatrice*, 1907, 9th edn., PUF, Quadrige, Paris, FR.

Bertalanffy, L. von (1964) The Mind-Body problem: A new view. *Psychosomatic Medicine* **26** (1): 29–45.

Castoriadis, C. (2005) *Figures of the Thinkable (including Passion and Knowledge).* Translated from the French and edited anonymously as a public service, eprint.

Caws, M. A. (1978) Review of *Paul Valery: An anthology* by Lawler J, R. (1977). In: *The French Review*, **51** (6), May: 903–904.

Cournot, A.-A. (1875) Matérialisme, vitalisme, rationalisme, Études sur l'emploi des données de la science en philosophie. Eprint made by Blachair A.

Cummins, R., Poirier, P. and Roth, M. (2004) Epistemological Strata and the Rules of Right Reason. *Synthese* **141**: 287–331.

Dennett, D. C. (2005) Sweet Dreams: Philosophical Obstacles to a Science of Consciousness. MIT Press, Cambridge, MA, USA.

De Waal, F. B. M. (2001) *Tree of Origin.* Harvard University Press, Cambridge, MA, USA.

Dhombres, J. and Kremer-Marietti, A. (2006) *L'épistémologie : état des lieux et positions.* Ellipse edn., Paris, FR.

Gardner, M. (1970) Mathematical Games: The fantastic combinations of John Conway's new solitaire game "life". *Scientific American* **223** (4): 120–123.

Gerdes, P. (2001) Origins of Geometrical Thought in Human Labor. *Nature, Society, and Thought*, 2003, **14**(4): 391–418.

Hacking, I. (2002) L'émergence de la probabilité, (French translation of The Emergence of Probability: A Philosophical Study of Early Ideas about Probability, Induction and Statistical Inference), Seuil edn., Paris, FR.

Höfer, T., Sherratt, J. A., and Maini, P. K. (1995), Dictyostelium discoideum: Cellular self-organisation in an excitable medium. *Proc. Roy. Soc. (Proceedings of the Royal Society)*, Biological Sciences **259** (1356): 249–257.

Jaeger, H. (2001a) The "echo state" approach to analysing and training recurrent neural networks. German National Research Center for Information Technology, Technical Report GMD, Report 148.

Jaeger, H. (2001b) Short term memory in echo state networks. German National Research Center for Information Technology, Technical Report GMD, Report 152.

Janmaat, K. R. L., Byrne, R. W. and Zuberbühler, K. (2006) Primates take weather into account when searching for fruit. *Current Biology* **16**: 1232–1237.

Karsai, I. and Pézes, Z. (2000) Optimality of cell arrangement and rules of thumb of cell initiation in Polistes dominulus: a modeling approach. *Behavioural Ecology* **11**: 387–395.

Kremer-Marietti, A. (2006) The constructal principle. *Dogma*, eprint.

Lima-de-Faria, A. (1988) Evolution without Selection: Form and Function by Autoevolution. Elsevier edn., Amsterdam, NL.

Maass, W. and Markram, H. (2002) Temporal integration in recurrent microcircuits. In: Arbib, M. A., *The Handbook of Brain Theory and Neural Networks*, 2nd edn, MIT Press, Cambridge, MA, USA, 1159–1163.

Maass, W., Natschläger, T., and Markram, H. (2002) Real-time computing without stable states: A new framework for neural computation based on perturbations. *Neural Computation* **14** (11): 2531–2560.

Marée, A. F. M. (2000) From Pattern Formation to Morphogenesis: Multicellular Coordination in Dictyostelium discoideum, Ph.D. thesis, Utrecht University, Utrecht, NL.

McCormick, D. A. (2001) Brain calculus: neural integration and persistent activity. *Nature Neuroscience* **4**: 113–114.

McCormick, D. A. (2005) Neuronal Networks: Flip-Flops in the brain. *Current Biology* **15**: 294–296.

Morowitz, H. J. (1987) The Mind Body Problem and the Second Law of Thermodynamics. *Biology and Philosophy* **2**: 271–275.

Mugur-Schächter, M. (1993) From Quantum Mechanics to Universal Structures of Conceptualization and Feedback on Quantum Mechanics, *Foundation of Physics* **23** (1): 37–122.

Mugur-Schächter, M. (1997) Les leçons de la mécanique quantique, *Le Débat*, Gallimard, Paris, FR, March/April, **94**: 169–192.

Mugur-Schächter, M. (2002) Objectivity and Descriptional Relativities, *Foundations of Science* **7**: 73–180.

Mugur-Schächter, M. (2005) But why...? But why the quantum mechanics? To reach the roots of the knowledge, eprint manuscript: http://www.mugur-schachter.net/maispourquoi.pdf

Mugur-Schächter, M. (2006) *Sur le tissage des connaissances*. Hermes edn., Science & Lavoisier, Coll. Ingénierie représentationnelle et constructions de sens, Paris, FR.

Nagarjuna, G. (2005) Muscularity of Mind: Towards an Explanation of the Transition from Unconscious to Conscious. *Cogprints: Cognitive sciences eprint archive*.

Natschläger, T., Maass, W., and Markram, H. (2002) The "liquid computer": A novel strategy for real-time computing on time series. *Special Issue on Foundations of Information Processing of TELEMATIK* **8**(1): 39–43.

Nimier, J. (1989) *Entretiens avec des mathématiciens. L'heuristique mathématique.* Institut de Recherche pour l'Enseignement des Mathématiques , Académie de Lyon, Villeurbanne, FR.

Oudeyer, P.-Y. (2003) L'auto-organisation de la parole. Ph.D. thesis, Paris VI University, Paris, FR.

Oudeyer, P.-Y. (2005) The self-organization of speech sounds, *Journal of Theoretical Biology* **233**: 435–449.

Pasti, L., Volterra, A., Pozzan, T., and Carmignoto, G. (1997) Intracellular calcium oscillations in astrocytes: A highly plastic, bidirectional form of communication between neurons and astrocytes in situ. *Society for Neuroscience* **17**(20): 7817–7830.

Pepperberg, I. M. and Lynn, S. K. (2000) Perceptual consciousness in Grey parrots. *American Zoologist* **40**: 893–901.

Pepperberg, I. M. and Gordon, J. D. (2005) Number Comprehension by a Grey Parrot (Psittacus erithacus), Including a Zero-Like Concept. *Journal of Comparative Psychology* **119**: 197–209.

Philippides, A., Husbands, P., Smith, T., and O'Shea, M. (2005) Flexible couplings: diffusing neuromodulators and adaptive robotics. *Artificial Life* **11**(1 and 2): 139–160.

Pirk, C. W. W., Hepburn, H.R., Radloff, S.E., and Tautz, J. (2004) Honeybee combs: construction through a liquid equilibrium process? *Naturwissenschaften* **91**: 350–353.

Reis, A. H., Miguel, A. F., and Bejan, A. (2006) Constructal theory of particle agglomeration and design of air-cleaning devices. *Journal of Physics D: Applied Physics* **39**: 2311–2318.

Simondon, G. (1964) L'individu et sa genèse physico-biologique: L'individuation à la lumière des notions de forme et d'information. Presses Universitaires de France, Paris, FR.

Simondon, G. (1989) Du mode d'existence des objets techniques, Aubier, Paris, FR.

Stepney, S., Braunstein, S. L., Clark, J. A., Tyrrell, A., Adamatzky, A., Smith, R. E., Addis, T., Johnson, C., Timmis, J., Welch, P., Milner, R., and Partridge, D. (2005) Journeys in non-classical computation I: A grand challenge for computing research. *International Journal of Parallel, Emergent and Distributed Systems* **20**(1): 5–19.

Sonnenschein, C. and Soto, A. M. (1999) The Society of Cells: Cancer and Control of Cell Proliferation. Back to the drawing board? BIOS Scientific Publishers, Oxford, UK.

Thom, R. (1977) *Stabilité structurelle et morphogenèse.* 2nd edn., InterEditions, Paris, FR.

Thom, R. (1994) Morphologie et individuation. In: *Gilbert Simondon: une pensée de l'individuation et de la technique*, Albin Michel edn., Paris, FR, 100–112.

Thom, R. and Noël, E. (1993) *Prédire n'est pas expliqué.* Flammarion, coll. Champs, Paris, FR.

Tort, P. (1996) *Dictionnaire du darwinisme et de l'évolution.* Presses Universitaires de France, Paris, FR.

Tort, P. (2002) La seconde révolution Darwinienne: Biologie évolutive et théorie de la civilization. Kimé, Paris, FR.

Valéry, P. (1919) The Crisis of the Mind. *The Athenaeum*, April 11 and May 2, London, UK. http://www.historyguide.org/europe/valery.html

Van Andel, P. (1994) Anatomy of the unsought finding: serendipity: origin, history, domains, traditions, appearances, patterns and programmability. *British Journal for the Philosophy of Science* **45**(2): 631–648.

Volterra, A. and Meldolesi, J. (2005) Astrocytes, from brain glue to communication elements: the revolution continues. *Nature Review Neuroscience* **6**: 626–640.
Whitehead, J. (1911) *Apocalypse Explained.* A posthumous publication of Emmanuel Swedenborg, 1757. Translation revised by J. Whitehead, Swedenborg Foundation, New York, NY, USA.

Index